The Global
Water Cycle

The Global Water Cycle

Geochemistry and Environment

Elizabeth Kay Berner
Robert A. Berner

Yale University

Prentice-Hall, Inc., Englewood Cliffs, New Jersey 07632

Library of Congress Cataloging-in-Publication Data

BERNER, ELIZABETH KAY, 1936–
    The global water cycle.

    Includes bibliographies and index.
    1. Hydrologic cycle.   2. Water chemistry.
3. Oceanography.   4. Geochemistry.
5. Environmental Chemisty.   I. Berner,
Robert A., 1935–   II. Title.
GB848.B47 1986   551.48   86-22603
ISBN 0-13-357195-5

Editorial/production supervision and
    interior design: Jean Hunter
Cover design: Diane Saxe
Manufacturing buyer: Barbara Kelly Kittle
Cover photographs by Robert A. Berner
(upper) Dordogne River, south of France
(lower) Big Sur, California

Printed in the United States of America

10 9 8 7 6 5 4 3 2 1

ISBN 0-13-357195-5 01

PRENTICE-HALL INTERNATIONAL (UK) LIMITED, *London*
PRENTICE-HALL OF AUSTRALIA PTY. LIMITED, *Sydney*
EDITORA PRENTICE-HALL DO BRASIL, LTDA., *Rio de Janeiro*
PRENTICE-HALL CANADA INC., *Toronto*
PRENTICE-HALL HISPANOAMERICANA, S.A., *Mexico*
PRENTICE-HALL OF INDIA PRIVATE LIMITED, *New Delhi*
PRENTICE-HALL OF JAPAN, INC., *Tokyo*
PRENTICE-HALL OF SOUTHEAST ASIA PTE. LTD., *Singapore*

To John, Susan, and James

Contents

Preface

This book came into being for two reasons. First, the junior author for some time had been teaching an undergraduate course at Yale University on the chemistry of natural waters and every year he had to explain to the students that there wasn't an appropriate text for the course. This led to a desire to write up the lecture notes, but there never seemed to be enough time and, besides, eventually some excellent potentially competing books on the subject, such as J. I. Drever's *Geochemistry of Natural Waters*, began to appear on the market. Enter the senior author. She wanted to see the book written, but also to broaden it to include much more atmospheric and environmental chemistry, and to attract an audience with less formal training in chemistry and mathematics than the audience at whom the junior author and most other already published authors were aiming. Thus, we decided to write a book together which could be understood by most anyone with an elementary knowledge of science, but which could still serve as a source of geochemical and environmental data for researchers in a variety of fields. The senior author ended up changing the junior author's approach to teaching the Yale University course, and also tested the idea of using this book as a textbook herself by teaching a one-semester undergraduate course based on it at Wesleyan University (in 1982) where results proved to be most successful.

The approach of this book, as the title indicates, is global. We attempt to quantify the rates by which water and its major dissolved constituents are transferred from one reservoir to another at the earth's surface and to track down the sources of these constituents, be they natural or human-produced. This has involved delving into the literature of many different fields including geology, geochemistry, environmental science, oceanography, meteorology, hydrology, limnology, physical chemistry, biology, ecology, soil science, agriculture, and economics. We try to bring all these fields to bear on the problem of the cycling of water and its principal dissolved and suspended constituents. Thus, the subject of this book does not fit neatly into any of the classical fields of instruction; nevertheless, it is mainly geological-geochemical in its approach.

Many people have been helpful by their criticisms both of lectures and earlier drafts of the manuscript. Notably we thank J. I. Drever, David Crerar, Peter Liss, H. D. Holland, J. C. G. Walker, and F. J. Millero for their reviews of the various chapters, and a variety of undergraduate and graduate students for their helpful comments on needed clarifications and numerous typographical errors. Considerable help in preparing the final manuscript was provided by Susan E. Berner. Finally, the junior author acknowledges the musical inspiration of Howard Hanson and Heitor Villa-Lobos.

New Haven, Connecticut

Elizabeth Kay Berner
Robert A. Berner

The Global
Water Cycle

1

Water:

An Important and Unique

Substance

INTRODUCTION

Earth is the only planet in the solar system having an abundance of liquid water on its surface; about 70% of the earth is covered by liquid water. Because of the particular combinations of temperature and pressure on the planet's surface, water can exist here in three states: as liquid water, as ice, and as water vapor. This is in sharp contrast to the surface of the planet Mars, for example, which is so cold and dry that water can exist there only as ice or as water vapor. Because liquid water on the earth's surface is so common, we tend to forget how unusual it really is. Moreover, water is unusual in *all* its physical and chemical properties. Its boiling point, its density changes, and its heat capacity, for example, are not what one would expect by comparison to other similar substances. As a result of its unique properties, the chemistry of water is unique also.

Water is essential to life; where there is no water there is no life. It is by far the most abundant substance in the part of the earth where life exists, the earth's surface or biosphere. The biological importance of water is manifested in several ways. First, in *photosynthesis*, which is the basis of almost all life on earth, water is combined with carbon dioxide by green plants in the presence of light to make carbohydrates. This process converts solar energy into chemical energy in the form of organic substances which can then be used by higher organisms in the food chain. Second, water is used for transport in plants, being drawn in by the plant roots and evaporated out through the plant leaves. Much more water is involved in this process than in photosynthesis. Similarly, water makes up a large part of the human body, where it transports dissolved and solid materials.

In this book we shall be concerned with the role of water in the global water cycle, or *hydrologic cycle*. Water moves from the atmosphere to the land surface as rain and snow. From there it passes downward through the soil into the underground and eventually makes its way into rivers where it flows to the ocean. From the ocean it is evaporated into the atmosphere and some of it is returned as rain to the oceans while some makes its way back to the continents where it falls again as rain and snow. As it passes around and around in the hydrologic cycle, water undergoes chemical reactions with atmospheric gases, rocks, plants, and other substances, resulting in changes in its chemical composition as well as profound changes in the substances with which it reacts. In fact, these changes ultimately act to maintain the overall chemical and physical conditions at the earth's surface.

This book is concerned with the origin of the principal constituents dissolved in and transported by natural waters. These include major components of both rocks and life: sodium, potassium, calcium, magnesium, silicon, carbon, nitrogen, sulfur, phosphorus, chlorine, and, of course, hydrogen and oxygen. It will be seen that global chemical cycles of these elements at the earth's surface are intimately interconnected with the hydrologic cycle. We shall also point out how the cycles of some of these elements, on a worldwide basis, have been perturbed by humans. Thus, environmental problems such as acid rain and eutrophic lakes are also tied up with the water cycle. Although they are of geochemical and environmental interest, we shall not be concerned with minor and trace elements (e.g., lead and mercury) or exotic synthetic chemicals (e.g., fluorocarbons and PCBs) since our goal is not that of all-inclusive environmental coverage. (The interested reader is referred to books such as those by Garrels, Mackenzie, and Hunt [1975], Horne [1978] and Raiswell et al. [1980] for detailed discussion of environmental problems.) Rather, we shall be concerned with water and how it interacts with life and with the major components of the land, the sea, and the air.

Water also plays an important role in heat transfer on the earth through the movement of water vapor in the atmosphere and its condensation and evaporation, and through currents in the oceans such as the Gulf Stream. In addition, water acts as a heat regulator in the atmosphere by absorbing outgoing earth radiation. Thus, water exerts a major influence on climate. These aspects of natural water are important, and so the present book will also discuss some physical aspects of the hydrologic cycle along with the chemical aspects.

Water, then, is an important substance with remarkable and unique properties. Before going on to discuss water in the hydrologic cycle, it is critical that we know something about these properties and how they arise. Thus, the remainder of this chapter is devoted to the fundamental physics and chemistry of water itself.

WATER STRUCTURE

What does water look like and why does it have such unusual properties? The reason why liquid water is so unusual lies in the way the molecules of liquid water associate with one another to form an ordered structure.

Structure of the Water Molecule

The individual water molecule, H_2O, consists of two hydrogen atoms and one oxygen atom. The hydrogen atoms are joined to the oxygen by means of two covalent bonds which, in terms of the simple dot structural formula, can be represented as

$$H:\ddot{O}: \quad \text{or} \quad H—\ddot{O}:$$
$$\dot{H} \qquad\qquad \phantom{H—\ddot{O}:}|$$
$$\qquad\qquad\qquad\qquad H$$

where the dots represent valence electrons and the short lines represent covalent bonds. Note that oxygen in water has two shared (or bonded) sets of electrons and two unshared sets.

In three dimensions, the geometry of the water molecule can be visualized as shown in Figure 1.1 (Horne 1969). The electron cloud of the water molecule forms a "jack," which can be visualized as being contained in a distorted cube with the oxygen in the center. The two lobes containing the hydrogen atoms point to diagonal corners of one face of the cube (hidden bottom face in Figure 1.1) and the two lobes containing the unbonded electrons point toward the opposite two corners of the opposite (top) side of the cube. In other words, the water molecule is shaped like the skeleton of a tetrahedron. However, the H—O—H angle is 104.5 °, which is less than the 109.5° tetrahedral angle that would result if the cube (and tetrahedron) were perfect. This structure of hydrogen-containing and hy-

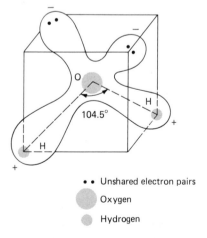

•• Unshared electron pairs

Oxygen

Hydrogen

Figure 1.1. Electron cloud of the water molecule showing hydrogen and oxygen nucleii and unshared electrons. The "side of the jack" with hydrogens is positive and the other "side" is negative, giving rise to the dipolar character of water. The H—O—H bond angle is 104.5°. (Modified from R. A. Horne, *Marine Chemistry*. Copyright © 1969. John Wiley & Sons, Inc. Reprinted by permission of John Wiley & Sons, Inc.)

drogen-free lobes pointing in opposite directions results in an uneven charge distribution. The oxygen "side" of the molecule with the unbonded electrons (top face) is more negative than the hydrogen "side" (bottom face). In this way the water molecule acts like an electrical *dipole* with a negative (oxygen) end and a positive (hydrogen) end.

Hydrogen Bonding

Because the water molecule acts as a dipole, it forms so-called *hydrogen bonds* with other water molecules. This hydrogen bonding is what gives water its unusual properties. A positively charged hydrogen atom in one water molecule is attracted to one of the unshared, negatively charged electron pairs in another water molecule to form a hydrogen bond. This bond is about 10 to 50 times weaker than the covalent bond between hydrogen and oxygen within each molecule, but it is still strong enough to cause water molecules to cluster together. In ice, or solid H_2O, this clustering is so strong that a well-ordered structure results with water molecules arranged around hydrogen bonds in tetrahedral, or fourfold, coordination. This tetrahedral coordination is illustrated in Figure 1.2

In liquid water the molecules are not perfectly ordered as they are in ice, but some ordering, due to hydrogen bonding, is still present. It is because of this tetrahedral ordering that water has such different properties from most other liquids which do not undergo hydrogen bonding. As ice melts, hydrogen bonds are broken or stretched and, as a result, the water molecules in the liquid pack more closely together and the average number of nearest neighbors for any given molecule is about 4.4 rather than 4 as in ice (Narten, Danford, and Levy 1967). This closer packing accounts for the well-known increase in density upon melting (see next section).

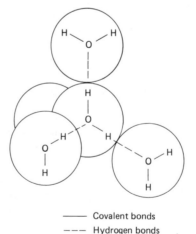

——— Covalent bonds
– – – Hydrogen bonds

Figure 1.2 Tetrahedral arrangement of the water molecules in ice. Each circle (sphere) represents a water molecule. (After R. A. Horne, *Marine Chemistry*. Copyright © 1969. John Wiley & Sons, Inc. Reprinted by permission of John Wiley & Sons, Inc.)

The exact structure of liquid water is a matter of some debate. (For a discussion, see Eisenberg and Kauzmann 1969 and Frank 1972.) According to one set of models, the so-called *mixture models*, the structure of water is a mixture of several distinguishable types of water molecules. In the simplest case, there are two types: non-hydrogen-bonded molecules and molecules in clusters. The cluster molecules are hydrogen-bonded to four neighboring molecules and the clusters are constantly breaking apart and reforming along the hydrogen bonds. Some mixture models call for as many as five different kinds of water molecules with from zero to four hydrogen bonds per molecule. In the mixture models, rising temperature leads to increases in the proportion of the non-hydrogen-bonded molecules.

An alternative model of water structure, the *interstitial model*, envisions one species of water molecule forming a hydrogen-bonded framework with cavities containing single non-hydrogen-bonded molecules. (This is actually a type of mixture model.) As ice melts, some of the water molecules break their hydrogen bonds with the lattice and move into neighboring cavities.

An additional set of water structure models, the so-called *distorted hydrogen-bond* or *uniformist* models, envision that the majority of hydrogen bonds are distorted or bent rather than broken. Tetrahedral hydrogen bonding to four other water molecules occurs but the bonds are distorted and form an irregular network of rings instead of an ordered lattice as in ice.

PHYSICAL AND CHEMICAL PROPERTIES OF LIQUID WATER

Table 1.1 lists the principal physical and chemical properties of liquid water. As can be seen from the table, water is anomalous in almost all of its properties when compared to other liquids. These unusual properties give to natural waters their distinctive behavior. In this section we shall discuss the different physical and chemical properties of water and see how these properties are derived from the structure of water and how they affect the behavior of water under natural conditions. (For further details on water properties and structure see Gymer 1977.)

Density

Looking at a graph of the density of pure water versus temperature, shown here in Figure 1.3, several important observations can be made: (1) at 0 °C, solid water (ice) is less dense than liquid water; (2) from 0° to 4° C, liquid water becomes heavier until it reaches its maximum density at 4° C; (3) above 4° C, the density decreases with increasing temperature. This is not normal behavior when compared to most other substances. Usually the solid form of a substance is denser than its liquid form and when the liquid is heated it continually becomes less dense. Why does water behave so strangely, and what are the implications of this behavior in our study of natural waters?

TABLE 1.1   Physical and Chemical Properties of Liquid Water

Property	Comparison with Other Substances	Importance to Environment
Density	Maximum density at 4° C, not at freezing point; expands upon freezing. Both properties unusual.	In lakes, prevents freezing up and causes seasonal stratification.
Melting and boiling points	Abnormally high.	Permits water to exist as a liquid at earth's surface.
Heat capacity	Highest of any liquid except ammonia.	Moderates temperature by preventing extremes.
Heat of vaporization	One of the highest known.	Important to heat transfer in atmosphere and oceans; moderates temperature extremes.
Surface tension	Very high.	Regulates drop formation in clouds and rain.
Absorption of radiation	Large in infrared and ultraviolet regions; less in visible regions.	Important control on biological activity (photosynthesis) in water bodies and on atmospheric temperature.
Solvent properties	Excellent solvent for ionic salts and polar molecules because of dipolar nature.	Important in transfer of dissolved substances in hydrological cycle and in biological systems.

Source: Modified from Sverdrup, Johnson, and Fleming 1942.

The reason for the strange density behavior of water lies in its structure and hydrogen bonding (Pauling 1953). As mentioned earlier, ice, due to complete hydrogen bonding, has an open structure which gives it a low density. When ice melts, some of the hydrogen bonds are broken or bent, and the resulting non-hydrogen-bonded water molecules crowd more closely together. In this way a given volume of liquid water contains more water molecules than the same volume of ice, resulting in a higher density for the liquid. Consequently, ice floats on water of the same temperature. In addition, upon freezing, water expands to assume the more open structure of ice.

As liquid water is heated from 0° to 4° C, hydrogen bonds break or bend and the water molecules pack together more closely, causing the water volume to contract. Less and less of the water has an ice-like structure. At 4° C (strictly speaking, 3.94° C) water attains its maximum density, and above 4° C it begins to behave like a normal liquid. In general, as a liquid is heated, its thermal energy is increased and the molecules bounce around more, occupying more space. In the case of water between 0° and 4° C, this effect is overshadowed by the breaking or bending of hydrogen bonds. Above 4° C, however, the tendency to move around overcomes the tendency to crowd together, and heating brings about an

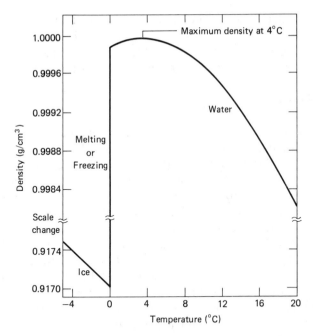

Figure 1.3 Density of water (and ice) as a function of temperature. Note maximum density of water at 4° C. (Data from Pauling 1953 and Hutchinson 1957: 204.)

overall drop in density. In this way the density of liquid water continuously decreases from 4° C up to its boiling point at 100° C.

The density characteristics of water are very important for freshwater lakes in mid-latitudes (see Chapter 6). If water became more dense when it froze, ice in lakes would sink to the bottom and the lakes would freeze up solid during the winter. This does not happen; instead, ice is less dense than liquid water and floats on it. As a result, ice at the surface of a lake acts as a thermal blanket, reducing heat loss from the underlying water during the winter and preventing total freeze-up of the lake. Also, because fresh water has a maximum density at 4° C, the attainment of temperatures both lower and higher than 4° in surface water can lead to stable stratification of lakes (layering of lighter over heavier water) in both winter and summer with accompanying phenomena such as loss of oxygen and build-up of nutrients at depth (see Chapter 6).

Seawater does not behave like pure water or fresh water. Because of its high salt content and high density (1.025), it undergoes only a continuous decrease in density with increasing temperature and there is no density maximum at 4° C (Sverdrup, Johnson, and Fleming 1942). Thus, it behaves like a "normal" liquid, and those phenomena that result from the presence of a 4° C density maximum, such as are found in lakes, are not present in the oceans.

The fact that water expands on freezing is important in the *physical weathering* of rocks (mechanical rock breakdown without chemical changes). Since the volume of water increases by 9% upon freezing, water that is confined in a crack in a rock tends, upon freezing, to force the rock apart. Repeated freezing and thawing are quite effective in shattering rock as is evidenced by the rubble that covers high mountaintops and by the potholes found in road pavement after the winter. The process of physical weathering is most effective in cool temperate climates because the temperature goes above and below freezing many times over the winter.

Melting and Boiling Points

The melting and boiling points of water are abnormal if compared with similar substances. Figure 1.4 shows the melting and boiling points of H_2O, H_2S, H_2Se, and H_2Te as a function of molecular weight. (Oxygen, sulfur, selenium, and tellurium all belong to group VI-A of the periodic table.) We would normally expect the melting and boiling points to occur at lower temperatures as molecular weight is decreased. This is because the normal attraction between molecules (the van der Waals force) is proportional to molecular weight and it is this attraction that holds a molecular solid or liquid together (Pauling 1953). This normal behavior is exhibited by H_2S, H_2Se, and H_2Te, but not by H_2O. If water behaved like the other substances, it would boil at $-80°$ C instead of $100°$ C! The large difference of $180°$ C is due to the need, before removal by boiling, to break the strong hydrogen bonds in liquid water that act to keep the H_2O molecules together. For similar reasons, the melting point of water is about $100°$ C higher than that predicted from the behavior of other substances.

The effect of the higher boiling point of water and the larger difference between melting and boiling points is to produce a liquid over most of the earth's surface. As pointed out earlier, this makes the earth unique when compared to other planets and permits life to exist.

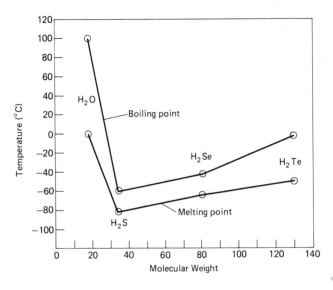

Figure 1.4 Plot of the boiling and melting points of H_2O, H_2S, H_2Se, and H_2Te versus molecular weight. Note the anomalous behavior of H_2O as compared to the other substances. (Adapted from R. A. Horne. "The Structure of Water and Aqueous Solutions," *Survey Progress in Chemistry*, 4, p. 3. © 1968 by Academic Press, reprinted by permission of the publisher.)

Heat Capacity

Heat capacity is the amount of heat required to raise the temperature of a specific amount of a substance by one degree. Water has the highest heat capacity of any known liquid except ammonia. This is because a lot of heat added to water is used to break hydrogen bonds and not to increase the average energy of motion of the constituent water molecules, the latter process constituting what is normally thought of as temperature change. Because of hydrogen bond breaking, water can absorb a lot of heat for a small temperature change.

The temperatures of large standing bodies of water such as lakes and the oceans are relatively constant, thanks partly to the high heat capacity of water. This thermal buffering serves to protect life from otherwise possibly lethal temperature fluctuations.

Heat of Vaporization

The heat of vaporization of water is 540 cal/cm^3 at 100° C, one of the highest of all liquids. (Hydrogen peroxide is higher but it also has hydrogen bonding.) This is the heat that must be supplied to convert liquid water to water vapor. It is the energy necessary for a water molecule to escape the attracting forces of the liquid, which involves breaking hydrogen bonds in addition to overcoming normal molecular attractions. When water condenses from vapor to liquid, it releases this same amount of heat.

The heat transfer of the earth, which drives the water cycle (see Chapter 2), is dependent upon the heat of vaporization and condensation of water. Liquid water absorbs heat at one place on the earth and evaporates; then the water vapor is transported by the atmosphere to another place where it is cooled and condenses, releasing heat. Thus, the high heat of vaporization of water means that a lot of heat is transported by this process. Evaporation also dissipates much of the heat of solar radiation (around 23%), thus moderating and stabilizing the earth's surface temperature without appreciably changing ocean temperatures.

Surface Tension

Water has a very high surface tension, much higher than most liquids. Surface tension is a measure of how strongly the molecules of a liquid are attracted to each other as compared to molecules of another substance across an interface. The hydrogen bonds in water cause it to be very cohesive and thereby to have a high surface tension. This plays a major role in the formation of drops in clouds and in rain.

Absorption of Radiation

Liquid water is fairly transparent to visible light except for some absorption at the red end of the spectrum. This allows light, necessary for photosynthesis, to penetrate to a considerable depth in large water bodies (up to more than 100 m in clear water), but with a shift in color at depth to blue-green due to the selective absorption of red light. By contrast, infrared and ultraviolet radiation (the latter being deleterious to life) are strongly absorbed, with the infrared absorption being due to the hydrogen bonds in the water. Water vapor in the atmosphere also absorbs strongly in the infrared region, which is important to the heat budget of the earth (see Chapter 2).

Water as a Solvent

Water is an excellent solvent. This property is essential in biological systems where water has a transport function. Two-thirds of the human body weight consists of water, which carries dissolved nutrients and eliminates wastes. In lakes and the ocean, life is maintained by the ability of water to supply food and oxygen to organisms and to remove wastes.

Water's solvent properties are also vital to transport in the geochemical cycle. Water, with its dissolved constituents (oxygen, carbon dioxide, humic acids, etc.), weathers rocks and carries the resulting ions via rivers to the ocean. Seawater owes its salinity to the ability of water to retain various ions in solution, particularly Na^+ and Cl^-.

Water is a good solvent for salts such as NaCl because it consists of *dipoles*. The hydrogen end of each water molecule has a net positive charge (see Figure 1.1) and the oxygen end a net negative charge. Na^+ and Cl^- in an ionic salt such as NaCl are very strongly held together by electrostatic attraction, yet NaCl will dissolve easily in water. This is because the attraction of the ends of the polar water molecules for the oppositely charged ions is greater than the attraction of the ions for one another. In addition to tearing apart the Na^+ and Cl^- ions when the salt dissolves, water is able to keep them as separate ions, despite the fact they would be expected to reunite.

The property of water that is an indication of its ability to keep ions in solution is the *dielectric constant*. This parameter is a measure of how much a substance reduces the attraction between charged particles. In the case of water, its highly dipolar nature reduces the electrical attraction between oppositely charged ions to 1/80 of what it would be in the absence of the water; in other words, water has a high dielectric constant (Pauling 1953). Surrounding each ion there is a layer of oriented water molecules that screens the interionic attracton; in this way ions in aqueous solution are said to be hydrated. Such hydration can be so strong that water molecules may accompany ions when they combine to form precipitated salts. A common example is the mineral gypsum, $CaSO_4 \cdot 2H_2O$ which contains waters of hydration within its crystal structure. Hydration also leads to a decrease in volume when salts are added to solution due to crowding of water molecules around each constituent ion.

Besides ions, water is also an excellent solvent for polar molecules. Many polar organic and inorganic substances are soluble in water because they can form hydrogen bonds with the water dipoles. Examples are sugars, alcohols, amino acids, and ammonia. The ability of water to dissolve polar organic molecules is essential both to life processes and to such geochemical processes as weathering.

With this (admittedly brief) description of the structure and properties of water and the importance of these properties to the behavior of natural waters, we are now ready to discuss the circulation of water near the surface of the earth, in other words, the hydrologic cycle.

REFERENCES

EISENBERG, D., and W. KAUZMANN. 1969. *The structure and properties of water.* Oxford: Clarendon Press. 296 pp.

FRANK, H. S. 1972. Structural Models. In *Water: A comprehensive treatise,* ed. F. Franks, vol 1: 1515–1543. New York: Plenum Press.

GARRELS, R. M., F. T. MACKENZIE and C. HUNT. 1975. *Chemical cycles and the global environment: Assessing human influences.* Los Altos, Calif.: William Kaufman, Inc., 206 pp.

GYMER, R. G. 1977. *Chemistry in the natural world.* Lexington, Mass.: D.C. Heath. 573 pp.

HORNE, R. A. 1968. The structure of water and aqueous solutions. In *Surv. Prog. Chem.,* vol. 4: 1–43.

—— 1969. *Marine chemistry.* New York: Wiley-Interscience. 568 pp.

—— 1978. *The chemistry of our environment.* New York: Wiley-Interscience. 869 pp.

HUTCHINSON, G. E. 1957. *A treatise on limnology,* vol. 1. New York: John Wiley. 1015 pp.

NARTEN, A. H., M. D. DANFORD and H. A. LEVY. 1967. X-Ray diffraction study of liquid water in the temperature range 4–200° C, *Discussion Faraday Soc.* 43: 97–107.

PAULING, L. 1953. *General chemistry,* 2nd ed. San Francisco: W. H. Freeman. 710 pp.

RAISWELL, R., P. BRIMBLECOMBE, D. L. DENT, and P. S. LISS. 1980. *Environmental chemistry.* London: Edward Arnold. 184 pp.

SVERDRUP, H. V., M. W. JOHNSON, and R. H. FLEMING. 1942. *The oceans.* Englewood Cliffs, N.J.: Prentice-Hall. 1087 pp.

2

The Water and Energy Cycles:

Oceanic and Atmospheric
Circulation and
the Greenhouse Effect

INTRODUCTION: MAJOR WATER MASSES

Water is by far the most abundant substance at the earth's surface. There are 1.4 billion km^3 of it in its three phases: liquid water, ice, and water vapor. As shown in Table 2.1, most of the earth's water (97.3%) is stored as seawater in the oceans. The remaining 2.7% is either on the continents or in the atmosphere. The amount of water in the atmosphere, in the form of water vapor, is very small in comparison with the other reservoirs, equivalent to only $0.013 \times 10^6 \ km^3$ of liquid water, or around 0.001% of the total. However, it plays a very important role in the water cycle as we shall see.

Of the fresh water stored on the continents, around three quarters is in the form of ice in polar ice caps and glaciers ($29 \times 10^6 \ km^3$). Most of the rest of the continental water is present either as subsurface groundwater ($9.5 \times 10^6 \ km^3$) or in lakes and rivers ($0.1 \times 10^6 \ km^3$). It is this small part of the earth's total water (0.7%) that humans draw on for their water supplies. In addition to the more accessible groundwater (above a depth of 4000 m), there is also at greater depths interstitial water in rock pores. Estimates vary for this deep water: Ambroggi (1977) gives $53 \times 10^6 \ km^3$ whereas Garrels, Mackenzie, and Hunt (1975) give $320 \times 10^6 \ km^3$. Whatever the volume, it will not be discussed here. Instead, we shall focus on water near the earth's surface and how it moves within and between the various reservoirs.

TABLE 2.1   Inventory of Water at the Earth's Surface

Reservoir	Volume (10^6 km^3)	Percent of Total
Oceans	1370	97.25
Ice caps and glaciers	29	2.05
Deep groundwater (750–4000 m)	5.3	0.38
Shallow groundwater (<750 m)	4.2	0.30
Lakes	0.125	0.01
Soil moisture	0.065	0.005
Atmosphere[a]	0.013	0.001
Rivers	0.0017	0.0001
Biosphere	0.0006	0.00004
Total	1408.7	100

Sources: Nace 1967; Peixoto and Kettani 1973; Turekian 1976; and Ambroggi 1977.

[a]As liquid equivalent of water vapor.

Because of their importance in the hydrologic cycle, emphasis will be placed in this chapter on the atmosphere and the oceans, and how their circulation takes place. This also includes a discussion of the energy cycle of the earth and its role in meteorology and oceanography. Finally, because of recent anthropogenic perturbations, the topic of CO_2 in the atmosphere and its effect on climate and the hydrologic cycle (the so-called atmospheric *greenhouse effect*) will be given special attention.

THE WATER CYCLE

Fluxes between Reservoirs

Water does not remain in any one reservoir, but is continually moving from one place to another in the *hydrologic cycle*. This is illustrated in Figure 2.1. (For a more detailed discussion see Penman 1970 and Baumgartner and Reichel 1975.) Water is evaporated from the oceans and the land into the atmosphere where it remains for only a short time, on the average about ten days, before falling back to the surface as snow or rain. Part of the water falling onto the continents runs off in rivers and, in some places, accumulates temporarily in lakes. Some also passes underground only to emerge later in rivers and lakes. (Little groundwater flows directly to the ocean.) The remaining portion of the precipitation on the continents is returned directly to the atmosphere via evaporation. Over the oceans, evaporation exceeds precipitation with the difference being made up by runoff from

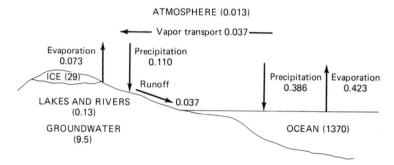

Figure 2.1   The hydrologic cycle.   Numbers in parentheses represent inventories (in 10^6 km³) for each reservoir.   Fluxes are in 10^6 km³ per year.   (Data from Tables 2.1 and 2.2).

the continents.   An idea of the sizes of these various fluxes of water (volumes transported per unit time) is given by the data of Table 2.2 which is also shown in Figure 2.1.

In order to conserve total water, evaporation must balance precipitation for the earth as a whole since the total mass of water at the earth's surface is believed to be constant over time (see section below on the origin of water).   The average global precipitation rate (Table 2.2), which is equal to the evaporation rate, is 496,000 km³/yr.   For any one portion of the earth, by contrast, evaporation and

TABLE 2.2   Fluxes in the Water Cycle with Methods of Determination

Process	Flux km³/yr	Flux cm/yr[a]	Source
Precipitation on land	110,300	74	Lvovitch 1973
Evaporation from land	72,900	49	(Precipitation on land minus runoff)
Runoff from land (river runoff and direct groundwater discharge to the ocean of ~ 6% of total)	37,400	25	Baumgartner and Reichel 1975; groundwater discharge, Meybeck 1986
Precipitation on oceans	385,700	107	(Total precipitation minus precipitation on land)
Evaporation from oceans	423,100	117	(Precipitation on oceans minus runoff)
Total precipitation on earth	496,000	97	Baumgartner and Reichel 1975
Total evaporation on earth	496,000	97	Baumgartner and Reichel 1975 (equal to total precipitation)

Note: Because of use of different areas, values in cm/yr do not balance between land and oceans.

[a] Fluxes in cm/yr calculated on the following basis: area of earth = 510×10^6 km² (total evaporation and precipitation); area of oceans = 362×10^6 km² (precipitation and evaporation over oceans); and area of land = 148×10^6 km² (runoff, precipitation and evaporation over land).

precipitation generally do not balance.   On the land, or continental part of the earth, the precipitation rate (110,300 km^3/yr) exceeds the evaporation rate (72,900 km^3 yr) whereas over the oceans evaporation (423,100 km^3/yr) dominates over precipitation (385,700 km^3/yr).   The difference in each case (37,400 km^3/yr) comprises water transported from the oceans to the continents as atmospheric water vapor, or that returned to the oceans as river runoff and a minor amount of direct groundwater discharge to the oceans.   Because the evaporation rates over many areas of the earth have not been measured, the values given in Table 2.2 are based, by necessity, on the difference between measured worldwide precipitation and river runoff values.   (Recently attempts have been made to obtain evaporation rates by other means, including atmospheric water vapor balance and heat balance; see section below on Radiation and Energy Balance.)

Assumption of constant volume of water in a given reservoir enables the use of the concept of *residence time*.   The residence time is defined as the volume of water in a reservoir divided by the rate of addition (or loss) of water to (from) it. It can be thought of as the average time a water molecule spends in a reservoir. For the oceans, the volume of water present (1370 $\times$ 10^6 km^3; see Table 2.1) divided by the rate of river runoff to the oceans (0.037 $\times$ 10^6 km^3/yr; Table 2.2) gives a residence time of 37,000 years.   This long residence time, which can also be thought of as a filling or replacement time, reflects the very large volume of water in the oceans.   By contrast, the residence time of water in the atmosphere, relative to evaporation from both the oceans and the continents, is only ten days. Lakes, rivers, glaciers, and shallow groundwater have residence times lying between these two extremes, but, because of extreme variability, no simple average residence time can be given for each of these reservoirs.

Geographic Variations in Precipitation and Evaporation

The values listed in Table 2.2 are only average values for precipitation and evaporation over the continents and oceans.   From one region to another there is considerable variation, as can be seen in Figure 2.2, which shows the mean annual precipitation for different areas of the continents.   In order to have rain or snow, there must be both sufficient water vapor in the atmosphere and rising air that can carry the water vapor up to a height where it is cold enough for condensation and precipitation to occur (see Chapter 3).   Net precipitation (precipitation minus evaporation), as shown in Figure 2.3, is highest near the equator (10°N to 10°S) and at 35° to 60° north and south latitudes, where there is frequent storm activity with its accompanying air motion.   Net precipitation is lowest in the subtropics (15° to 30° N and S), where the air is stable, and near the poles, which have both stable air and a very low moisture content due to low temperatures.   (However, since there is also very low evaporation near the poles, precipitation can exceed evaporation in certain places, resulting in the formation of the ice caps of Greenland and Antarctica.)

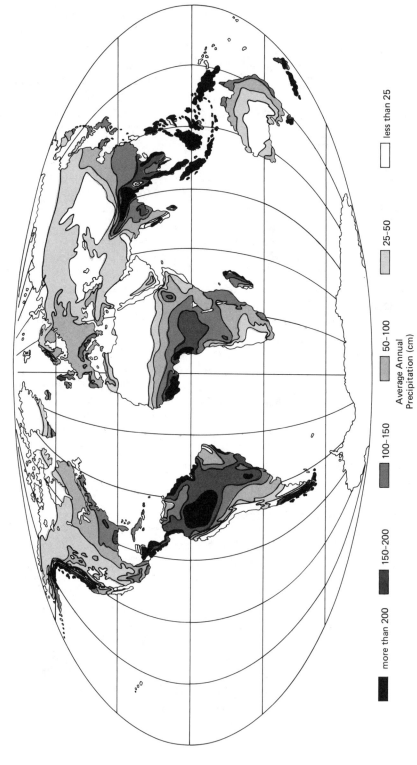

Figure 2.2 Global average annual precipitation. (From H. L. Penman, "The Water Cycle." Copyright © September 1970 by Scientific American, Inc. All rights reserved.)

Average Annual
Precipitation (cm)

more than 200   150–200   100–150   50–100   25–50   less than 25

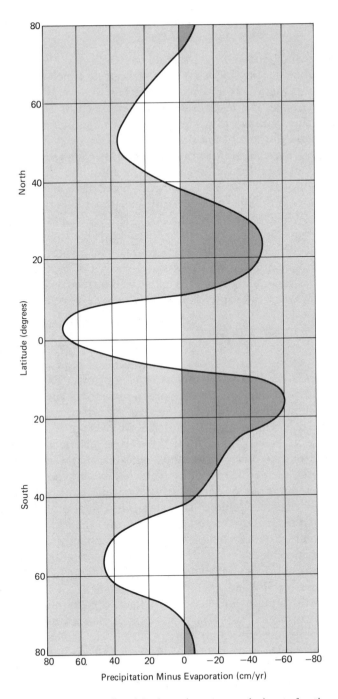

Figure 2.3 Net precipitation (precipitation minus evaporation) as a function of latitude. Positive values represent net precipitation while negative values represent net evaporation. (From J. P. Peixoto and M. A. Kettani, "The Control of the Water Cycle." Copyright © April 1973 by Scientific American, Inc. All rights reserved.)

17

In continental areas of high rainfall, runoff is also high.   Examples are the large rivers of equatorial regions (Amazon, Zaire, Orinoco) and those of middle latitudes such as the Mississippi and Hwang Ho.

Evaporation also varies considerably over the earth's surface.   For net evaporation to occur, there must be a heat source (i.e., radiation from the sun), a low moisture content in the air, and the presence of water available for evaporation. In arid regions the evaporation rate is high but is limited by the availability of water.   As shown in Figure 2.3, overall, evaporation exceeds precipitation at subtropical latitudes (15° to 30° N and S).   Over the continents this leads to the formation of large deserts at these latitudes (for example, the Sahara Desert in Africa and the Great Desert of Australia).   The greatest evaporation rates on the earth (more than 200 cm/yr), however, occur over the subtropical oceans such as over the Gulf Stream in winter where warm water, which is carried northward, encounters cooler, drier air and evaporates.   High evaporation rates can lead to locally higher ocean salinities, such as in the Mediterranean Sea, due to the removal of almost pure water in evaporation, which leaves excess sea salt behind.

Origin of Water on the Earth

As we noted above, it is believed that the total amount of water at the earth's surface has remained fairly constant over geologic time.   This is based on our ideas about the history of the earth and the origin of water.   The earth is thought to be around 4.6 billion years old, based on the ages of meteorites that presumably formed at the same time.   However, the oldest known rocks on earth have been dated as only 3.8 billion years old (Moorbath 1977).   These rocks include sedimentary rocks, which were deposited under water, thus indicating that water existed on the earth at that time.   Of the period between 4.6 and 3.8 billion years ago, some 800 million years, we have no direct evidence for water because no rocks have been found, nor is it likely that they will be, since there is a high probability that they have been destroyed by erosion or metamorphism since that time.   Thus, although we know that water existed 3.8 billion years ago, we can only make an educated guess about its formation based on our ideas about the origin of the earth.   (The following discussion is based on Turekian 1976 and Arrhenius, De, and Alfven 1974; see these references or Holland 1984 for further details.)

The earth formed from a cloud of ionized gas around the sun.   The gas condensed to form small globules or pieces and, once a number of these coalesced to form a protoplanet, more and more of them were drawn into it by gravitational attraction.   The energy of motion of these pieces was changed into heat energy when they collided with the earth and, as a result, the earth heated up.   A hot earth could not retain water or water vapor, or also a number of other gases such as carbon dioxide, nitrogen, ammonia or methane, because these gases are all volatile, are easily evaporated, and thus, would escape into space.   Thus, at this hot early stage, the earth did not have water or water vapor at its surface.

Because rocks of the earth's crust (outer layers) do not contain appreciable amounts of water, scientists were originally at a loss to explain the origin of water. However, it has been discovered that one group of stony meteorites, the *carbonaceous chondrites*, do contain up to 20% water. These meteorites have water locked up in the structure of minerals, particularly in clay minerals in the form of hydroxyl ion (OH^-). Since the carbonaceous chondrites are debris left over from when the solar system and the earth originally formed, they are believed to hold the clue as to how volatiles like water were added to the earth.

Probably some of the bodies hitting the early earth had a composition like the carbonaceous chondrite meteorites (containing water). Due to impact, and the internal generation of heat by the decay of radioactive elements within the earth, the accumulated chondritic material may have been heated to the point where water vapor was released from the clay minerals and other water-containing silicates (in the so-called *degassing* process). This water vapor would escape to the earth's surface where it would cool and condense, forming liquid water that could then accumulate on the earth. The ability of water vapor and liquid water to accumulate on the earth is determined by the balance between the surface temperature of the earth (the hotter the surface, the more likely molecules of water vapor are to escape) and the gravitational attraction of the earth on water vapor (the larger the earth, the stronger its attraction). Thus, a large, cool earth would serve as a better trap for water.

The formation of water, according to the process we have described, took place mainly during the formation of the earth itself, and the ocean would have had essentially its present volume early in earth history. (Henderson-Sellers and Cogley [1982] reach a similar conclusion from models of the radiation balance of the early earth of the type shown in Figure 2.4.) Since that time, only very small additions and losses of water have occurred. Water is lost slowly from the earth by photodissociation to hydrogen and oxygen atoms in the upper atmosphere. This is brought about by ultraviolet radiation, which is more intense at high altitudes where it has not been absorbed by the earth's atmosphere. After the water breaks down, the hydrogen escapes to space while oxygen remains behind. Almost all water vapor is prevented from escaping from the atmosphere via this process by the so-called *cold trap* at about 15 km above the ground. At this altitude the atmosphere becomes cold enough that practically all water vapor is condensed and thereby returned to lower altitudes where it cannot undergo photodissociation. As a result, only about 4.8×10^{-4} km^3 of water (as water vapor) are destroyed each year by photodissociation (Walker 1977).

Very small amounts of water (juvenile water) are also added to the earth's surface by continued degassing from deep within the earth. The amount cannot be very great; Garrels and Mackenzie (1971) and Walker (1977) estimate that the *extreme maximum* rate of degassing is only 0.3 km^3/yr, which is negligible compared to other fluxes in the hydrologic cycle (see Table 2.2). At most, degassing of water vapor can account for no more than 0.001% of the water transported each year from the continents to the oceans and back to the continents.

Relative constancy of total water at the earth's surface through geologic time does not imply constancy of the volume of water in any given reservoir. The distribution of water between land and sea has certainly varied with time. During glacial periods much more water was present on the continents, in the form of ice caps, than at present, whereas during warmer periods practically all of the continents were ice-free. For example, during the last Pleistocene glacial maximum 18,000 years ago, sea level was lowered by around 130 m (Bloom 1971), accounting for a transfer of about 47×10^6 km^3 of water (equal to about 3.5% of the oceanic volume) from the oceans to the land. This is a very large amount of water and constitutes a doubling of the continental reservoir.

ENERGY FOR THE WATER CYCLE: THE ENERGY CYCLE

Introduction

Before considering the atmospheric and oceanic parts of the water cycle, in other words, the fluxes of water vapor in the atmosphere and water in the oceans, we shall take a look at the energy cycle of the earth. This is what drives the water cycle, especially the movement of water vapor in the atmosphere. The atmosphere contains 0.013×10^6 km^3 of water in the form of water vapor at any one time (Table 2.2) and this water remains in the atmosphere for only about ten days on the average. However, during this time it has travelled a mean distance of 1000 km. (Peixoto and Kettani 1973), and this transport is controlled by the energy cycle of the earth. The energy cycle, in turn, is greatly influenced by the presence of water vapor in the atmosphere. Thus, the earth's energy and water cycles are intimately interconnected and they exert strong influences on one another. (For more information on the earth's energy cycle see Oort 1970, Sellers 1965, and Ingersoll 1983.)

Radiation and Energy Balance

The primary energy sources for the earth's surface are summarized in Table 2.3. As can be seen, radiation from the sun is by far the most important source of energy (99.98%) and, consequently, it is the dominant influence on the circulation

TABLE 2.3 Primary Energy Sources for the Surface of the Earth

Source	Energy Flux (cal/cm²/min)	Percent of Total Energy Flux
Solar radiation	0.5	99.98
Heat flow from interior of earth	0.9×10^{-4}	0.018
Tidal energy	0.9×10^{-5}	0.002

Sources: Hubbert 1971; and Flohn 1977.

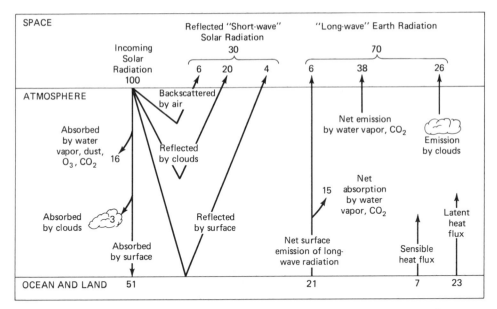

Figure 2.4 The mean annual radiation and heat balance of the atmosphere and earth. Units are assigned so that incoming solar radiation (0.5 cal/cm²/min) is set equal to 100. ("Short-wave" solar radiation is that with <4μm wave length; "long-wave" earth radiation is >4μm). (Adapted from U.S. Committee for the Global Atmospheric Research Program 1975).

of the atmosphere and oceans. It is the only energy source that will be discussed here. The incoming solar radiation impinges on the top of the atmosphere, below which it is reflected, absorbed, and converted into other forms of energy. Apportionment of this energy into different forms is described in terms of the *radiation or energy balance* of the earth (Figure 2.4). (To simplify terminology, we shall express radiation in relative units with 100 units set equal to the solar flux of 0.5 cal/cm²/min.).

As shown in Figure 2.4, incoming solar radiation at the top of the atmosphere is balanced by an equivalent outgoing energy flux from the earth to avoid heating up or cooling of the earth. The assumption of a radiation or energy *balance* is based on the fact that over geologically short periods of time (a hundred to several thousand years) the average temperature (and climate) of the earth have remained constant. Thus, no net energy as heat is gained by the earth, and incoming solar radiation is balanced by outgoing energy loss. Since the earth's climate has changed over longer periods of time, as evidenced by glaciation in the mid-latitudes some 15,000 years ago, the incoming and outgoing earth radiation are not necessarily perfectly balanced over larger time intervals.

The earth's outgoing energy flux is divided into two basic components. About 30 units represent short-wave (<4 μm) solar radiation that has simply been reflected back to space from air, clouds, and the land and oceans, without change in wave-

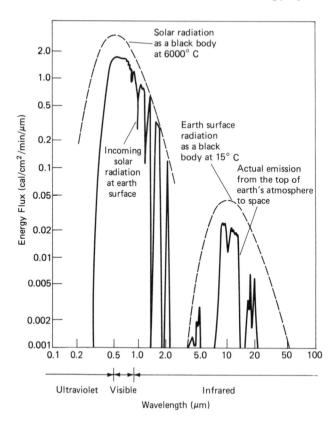

Figure 2.5   Incoming radiation from the sun at the earth's surface and outgoing radiation from the top of the earth's atmosphere as a function of wavelength. Also shown are the energy spectra expected for black bodies at the same temperature as the sun and earth. (A black body emits the maximum radiation for its temperature.) (Adapted from W. D. Sellers, *Physical Climatology*. Copyright © 1965 by The University of Chicago Press. All rights reserved.)

length. The fraction of incoming solar radiation that is reflected, 0.3, or 30%, is known as the earth's *albedo*. It is a measure of how bright the earth would appear if viewed from outer space. The remaining 70 units represent short-wave solar radiation that has been absorbed and reradiated as longer-wave (>4 μm) infrared radiation by the ground and various components of the atmosphere. This 70 units is what is of primary interest to the present discussion because of the interaction between this radiation and the water cycle.

Because the sun is very hot (6000° C), most of the solar radiation is in shorter wavelengths (less than 4 μm) with the peak radiation in the visible wavelengths (0.4–0.7 μm). Much of this visible radiation, or light, penetrates the atmosphere and ultimately reaches the ground. This is important because life is dependent on the absorption of light by photosynthetic organisms. By contrast almost all of the ultraviolet solar radiation (<0.4 μm) is absorbed in the upper atmosphere by ozone and O_2, which protects life from its harmful effects. The sun's incoming long-wave, or infrared radiation is also absorbed at certain wavelengths by atmospheric water vapor and CO_2 and by water droplets in clouds. Thus, the incoming solar radiation reaching the earth's surface has gaps at those wavelengths where atmospheric absorption occurs (see Figure 2.5). Overall, due to the absorption of 19 units by atmospheric ozone and water vapor combined with reflection of 30 units, only 51 units or 51% of the incoming solar radiation actually reaches the earth's surface (Figure 2.4) to be absorbed there.

The surface of the earth reradiates part of its absorbed solar energy back into space and the atmosphere, but since it is cooler than the sun (the average earth surface temperature is 15° C according to Hansen et al. 1981), the earth radiates at longer wavelengths (>4 μm) with a maximum in the infrared at about 10 μm (see Figure 2.5). Atmospheric water vapor is a very good absorber of energy in the infrared wavelengths. Therefore, the earth's infrared radiation that manages to pass through the atmosphere into space differs considerably from that expected for a black body at 15° C (see Figure 2.5) because atmospheric water vapor and carbon dioxide absorb most of the outgoing radiation from the earth's surface and stop it from leaving the atmosphere. This can be seen in Figure 2.4. Of the *net* of 21 units of infrared radiation emitted by the earth's surface, only 6 units are emitted directly to space with the remaining 15 units being absorbed by the atmosphere.

The net absorption of 15 units of outgoing earth radiation by atmospheric water vapor and carbon dioxide actually reflects a more complicated situation, which is important in maintaining the earth's surface temperature. The surface of the earth actually emits approximately 108 units of infrared radiation to the atmosphere where it is absorbed with all but 15 units of this (or 93 units) reradiated back again to the earth's surface where it is reabsorbed, thus keeping the earth warm. The 15 units not returned to the earth's surface are ultimately lost to space. Overall, then, only 21 (15 + 6) units of a much larger total amount of long-wave radiation arriving and leaving the earth's surface are ultimately lost to space. (For further details, see Sellers 1965 and Ingersoll 1983.) To look at this another way, the total amount of long-wave radiation given off by the surface of the earth (108 + 6 = 114 units) is determined by its temperature. However, in order to maintain a *constant* earth temperature, the amount of incoming solar radiation received at the earth's surface (51 units; see Figure 2.4) must be balanced by the *net* loss of long-wave radiation by the earth's surface to space (21 units) combined with the sensible and latent heat given off to the atmosphere (30 units).

The role of atmospheric water vapor (and, to a lesser extent, carbon dioxide) in allowing the incoming short-wave solar radiation to pass through to the earth, while absorbing and reradiating to the earth's surface most of the earth's outgoing long-wave radiation, is referred to as the *atmospheric greenhouse effect* by comparison to the glass in a greenhouse. A greenhouse lets solar radiation in, but keeps the greenhouse warm by preventing long-wave back-radiation from leaving.[1] The greenhouse effect makes the earth much warmer (around 30° C warmer, according to Budyko 1974) than it would be otherwise. Lately, there has been much concern about the atmospheric buildup of carbon dioxide released from fossil fuel

---

[1]Actually a greenhouse is heated more by the trapping of warm air than by the "greenhouse effect."

burning in that it may result in an increase of the earth's temperature by absorbing more than the normal amount of the earth's outgoing radiation. Because of the importance of this problem, we have devoted a special section at the end of the present chapter to the greenhouse problem as it relates to CO_2.

When liquid water is evaporated to form atmospheric water vapor, heat (energy) is absorbed. This is so-called *latent heat* since upon subsequent condensation of the water vapor into rain and snow, the previously added energy is released as heat. Since condensation can occur at great distances from the original site of evaporation, the transport of water vapor in the atmosphere also involves the transport of heat.

Of the 51 units of solar energy absorbed at the earth's surface, 23 units are used to evaporate water, giving rise to a *latent heat flux* from the land and oceans to the atmosphere (see Figure 2.4). Since it takes 588 calories to convert 1 cm³ of liquid water to water vapor at the average earth surface temperature of 15° C, the rate of evaporation corresponding to the absorption of 23 radiation units over the earth's surface (510 × 10⁶ km²) is calculated as follows (after Miller et al. 1983; see also Budyko and Kondratiev 1964):

$$\frac{(0.23) \cdot (0.5 \text{ cal/cm}^2/\text{min}) \cdot (510 \times 10^{16} \text{ cm}^2)}{588 \text{ cal/cm}^3} = 10^{15} \text{ cm}^3/\text{min}$$

This is equivalent to 525,000 km³ of water per year, which agrees quite well with estimates of total annual evaporation from the earth (495,000 km³ from Table 2.2) based on setting the total evaporation equal to measured total precipitation in the global water balance.

The remaining process that transports energy from the earth's surface to the atmosphere is the flux of *sensible heat* by conduction and convection. Heat flows from the ground and sea surface to the air simply because, on the average, they are warmer than the air. This constitutes *conduction*. The heated air tends to rise and be replaced by cooler sinking air and this overall turnover process is known as *convection*. Together conduction and convection comprise 7 units of energy transfer from the earth's surface to the atmosphere (Figure 2.4).

Variations in Solar Radiation: The Atmospheric and Oceanic Heat Engine

The amount of solar radiation absorbed by the earth decreases with latitude from the equator to the poles. It is this variation in the earth's heating that drives the circulation of the ocean and atmosphere and, thus, the hydrologic cycle.

Latitudinal variations in the input of solar energy are due to two factors. First, the earth is a sphere and the angle at which the sun's rays hit its surface varies from 90° (or vertical) near the equator to 0° (or horizontal) near the poles. This is shown in Figure 2.6. Less energy is received at the poles because the same amount of radiation is spread out over a much larger area at high latitudes (compare situation C with situations A and B in Figure 2.6) and because at high latitudes

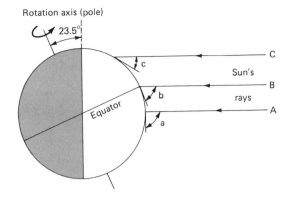

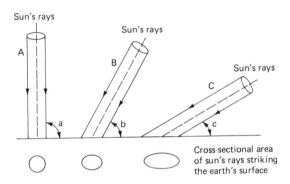

Figure 2.6 Schematic diagrams showing the variations of solar intensity (energy per unit area) with angle of incidence to the earth's surface. Lower angles (higher latitudes) result in the same energy spread out over a larger area and, thus, in a lower intensity of radiation. Scene depicted is for Northern Hemisphere winter. (Modified from A. Miller, et al. *Elements of Meteorology, 4th ed.* Copyright © 1983 by Charles E. Merrill Publ. Co. Reprinted by permission of the publisher.)

the sun's rays must travel through a much greater thickness of atmosphere where more absorption and reflection occur.

The second factor affecting latitudinal variations in heating is the duration of daylight. Because the polar axis of the earth is tilted at an angle of 23.5° with respect to the ecliptic (the plane of the earth's orbit about the sun), we have a progression of *seasons* where the angle of the sun's rays striking any given point varies over the year. This is shown in Figure 2.7.

In the Northern Hemisphere winter (Figure 2.7), no sunlight strikes the area around the North Pole, during a full day's rotation of the earth, because it is in the earth's shadow. Thus, little or no solar heating occurs in this area at this time. Conversely, at the South Pole there is continual daylight, but at a very low sun angle, during the Northern Hemisphere winter. As the seasons shift, the South Polar region eventually becomes plunged into 24-hour darkness (during the Southern Hemisphere winter and Northern Hemisphere summer) just as the North Pole had been earlier. Low-latitude regions near the equator, by contrast, undergo little seasonal change in the duration of daylight whereas intermediate latitudes are subjected to changes intermediate between those of the poles and the equator.

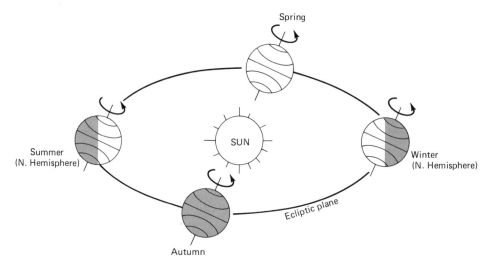

Figure 2.7 Revolution of the earth in its orbit around the sun, showing the changing seasons (also length of day). The seasons given are for the Northern Hemisphere; they are reversed in the Southern Hemisphere. (Modified from A. Miller, et al. *Elements of Meteorology, 4th ed.* Copyright © 1983 by Charles E. Merrill Publ. Co. Reprinted by permission of the publisher.)

Thus, because of seasonality, more annual radiation is received per unit area at lower, as compared to, higher latitudes.

The effects of variation in angle of the sun's rays with latitude and seasonal changes in the amounts of daylight result in strong variation with latitude in the total solar radiation received during a year. Since the long-wave radiation leaving the earth varies little with latitude, there are radiation *imbalances* over the surface of the earth (Figure 2.8). From 40° north and south latitudes to the poles there is a net deficit of radiation (more leaves than enters), whereas from 40° to the equator there is a net surplus (more solar radiation enters than the earth radiates back). To keep the poles from getting colder and the tropics from getting warmer, heat must be transported from lower to higher latitudes. This is accomplished by the circulation of the atmosphere and oceans. Thus the atmosphere and oceans act like a "heat engine" driven by latitudinal variations in solar radiation.

Heat is transported from the equator to the poles in three ways: (1) by ocean currents carrying warm water, (2) by atmospheric circulation (wind) carrying warm air, and (3) by atmospheric circulation carrying latent heat in the form of water vapor. The relative contribution of each of the three processes to total heat transport (Sellers 1965) is shown in Figure 2.9 as a function of latitude.

Warm, wind-driven ocean currents move poleward from the zone between 20°N and 20°S carrying 20%–25% of the heat surplus; they are well known, examples being the Gulf Stream in the North Atlantic and the Kuroshio Current in the North Pacific. The poleward transport of warm water by the oceans tends to warm the overlying atmosphere at higher latitudes, especially during the winter.

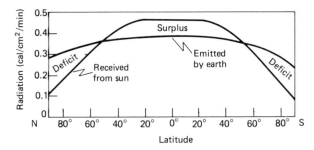

Figure 2.8 Mean annual radiation absorbed from the sun and radiated from the earth to space, as a function of latitude. (Adapted from A. Miller, et al. *Elements of Meteorology, 4th ed.* Copyright © 1983 by Charles E. Merrill Publ. Co. Reprinted by permission of the publisher.)

This provides a moderating influence on climate and helps to explain the relatively mild winters experienced, for example, by western Europe (Gulf Stream) and by Japan (Kuroshio Current).

Most heat (75%–80%) is carried poleward by the atmosphere, as warm air and latent heat. A major source of warm air is the tropical region between 10°N and 10°S, which is the zone of maximum surplus of solar radiation. The tropical air is heated both by sensible heat and by the release of latent heat upon condensation of moisture. (The hot tropics are a zone of both high evaporation and high rainfall with the latter predominating). Latent heat in the form of water vapor is injected into the atmosphere mainly in the subtropic zones (15°–30° N and S) where evaporation exceeds precipitation. From there, poleward transport of latent heat takes place (see Figure 2.9). The latent heat is subsequently released by condensation which warms the atmosphere in the mid-latitude zones of intense storm activity at 30°–50°N and S.

A small part (about 0.7%) of the incoming solar radiation is converted into the energy of motion (*kinetic energy*) of ocean currents, winds, and waves. Al-

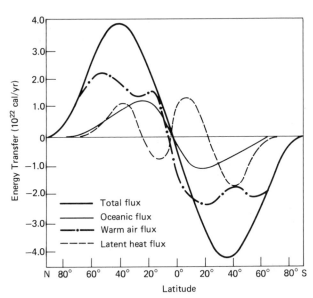

Figure 2.9 Rates of poleward energy transport by various mechanisms. Positive energy transfer values refer to northward transport and negative values to southward transport. (Adapted from W. D. Sellers, *Physical Climatology.* Copyright © 1965 by The University of Chicago Press. All rights reserved.)

though this is a small number, it represents an energy of major interest to the hydrologic cycle, that associated with the circulation of the atmosphere and oceans. This circulation will be discussed next.

CIRCULATION OF THE ATMOSPHERE

The atmosphere circulates as a consequence of the latitudinal heat imbalance discussed above.   If the circulation were due solely to heating, hot air would rise at the equator and flow poleward at high levels.   As the air was cooled in transit and piled up at the poles, it would tend to sink at the poles.   To complete the cycle there would be a return flow near the earth's surface of cool air toward the equator where heating would produce two symmetrical closed circuits, or *cells*, one in the Northern and one in the Southern Hemisphere.   Such a circulation (incorporating the earth's rotation) was originally proposed in 1735 by George Hadley but, because

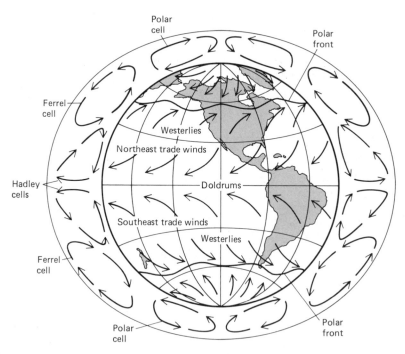

Figure 2.10 Schematic representation of the general circulation of the atmosphere. (Modified from A. Miller, et al. *Elements of Meteorology, 4th ed*.   Copyright © 1983 by Charles E. Merrill Publ. Co.   Reprinted by permission of the publisher.)

of a number of factors, the actual circulation has turned out to be considerably more complicated.

The *general circulation* of the atmosphere, showing the mean annual winds, is depicted in Figure 2.10. Note that the winds do not simply blow along north-south, or meridional lines. This is because they are deflected by the rotation of the earth. The force that deflects moving objects—in this case moving air masses—to the right in the Northern Hemisphere, and to the left in the Southern Hemisphere, is referred to as the *Coriolis force*. The general circulation differs from the simple Hadley circulation by being broken up into several latitudinal zones; however, there is still symmetry more or less between the Northern and Southern Hemispheres. For the sake of brevity, we will discuss only the Northern Hemisphere but what is said applies equally to the Southern Hemisphere (with a leftward instead of rightward directed Coriolis force). (For details of circulation not discussed here, the interested reader is referred to books on meteorology and climatology such as Sellers 1965, Goody and Walker 1972, and Miller et al. 1983.)

Hot moist air rises at the equator and, as it rises, the moisture condenses and intense precipitation results. Here there are only very weak surface winds giving rise to the equatorial *doldrums*. After rising, the air flows northward at high levels, cools, and eventually sinks around 30°N. The descending air is very dry (having lost most of its moisture in the tropics) and, when it is warmed, its capacity to take up moisture is further increased. The resulting hot dry air causes intense evaporation at the earth's surface and this gives rise to the subtropical belt of deserts centered between 15° and 30°N. After reaching the surface, the air flows southward, picking up moisture as it flows over the ocean, and being deflected to the right by the Coriolis force. This surface flow is known as the northeast *trade winds*. Upon reaching the equator the northeast trades converge with the southeast trades from the Southern Hemisphere, and the air rises at the equator to complete the low-latitude cycle known as the *Hadley cell* (which behaves rather as Hadley expected the whole atmospheric circulation to behave).

At around 30°N, additional air descends and then flows north at the surface rather than south. This is the beginning of the *Ferrel cell*. The northward flowing air is deflected to the right, forming the prevailing *westerlies*, which flow from southwest to northeast in the Northern Hemisphere. The westerlies continue until they encounter a cold mass of air moving south from the North Pole at about 50°N. This zone where the air masses meet is known as the *polar front* and it is a region of unstable air, storm activity, and abundant precipitation. The warmer air from the south rises over the polar air and then turns south at high altitude to complete the Ferrel cell. Meanwhile the southward flowing polar air (*polar easterlies*) becomes warmed by condensation at the polar front and by contact with the southern air. As a result, it too rises and then flows northward at high altitude to the pole where it sinks, thus completing the *polar cell*.

In the mid-latitudes, the west-to-east flow (Northern Hemisphere westerlies) is subject to considerable turbulence because of the earth's rotation. At higher levels of the atmosphere, the flow forms waves (so-called *planetary waves*) that transport warm air from the surface to the top of the atmosphere (Goody and Walker 1972; Ingersoll 1983). These waves are expressed, in the lower atmosphere, in a series of storms that travel west to east around the globe, transporting warm air poleward and cool air equatorward and releasing heat by precipitation.

OCEANIC CIRCULATION

Introduction

The oceans can be divided into two portions for the purpose of discussing circulation. The top 50-300 m, or surface layer, is stirred by the wind and is well mixed from top to bottom (Figure 2.11). Below the surface layer the remaining deeper water is colder, less well mixed, and divided into a number of roughly horizontal layers of increasing density. The uppermost portion of the deep water is marked by a region of steeply decreasing temperature gradient, known as the *thermocline* (Figure 2.11). In the surface ocean, lateral circulation is predominantly driven by the wind; in the deep ocean, circulation is driven by density variations due to differences in temperature and salinity, giving rise to the term *thermohaline circulation*. In this section the wind-driven and thermohaline circulations are discussed separately.

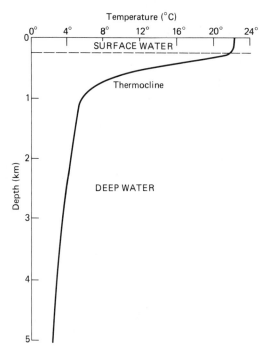

Figure 2.11 Generalized temperature-versus-depth profile for the oceans (at low- to- mid-latitudes) showing vertical stratification into surface and deep-water masses.

Wind-Driven (Shallow) Circulation

The circulation of the shallow ocean is driven by prevailing winds, which are in turn caused by uneven heating of the earth's surface. The circulation pattern can be summarized as a number of current rings or gyres that flow clockwise in the Northern Hemisphere and counterclockwise in the Southern Hemisphere due to the stresses imparted by the prevailing winds. This is shown in Figure 2.12. Each gyre has a strong narrow poleward current on the western side with much weaker currents to the east. The most pronounced of these "westward intensification" currents are found in the Northern Hemisphere as the Gulf Stream in the Atlantic Ocean and the Kuroshio Current in the Pacific.

The circulation pattern shown in Figure 2.12 is not that expected if water were simply carried downwind. Transport of surface water, on the scale of the oceans, involves factors such as the earth's rotation and friction against continents, as well as wind stresses on the surface. The interaction of wind and water is complicated, and only a brief summary will be given here. (For details, the interested reader should consult references on physical oceanography such as Munk 1955, Stommel 1965, von Arx 1962, or Stewart 1969.)

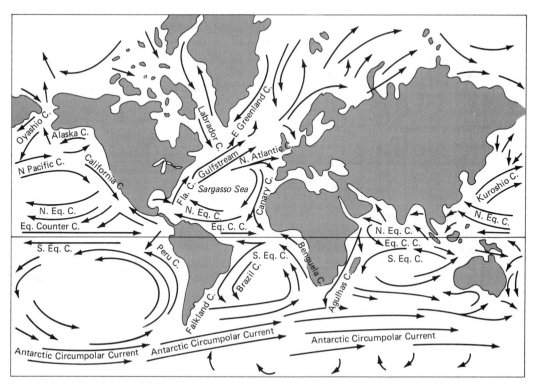

Figure 2.12 Surface currents of the oceans. (After Drake et al. 1978.)

The origin of prominent gyres can best be understood by reference to the North Atlantic. Here the prevailing winds are westerlies (from the west) at 40°–50°N and trade winds (from the east) at 15°–30°N. Because of the Coriolis force (see previous section), water in the top layer does not simply move downwind, but instead is moved to the right of the wind direction (*Ekman flow*). This brings about a convergence or piling up of water from both the north and south into the central portion of the North Atlantic. The piled-up water then sinks and begins to return just below the surface, back to the north and south. As it does, it is turned to the right by the Coriolis force, which results in a strong east-flowing current on the north, and a strong west-flowing current on the south, just below the surface. These so-called *geostrophic* currents run parallel to the wind direction until they encounter the European continent at the northeast portion of the gyre and the North American continent along the southwest portion. Here they must turn. Since friction is strong along the continents and reduces the effect of the Coriolis force, the currents will flow from high pressure (where water is accumulating due to the current) to low pressure (where it is being removed), resulting in a north-to-south-flowing current on the European side and a south-to-north-flowing current (Gulf Stream) on the North American side. Thus a clockwise-flowing gyre results. The current on the west side of the Atlantic becomes intensified because of the super-imposition of an additional pressure gradient (decreasing pressure from south to north due to the variation in Coriolis force) that reinforces the western current and opposes the eastern current.

This type of explanation for the North Atlantic circulation can be applied to the other oceans except that in the Southern Hemisphere the Coriolis force is to the left of the direction of motion and, as a result, the gyres are counterclockwise (see Figure 2.12).

Coastal Upwelling

A special case of the wind-driven circulation is coastal upwelling, which has an important effect on biological productivity in the ocean (see Chapter 8). Major upwelling occurs along the western boundaries of the continents where the surface currents flowing toward the equator are broad and relatively weak. Winds blowing equatorward along the coasts bring about a transport of surface water offshore. This is because the net transport of water (Ekman drift) is to the right of the wind in the Northern Hemisphere and to the left of the wind in the Southern Hemisphere. Transport of surface water away from shore leaves a nearshore deficit that is replaced by deeper water flowing up from below (see Figure 2.13). This deeper water is enriched in nutrients, which results in high planktonic productivity and teeming life, including abundant fish. Some classic examples of upwelling areas are those off Peru and Chile, the bulge of West Africa, southwest Africa, and the California coast.

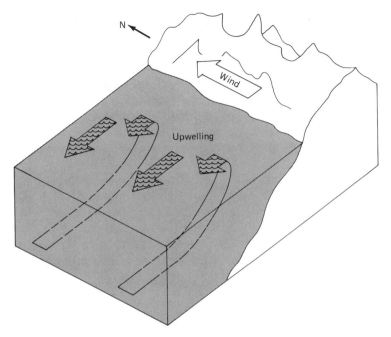

Figure 2.13   Upwelling, or the result of Ekman drift, in response to a north-blowing wind in the Southern Hemisphere. (After K. K. Turekian, *Oceans, 2nd ed.* Copyright © 1976, p. 35. Reprinted by permission of Prentice-Hall, Inc., Englewood Cliffs, N.J.)

Thermohaline (Deep) Circulation

Below the top few hundred meters, the oceans are not directly affected by the winds. Here circulation is brought about by density differences arising from differences in temperature and salinity. (The following discussion of deep ocean circulation is from Pickard and Emery 1982; see also Drake et al. 1978). In seawater, density increases continuously with decrease of temperature and no density maximum, as is found in fresh water at 4° C (see Chapter 1), is encountered. Density also increases with increasing salt content, or salinity. Overall the deep ocean is vertically *stratified* into various *water masses*, with the densest water at the bottom and the lightest at the top, and the density stratification is due primarily to the decrease of temperature with depth. The stratification severely inhibits vertical motion; in other words, it is much easier to move water *along* surfaces of constant density than it is to move it *across* them. Thus, the deep-water circulation can be viewed as primarily horizontal.

The density stratification and density differences between water masses of the deep sea owe their origin to surface processes. Here changes in density are brought about by heating and cooling, evaporation, addition of fresh water, and freezing out of sea ice.   At high latitudes, surface waters become more dense than those at lower latitudes due to cooling and sea-ice formation.   (When sea ice forms, dissolved salts are excluded from the ice and the remaining water becomes more saline and, thus, more dense).   In certain locations in the far North and far South Atlantic, this surface density increase becomes so great that the cold salty surface water occasionally becomes denser than the underlying water and sinks downward to replace it.   In this way deep ocean water originates, and is replenished from above.   Once it reaches great depths, the water tends to conserve its temperature and salinity as it flows laterally throughout the oceans, bringing about the deep-water circulation.   As the bottom water traverses the ocean floor, there is very slow, diffuse upwelling (at the rate of about 1m/yr) of deep water into surface layers.   This slow upwelling is supplemented by the more intense but localized coastal upwelling discussed in the previous section.

An example of stratification and deep circulation for the Atlantic Ocean, is shown in Figure 2.14.   Here each water mass is identified by its characteristic

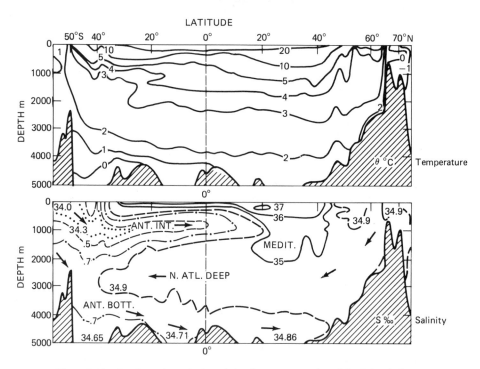

Figure 2.14   South-north vertical sections of water properties of the Atlantic Ocean along the western trough as delineated by lines of constant temperature and salinity. N. Atl. Deep = North Atlantic Deep Water; Ant. Bott. = Antarctic Bottom Water; Ant. Int. = Antarctic Intermediate Water; Medit. = Mediterranean Water. (After Pickard and Emery 1982, based on data from Bainbridge, 1976.)

temperature and salinity. The deep water is dominated by two cold water masses, the North Atlantic Deep Water and the Antarctic Bottom Water. The North Atlantic Deep Water originates in the Norwegian Sea off Greenland from the cooling of Gulf Stream surface water. This water sinks and flows at depth, southward across the equator. In the Antarctic an even denser water, the Antarctic Bottom Water, is formed by cooling and the freezing out of sea ice in the winter. This sinks to the bottom and flows north. In the Antarctic region further north, the Antarctic Intermediate Water is formed which also sinks but not to the bottom because it cannot displace the underlying denser North Atlantic Deep or Antarctic Bottom waters. This water thus occupies intermediate depths as implied by its name. Another intermediate-type water is formed in the Mediterranean Sea. Here intense evaporation causes the water to be sufficiently saline and dense that, upon passing out into the Atlantic through the straits of Gibraltar, it sinks and fills intermediate depths. It does not sink to the bottom because, although it is more saline than North Atlantic Deep Water, it is also warmer and the temperature difference counteracts the salinity effect making it less dense than North Atlantic Deep Water. For the same reason, surface water at lower latitudes remains at the surface, even though it is more saline (see Figure 2.14, bottom); that is, the density-lowering effect of higher temperature overpowers the density-raising effect of higher salinity.

The lateral deep-water circulation over the entire ocean, according to the model of Stommel (1958), is shown in Figure 2.15. (Because it is so slow, deep-water circulation is not well documented and maps, such as that shown in Figure 2.15, are based largely on theoretical models). North Atlantic Deep Water flows away southward from its source as an intense bottom current on the western side of the North Atlantic. This meets a strong northward flowing current of Antarctic Bottom Water in the South Atlantic, and as a result they merge and flow east into the Indian Ocean and ultimately into the Pacific Ocean. Thus, the deep water originates only at two places, and both are in the Atlantic Ocean. Besides the strong currents there are slower deep return flows spiralling around gyres in each ocean basin (Figure 2.15) and diffuse upwelling everywhere as mentioned above.

The deep-water circulation is slow, with average flows of a few kilometers per month, as compared to surface currents of a few kilometers per hour. Thus, dissolved substances at depth remain out of contact with the atmosphere for long periods of time which can result in appreciable changes in the chemical composition of the water (see Chapter 8). The residence time of deep water, or average time it spends out of contact with the atmosphere, is about 200–500 years for the Atlantic and 1000–2000 years for the Pacific (see, e.g., Turekian 1976).

CARBON DIOXIDE AND THE ATMOSPHERIC GREENHOUSE

In this chapter we have discussed both the water cycle and the energy cycle of the earth and how they are interconnected. We shall conclude by considering the rise in the atmospheric concentration of carbon dioxide gas (and certain other gases),

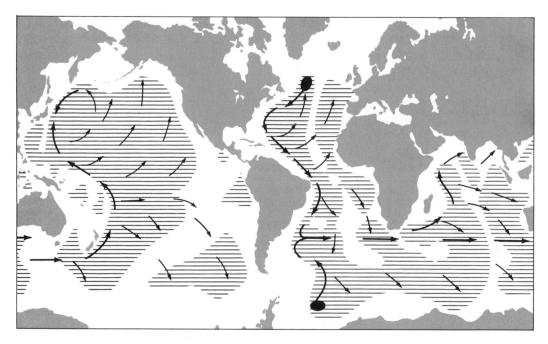

Figure 2.15   Deep (thermohaline) circulation of the ocean.   Deep water originates by sinking at the two locations shown as large black dots and then flows at depth as intensified currents along the western sides of the ocean basins. (After H. Stommel, "Circulation of the Abyss." Copyright © July 1958 by Scientific American, Inc.   All rights reserved.)

which has been caused by humans, and which has a strong potential to alter the earth's energy balance by raising the earth's surface temperature and altering the water cycle.

Although carbon dioxide is the fourth most abundant gas in the atmosphere after nitrogen, oxygen, and argon, it constitutes only some 0.03% (by volume) with nitrogen and oxygen together accounting for 99%.   Atmospheric CO_2 is important for two reasons: (1) it strongly absorbs infrared (long-wave) radiation given off by the earth and reradiates energy back to the earth, thus helping to maintain the earth's surface temperature (the so-called greenhouse effect discussed earlier); and (2) it is a source of carbon, which is the dominant element in life and in the biological cycle of the earth.

Anthropogenic Carbon Dioxide— Present, Past, and Future

Beginning in 1958, the atmospheric concentration of CO_2 has been measured at Mauna Loa in Hawaii (Figure 2.16).   There is an obvious annual oscillation in atmospheric CO_2 concentration of around 6 ppm which is a result of biological cycling, with uptake of CO_2 by plants during the spring and summer due to excess photosynthesis, and its release during the winter due to excess respiration. However, the *yearly average* value of CO_2 has clearly increased from 315 ppm in 1958 to ~339 ppm in 1980 and the CO_2 concentration has been rising faster in the last ten years (at the average rate of 0.4%/year) than it did earlier (NRC 1983).

The rise in atmospheric CO_2 has been attributed to the burning of fossil fuels (coal and oil), which release CO_2 to the atmosphere. The rate of fossil fuel CO_2 release has risen 3.5% per year over the last 120 years and is estimated to rise 1%–2% per year over the next 100 years; thus, this source of CO_2 is obviously increasing (NRC 1983). Deforestation by humans, that is, the burning or organic breakdown of carbon-containing plant material, is another anthropogenic source of atmospheric CO_2, the magnitude of which, both now and in the future, is less well agreed upon.

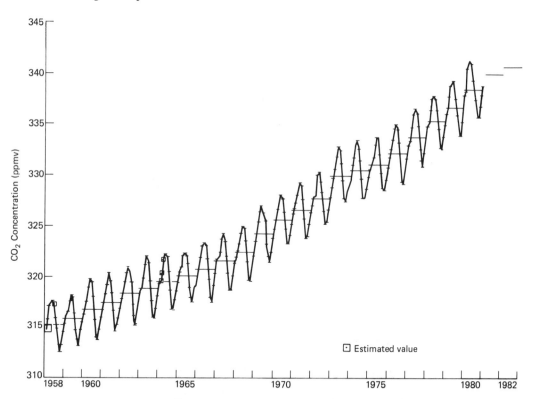

Figure 2.16 Mean monthly concentrations of atmospheric CO_2 at Mauna Loa, Hawaii. The horizontal bar represents the mean annual concentration of CO_2 for each year. The yearly oscillation is explained mainly by the annual cycle of photosynthesis and respiration of plants in the Northern Hemisphere. (Note 1 ppmv CO_2 = 2.12 Gt C, where 1 Gt C = 10^9 tons C). (After NRC 1983, based on measurements of C. D. Keeling and the National Oceanic and Atmospheric Administration. Average annual CO_2 concentrations for 1981 and 1982 from Komhyr et al. 1985.)

Since the amount of fossil fuel CO_2 released from 1959 through 1980 has been estimated as 84.5 Gt C (Marland and Rotty 1984; the symbol Gt refers to gigatons or 10^{15}g), and the atmospheric CO_2-C increase in the same period was only 51 Gt C, about 60% of the fossil fuel carbon added has remained in the atmosphere. Where has the rest of the CO_2 gone? The answer to this question lies in the carbon cycle (see Figure 2.17).

There are two major carbon reservoirs which could take up the missing CO_2 and which are both much larger than the atmosphere and exchange rapidly with it on a time scale of years to hundreds of years. These reservoirs are the oceans and the terrestrial biosphere plus soils. (Carbonate rocks and buried organic matter represent an even larger carbon reservoir, which is important over geologic time but not on a human time scale; see, e.g., Berner, Lasaga, and Garrels 1983). The oceans represent the largest of the rapidly exchanging reservoirs and most of the missing atmospheric fossil fuel CO_2-C has probably been stored in them. Carbon is present in the oceans primarily as inorganic carbon, in the form of dissolved bicarbonate ion (HCO_3^-), and carbonate ion (CO_3^{--}). When atmospheric CO_2 is added to ocean surface water, the following reaction occurs:

$$CO_2 + CO_3^{--} + H_2O \rightarrow 2HCO_3^-$$

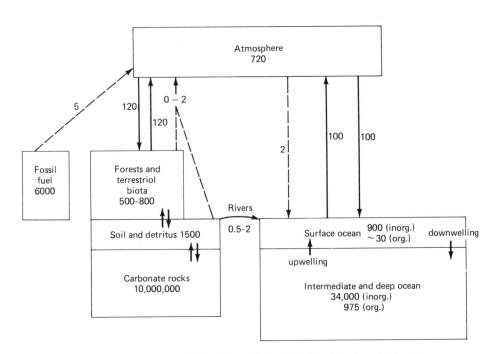

Figure 2.17 The Carbon Cycle. Reservoirs in 10^{15} g C = 10^9 t C = Gt C. Fluxes in Gt C/yr (dashed fluxes due to human activity; solid fluxes, natural). (Modified from NRC 1983; additional data from Bolin et al. 1979.)

Thus, atmospheric CO_2 is converted to HCO_3- ion and stored in the oceans. (For a more thorough discussion of ocean chemistry, see Chapter 8.)

The surface oceans (top 75 m) are well mixed with the atmosphere. Below this lies the thermocline (1000 m thick) where exchange with atmospheric CO_2 is on a time scale of decades, and below the thermocline lies the deep sea which is isolated from the atmosphere and mixes very slowly with it on a time scale of a thousand years (Broecker et al. 1979; for further details on ocean structure see the oceanic section of the present chapter). Thus, on a time scale of years, only the surface ocean takes up CO_2. Because of the solubility of CO_2 and the large amount of carbon in the oceans, the recent increase in atmospheric CO_2 of about 10% should have been accompanied by a rise of only about 1% in the CO_2 stored in the *surface* oceans (Broecker et al. 1979). This is difficult to detect. Since 1958 the oceans are believed to have taken up about 40% of the fossil fuel CO_2 released (Brewer 1983). The amount of CO_2 removed over longer time periods (decades to hundreds of years) depends upon how fast it has been mixed to greater depths.

Some atmospheric CO_2 is also removed by marine organisms as organic matter during photosynthesis. However, most of this CO_2 is released fairly rapidly by organic matter decay and only a very small amount is eventually buried at the ocean bottom and permanently removed. It is probably not a major mechanism for the removal of anthropogenic CO_2 (see Berner 1982).

The other large carbon reservoir and possible missing CO_2 sink is forests and the terrestrial biosphere (500–800 Gt C) plus soils (1500 Gt C). There is rapid annual exchange between the terrestrial biosphere and the atmosphere of ~120 Gt C/yr and this causes the annual oscillation in atmospheric CO_2 concentration mentioned previously. The amplitude of the annual oscillation of atmospheric CO_2 seems to be increasing, which leads some to suspect that the size of the biosphere may be growing in response to increasing atmospheric CO_2 concentration. This would represent increased CO_2 storage in the biosphere. However, Woodwell et al. (1983) and others believe that the rate of net deforestation (deforestation minus forest regrowth) leads to an *addition* of CO_2 to the atmosphere of 1.8–4.7 Gt C/yr. Whether the biosphere represents net addition or loss of CO_2 from the atmosphere is important because it affects projections of atmospheric CO_2 increases in the future.

The annual balance for anthropogenic CO_2 can be expressed as follows (Woodwell 1983):

$$A = F - S - B$$

where A = increase in atmospheric CO_2 per year; F = fossil fuel addition per year; S = oceanic storage per year; and B = loss to the biosphere ($B < 0$ means that there is net gain from deforestation). From the estimates in 1980 (in Gt C/yr) of $A = 2.5 \pm 0.2$ (Bacastow and Keeling 1981); $F = 5.2 \pm 0.5$ (Marland and Rotty 1984); $S = 2.1 \pm 0.4$ (Brewer 1983), we obtain, from the above equation, $B = 0.6 \pm 1.1$ Gt/yr. This means that errors in the calculation allow for a range from a net loss to the biosphere (storage) of 1.7 Gt/yr to a net gain to the atmosphere from deforestation of 0.5 Gt/yr. Elliott, Machta, and Keeling (1985) suggest that net loss to the biosphere is roughly zero, with 0.5 Gt/yr as a maximum storage value, and that changes in atmopheric CO_2 concentration are due solely to fossil fuel burning. By contrast, Peng (1985), based on ^{13}C in tree rings, suggests a net input to the atmosphere from the biosphere of 1 Gt C/yr, which is the value used by Nordhaus and Yohe (NRC 1983). At any rate, the problem of CO_2 uptake or loss from the biosphere is an unsettled issue but it most likely is a smaller flux than the other three.

In order to make projections of future rises in temperature with increasing CO_2, and in order to estimate the size of the biosphere and fossil fuel contributions of CO_2 over the past century, it is necessary to know what the concentration of CO_2 in the atmosphere was *before* human intervention. Estimates of the concentration of CO_2 around 1850, assumed to be the preindustrial or preanthropogenic concentration, range from 250 to 295 ppm, with 260–280 ppm the preferred range (Machta 1983). Back-extrapolation of CO_2 concentration based on fossil fuel emissions alone suggests a value of $\sim$ 290 ppm in the late nineteenth century. Measurements of CO_2 concentrations in ice cores from $\sim$1850 give a lower value of approximately 265 ppm (Neftel et al. 1982; Raynaud and Barnola 1985), as do ocean measurements (Brewer 1983) and studies of ^{13}C in tree rings (Peng 1985). If these measurements of $\sim$265 ppm for 1850 are correct, this implies that the difference of about 25 ppm between these observations and the back-extrapolated concentration from fossil fuel additions alone (290 ppm) represents CO_2 added from the terrestrial biosphere via deforestation during the late nineteenth or early twentieth century (Brewer 1983). Ice core data (Raynaud and Barnola 1985) also suggest that the so-called "preindustrial" CO_2 level of $\sim$260 ppm may not have been constant during the several hundred years before large human perturbations. The data suggest that there may have been a decrease in CO_2 of approximately 10 ppm from 1350 to 1700 due to *natural* changes in the oceans and biosphere (possibly related to the little Ice Age), followed by a CO_2 increase after 1700.

Other trace gases, such as methane, nitrous oxide, and chlorofluorocarbons, whose concentrations are increasing in the atmosphere, can also cause a greenhouse effect by absorbing the earth's long-wave radiation. The combined effect of these gases in raising the earth's surface temperature (which is less well known than CO_2) may be about equal to the effects of CO_2 (Ramanathan et al. 1985). Together, the trace gases and CO_2 would cause the earth's temperature to rise more rapidly than CO_2 alone.

Climatic and Hydrologic Effects of CO_2 (and Trace Gas) Increase

The climatic effects of CO_2 are often modelled for a doubling of the CO_2 concentration in the atmosphere to 600 ppm. Estimates of *when* atmospheric CO_2 might be doubled involve projections of future fossil fuel use. According to a fairly conservative recent estimate (NRC 1983), the atmospheric concentration of CO_2

will probably pass 600 ppm in the third quarter of the twenty-first century ($\sim$2065). A source of error is whether deforestation has been a large net CO_2 source over the last several decades. If large amounts of deforestation CO_2 have also been added to the atmosphere, then the percentage of fossil fuel CO_2 that is airborne is lower, and CO_2 will increase more slowly in the future. In addition, greenhouse effects from increases in trace gases would cause climatic changes to occur sooner than would be expected from CO_2 increases alone.

For a doubling of atmospheric CO_2, the worldwide warming of the mean earth surface temperature predicted by climatic greenhouse models (Manabe and Stouffer 1980; Hansen et al. 1981) is 1.5°–3.0° C. The most realistic models give results in the range 2.0°–3.0° C (NRC 1983). (The value of 2° C is supported by modelling of data for the worldwide air temperature and CO_2 increases over the past 100 years; NRC 1983.) This warming of the earth's surface is due to several effects: (1) increased CO_2 concentrations cause more absorption of the earth's outgoing long-wave radiation, thereby increasing surface temperatures: (2) increased surface temperatures also cause an increase of atmospheric water vapor, which absorbs earth radiation even more strongly than CO_2; and (3) increased surface temperatures cause melting and decreased areal extent of surface ice and snow at high latitudes, which reduces the earth's *albedo*—the amount of solar radiation reflected from the earth (see above). Ice and snow are much stronger reflectors of solar radiation than vegetation or bare ground so that increased absorption of solar radiation accompanying a lower albedo further increases surface temperature.

Other factors need to be taken into consideration. Any changes in cloudiness (not predictable by the models) can cause either a positive or negative effect. The net effect of high-level clouds is to reduce the terrestrial back-radiation leaving the top of the atmosphere (warming effect), while low-level clouds reflect more entering solar radiation (cooling effect). The oceans tend to slow down a rise in earth surface temperature by 20–30 years because of their ability to absorb heat (NRC 1983). Thus, even if the earth's energy balance has already been changed at present, changes in the earth's surface temperature may be delayed.

Although the overall rise in earth surface temperature predicted for an atmospheric CO_2 doubling is probably 2.0° to 3.0° C, the change varies both with latitude and with season (NRC 1982). The predicted temperature rise is two to three times greater at the poles than near the equator, and the Arctic (North Pole) will be warmed more than the Antarctic (South Pole). The amount of warming will be greater in winter than in summer over the Arctic whereas at low latitudes ($<$45°), there will be little seasonal differences in warming (NRC 1983).

The overall effect of a rise in surface temperature will be to speed up the hydrologic cycle—with increases in both the global mean evaporation and precipitation rates of 4%–7% for a doubling of CO_2 (NRC 1982). However, the hydrologic changes are variable in different zones on the earth's surface. In middle latitudes of the Northern Hemisphere, in a zone from 35°N to 50°N with a peak at 45°N, there will be a tendency to greater summer dryness with a maximum of

10% less soil moisture for a doubling of CO_2 (Manabe, Wetherald, and Stouffer 1981). Areas such as the north central and western United States and central Asia would be particularly affected. The cause of the reduction in soil moisture in summer is primarily stronger evaporation coupled with an earlier occurrence of the summer period of low rainfall and an earlier end of the snowmelt season. In areas of the United States where agriculture is dependent upon rain, the positive effects of increased CO_2 and temperature (i.e., increased photosynthesis, improved plant hydration, and longer growing season) are less important than the negative effects of a lack of water (Waggoner 1983). In arid western areas of the United States dependent upon irrigation, decreases in runoff could cause severe water shortages (Revelle and Waggoner 1983).

Models also predict that precipitation and runoff, except in summer, would increase at high northern latitudes ($>60°N$) (Manabe and Stouffer 1980; Manabe, Wetherald, and Stouffer 1981) due to greater poleward transport of moisture. In summer there would be greater dryness in a zone centered at 60° N, including northern European Russia, due mainly to an earlier end to snowmelt.

Global warming could cause a *gradual* rise in sea level through two processes: (1) slow partial melting of ice and snow in the Greenland and Antarctic ice caps, along with mountain glaciers, which transfers water to the ocean; and (2) expansion of ocean water due to heating. Revelle (1983) predicts a gradual rise in sea level over the next century of 0.7 m from a combination of these effects.

The potential disintegration of the West Antarctic Ice Sheet would cause a much larger additional rise of 5–6 m over the next several hundred years (Revelle, 1983). Disintegration of the West Antarctic Ice Sheet would cause an abrupt addition of large masses of floating ice to the oceans bringing about a rapid rise of sea level.

A reduction in the areal extent and thickness of sea ice in the Arctic and circum-Antarctic oceans is another hydrologic change expected from rising temperatures. (This change would *not* effect sea level, however.) There have been suggestions that the Arctic sea ice might completely melt in summer as a result of warming from doubled CO_2. This could cause a northward shift in storm tracts during fall and spring and a reduction in precipitation in the Northern Hemisphere mid-latitudes (Revelle, 1983). Also, a worldwide reduction in sea ice could lead to more evaporation from the ocean and increased low-altitude cloudiness which would reflect solar radiation and cause cooling (NRC 1982).

Thus, to summarize, a rise in atmospheric CO_2 concentration from anthropogenic CO_2 release alters the earth's energy balance through the greenhouse effect. Since the hydrologic cycle is so intimately connected with the earth's energy cycle, increases in the earth's surface temperature cause modification in the hydrologic cycle. Although the rates of evaporation and precipitation are increased overall, in some areas a drier climate results. In addition, the amount of ice stored on the earth decreases resulting in a rise in ocean levels. Although the percentage change is not great, the human consequences of such changes are large because humans are so sensitive to changes in the hydrologic cycle.   (For further discussion of this last point see NRC 1983.)

REFERENCES

AMBROGGI, R. P. 1977.   Underground reservoirs to control the water cycle, *Sci. Amer.* 236 (5): 21–27.

ARRHENIUS, G., B. R. DE, and H. ALFVEN.   1974. Origin of the ocean.   In *The sea*, vol. 5, ed. E. Goldberg, pp. 839–861.   New York: Wiley-Interscience.

BACASTOW, R., and C. D. KEELING.   1981.   Atmospheric carbon dioxide concentration and the observed airborne fraction.   In *Carbon cycle modelling*, ed. B. Bolin, pp. 103–112.   SCOPE Report no. 16.   New York: John Wiley.

BAINBRIDGE, A. E.   1976.   *GEOSECS Atlantic Expedition*, v. 2: (198).   *Sections and Profiles*.   Washington, D.C.: National Science Foundation.

BAUMGARTNER, A., and E. REICHEL.   1975. *The world water balance*.   Munich and Vienna: R. Olenburg.   179 p.

BERNER, R. A.   1982.   Burial of organic carbon and pyrite sulfur in the modern ocean and its geochemical and environmental significance, *Amer. J. Sci.* 282: 451–473.

BERNER, R. A., A. C. LASAGA, and R. M. GARRELS.   1983.   The carbonate-silicate geochemical cycle and its effect on atmospheric carbon dioxide over the past 100 million years, *Amer. J. Sci.* 283: 641–683.

BLOOM, A. L.   1971.   Glacial-eustatic and isostatic controls of sea level since the last glaciation.   In *The late Cenozoic ice ages*, ed. K. K. Turekian, pp. 355–379.   New Haven, Conn: Yale Univ. Press.

BOLIN, B., E. T. DEGENS, P. DUVIGNEAUD, and S. KEMPE.   1979.   The global biogeochemical carbon cycle.   In *The global carbon cycle*, ed. B. Bolin, pp. 1–56.   SCOPE Report no. 13.   New York: John Wiley.

BREWER, P. G.   1983.   Carbon dioxide and the oceans. In *Changing climate*, pp. 188–215. Report of the Carbon Dioxide Assessment Committee, NRC Board on Atmospheric Sciences and Climate.   Washington D.C.: National Academy Press.

BROECKER, W. S., T. TAKAHASHI, H. J. SIMPSON, and T.-H. PENG.   1979.   Fate of fossil fuel carbon dioxide and the global carbon budget, *Science* 206: 409–418.

BUDYKO, M. I.   1974.   *Climate and life*, English ed., ed. D. H. Miller.   New York: Academic Press.   508 pp.

BUDYKO, M. I., and K. Y. KONDRATIEV.   1964.   The heat balance of the earth. In *Research in geophysics*, vol. 2: 529–554.   Cambridge, Mass.: MIT Press.

DRAKE, C. L., J. IMBRIE, J. A. KNAUSS, and K. K. TUREKIAN.   1978.   *Oceanography*. New York: Holt, Rinehart & Winston.   447 pp.

ELLIOTT, W. P., L. MACHTA, and C. D. KEELING.   1985.   An estimate of the biotic contribution to the atmospheric CO_2 increase based on direct measurements at Mauna Loa Observatory, *J. Geophys. Res.* 90 (D2): 3741–3746.

FLOHN, H. 1977. Man-induced changes in the heat budget and possible effects on climate. In *Global chemical cycles and their alterations by man*, ed. W. Stumm, pp. 207–224. Berlin: Dahlem Konferenzen.

GARRELS, R. M., and F. T. MACKENZIE. 1971. *Evolution of sedimentary rocks*. New York: W. W. Norton, 397 pp.

GARRELS, R. M., F. T. MACKENZIE, and C. HUNT. 1975. *Chemical cycles and the global environment: Assessing human influences*. Los Altos, Calif: Wm. Kaufman, Inc., 206 pp.

GOODY, R. M., and J. C. G. WALKER. 1972. *Atmospheres*. Englewood Cliffs, N.J.: Prentice-Hall. 150 pp.

HANSEN, J., D. JOHNSON, A. LACIS, S. LEBEDEFF, P. LEE, D. RIND, and G. RUSSELL. 1981. Climate impact of increasing atmospheric carbon dioxide, *Science* 213 (4511): 957–966.

HENDERSON-SELLERS, A., and J. G. COGLEY. 1982. The Earth's early hydrosphere, *Nature* 298: 832—835.

HOLLAND, H. D. 1984. *The chemical evolution of the atmosphere and oceans*. Princeton, N.J.: Princeton Univ. Press. 582 pp.

HUBBERT, M. K. 1971. The energy resources of the earth, *Sci. Amer.* 224 (3): 60–70.

INGERSOLL, A. P. 1983. The atmosphere, *Sci. Amer.* 249 (3): 162–175.

KOMHYR, W. D., R. H. GAMMON, T. B. HARRIS, L. S. WATERMAN, T. J. CONWAY, W. R. TAYLOR, and K. W. THONING. 1985. Global atmospheric CO_2 distributions and variations from 1968–1982 NOAA/GMCC flask sample data, *J. Geophys. Res.* 90 (D3): 5567–5596.

LVOVITCH, M. I. 1973. The global water balance, *Trans. Amer. Geophys. Union* 54 (1): 28–42.

MACHTA, L. 1983. The atmosphere. In *Changing climate*, pp. 242–251. Report of the Carbon Dioxide Assessment Committee, NRC Board on Atmospheric Sciences and Climate. Washington D.C.: National Academy Press.

MANABE, S., and R. L. STOUFFER. 1980. Sensitivity of a global climate model to an increase of CO_2 concentration in the atmosphere, *J. Geophys. Res.* 85: 5529–5584.

MANABE, S., R. T. WETHERALD, and R. S. STOUFFER. 1981. Summer dryness due to an increase of atmospheric CO_2 concentrations, *Climatic Change* 3: 347–386.

MARLAND, G., and R. ROTTY. 1984. Carbon dioxide emissions from fossil fuels: A procedure for estimation and results for 1950–1982, *Tellus* 36B: 232–261.

MEYBECK, M. 1986. Origin of riverbone elements derived from continental weathering, *Am. J. Sci.* (in press).

MILLER, A., J. C. THOMPSON, R. E. PETERSON, and D. R. HARAGAN. 1983. *Elements of meteorology*, 4th ed. Columbus, Ohio: Chas. E. Merrill.

MOORBATH, STEPHEN. 1977. The oldest rocks and the growth of continents, *Sci. Amer.* 236 (3): 92–104.

MUNK, W. H. 1955. Circulation of the ocean, *Sci. Amer.* 193 (34): 96–102.

NACE, R. L. 1967. Water resources: A global problem with local roots, *Environ. Sci. Technol.* 1: 550–560.

NATIONAL RESEARCH COUNCIL (NRC). 1982. *Carbon dioxide and climate: A second assessment*. Report of the CO₂/Climate Review Panel, J. Smagorinsky, Chairman. Washington, D.C.: National Academy Press. 72 pp.

NATIONAL RESEARCH COUNCIL BOARD ON ATMOSPHERIC SCIENCES AND CLIMATE (NRC). 1983. *Changing climate*. Report of the Carbon Dioxide Assessment Committee. Washington, D.C.: National Academy Press. 496 pp.

NEFTEL, A., H. OESCHGER, J. SCHWANDER, B. STAUFFER, and R. ZUMBRUNN. 1982. New measurements on ice core samples to determine the CO_2 content of the atmosphere during the last 40,000 years, *Nature* 295: 220–223.

OORT, A. H. 1970. The energy cycle of the earth, *Sci. Amer.* 223 (3): 54–63.

PEIXOTO, J. P., and M. KETTANI. 1973. The control of the water cycle, *Sci. Amer.* 228 (4): 46–61.

PENG, T.-H. 1985. Atmospheric CO_2 variations based on the tree-ring ^{13}C record. In *The carbon cycle and atmospheric CO_2: Natural variations, Archaean to present*, ed. E. T. Sunquist and W. S. Broecker, pp. 123–131. AGU Geophysical Monograph no. 32. Washington, D.C.: Amer. Geophys. Union.

PENMAN, H. L. 1970. The water cycle, *Sci. Amer.* 223 (3): 98–108.

PICKARD, G. L., and W. J. EMERY. 1982. *Descriptive physical oceanography*, 4th ed. New York: Pergamon Press. 249 pp.

RAMANATHAN, V., R. J. CICERONE, H. B. SINGH, and J. T. KIEHL. 1985. Trace gas trends and their potential role in climate change, *J. Geophys. Res.*, 90 (D3): 5547–5566.

RAYNAUD, D., and J. M. BARNOLA. 1985. An Antarctic ice core reveals atmospheric CO_2 variations over the past few centuries, *Nature*, 315: 309–311.

REVELLE, R. 1983. Probable future changes in sea level resulting from increased atmospheric CO_2. In *Changing Climate*, pp. 433–448. Report of the Carbon Dioxide Assessment Committee, NRC Board on Atmospheric Sciences and Climate. Washington, D.C.: National Academy Press.

REVELLE, R. R., and P. E. WAGGONER. 1983. Effects of a carbon dioxide induced climatic change on water supplies in the western United States. In *Changing Climate*. pp. 419–432. Report of the Carbon Dioxide Assessment Committee, NRC Board on Atmospheric Sciences and Climate. Washington, D.C.: National Academy Press.

SELLERS, W. D. 1965. *Physical climatology*. Chicago: Univ. of Chicago Press. 272 pp.

STEWART, R. W. 1969. The atmosphere and ocean. *Sci. Amer.* 221 (3): 76–86.

STOMMEL, H. 1958. Circulation of the abyss, *Sci. Amer.* 199 (1): 85–90.

———. 1965. *The Gulf Stream*, 2nd ed. Berkeley, Calif.: Univ. of California Press. 248 pp.

TUREKIAN, K. K. 1976. *Oceans*, 2nd ed. Englewood Cliffs, N.J.: Prentice-Hall. 149 pp.

U.S. COMMITTEE FOR THE GLOBAL ATMOSPHERIC RESEARCH PROGRAM. 1975. *Understanding climatic change—A program for action*. Washington, D.C.: National Academy of Sciences, National Research Council. 239 pp.

VON ARX, W. S. 1962. *An introduction to physical oceanography*. Reading, Mass.: Addison-Wesley. 422 pp.

WAGGONER, P. E. 1983. Agriculture and a climate changed by more carbon dioxide. In

Changing Climate, pp. 383–418. Report of the Carbon Dioxide Assessment Committee. NRC Board on Atmospheric Sciences and Climate. Washington, D.C.: National Academy Press.

WALKER, J. C. G. 1977. *Evolution of the Atmosphere*. New York: Macmillan. 318 pp.

WOODWELL, G. M. 1983. Biotic effects on the concentration of atmospheric carbon dioxide: A review and projection. In *Changing Climate*, pp. 216–241. Report of the Carbon Dioxide Assessment Committee, NRC Board on Atmospheric Sciences and Climate. Washington, D.C.: National Academy Press.

WOODWELL, G. M., J. E. HOBBIE, R. A. HOUGHTON, J. M. MELILLO, B. MOORE, B. J. ROBERTSON, and G. R. SHAVER. 1983. Global deforestation: Contribution to atmospheric carbon dioxide, *Science* 222: 1081–1086.

3

Rainwater and Atmospheric Chemistry

INTRODUCTION

Most of the water in the atmosphere (more than 95%) is present in the form of water vapor. How does this water become liquid rain or snow and what chemical changes have occurred in it in the meantime? Even though we often think of rainwater as being very pure, it is, in fact, no longer just H_2O. It has become a dilute solution of a number of substances picked up during its trip through the atmosphere. We shall discuss these "impurities" in detail in this chapter—how they get into the atmosphere and how they become dissolved in rainwater.

In addition to its intrinsic importance, the chemistry of rainwater is also interesting from other points of view. Rainfall provides a major input of several elements to the earth's surface and the importance of the rain input can be determined only if its composition is well known. Likewise, in attempting to determine the effects of rock weathering or biological processes on the concentration of a given element in a lake, river, or groundwater, one must first correct for the concentration of this element in rainwater arriving at the ground.

From a practical viewpoint, rainwater composition is of interest to those concerned with air pollution and the role of humans in altering the chemistry of the atmosphere. An outstanding example of this is the formation, in recent years, of acid rain downwind from industrial areas. Because of intense public interest in acid rain and air pollution in general, much attention will be given in the present chapter to these subjects.

FORMATION OF RAIN (AND SNOW)

How is atmospheric water vapor transformed into the precipitation (rain and snow) that arrives at the earth's surface? In this section we shall address this problem, starting with a discussion of water vapor itself.

Water Vapor in the Atmosphere

The amount of water vapor that is present in any given volume of air varies from place to place. One way of expressing the quantity of water vapor in air is the *absolute humidity* or density of water vapor as grams of water vapor in a unit volume of air (g/m³). (A similar measure, which can be derived from the absolute humidity, is the *mixing ratio* or grams of water vapor per kilogram of dry air.) The other commonly used expression for the water vapor content of the atmosphere is the *water vapor pressure*, that part of the total atmospheric pressure that is due to water vapor. Since water vapor is a minor constituent of air, its pressure is much lower than that of the atmosphere (about 2% on the average of the atmospheric pressure at the earth's surface).

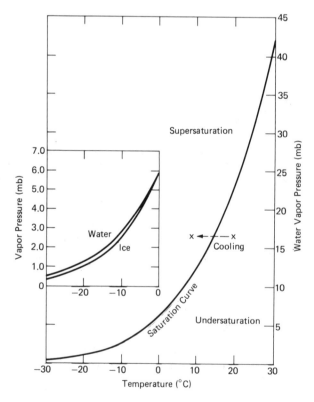

Figure 3.1 Saturation vapor pressure in millibars (1 mb = 10^3 dynes/cm²) of pure water as a function of temperature (°C). The dashed arrow represents the change from undersaturation to supersaturation upon cooling of air with a constant water vapor pressure. Inset: Saturation vapor pressure (mb) over water and ice at temperatures below 0°C. (After H. R. Byers, *Elements of Cloud Physics*. © 1965 by The University of Chicago Press. All rights reserved.)

A volume of air can hold just so much water vapor before the air becomes saturated and the water vapor condenses as a liquid or sublimates as ice. The saturation vapor pressure, or maximum amount of water that air can hold before condensation or sublimation, is a function of temperature as shown in Figure 3.1. As can be seen, if air that is saturated with water vapor at a certain temperature is cooled, it becomes supersaturated and water condenses (or sublimates). A measure of how close a given air mass is to saturation is given by the familiar term *relative humidity*, which is often mentioned in weather reports. Relative humidity is the ratio between the actual water vapor pressure and the saturation vapor pressure for the same temperature. It is usually expressed in terms of percent. (For further discussion see Miller et al. 1983, or Neiburger, Edinger, and Bonner 1973.)

Because temperature decreases with height in the atmosphere (at an average rate of $-6.5°$ C/km), water vapor content, due to condensation and sublimation, also decreases. This is shown in Figure 3.2. From the ground to an altitude of about 10 km, water vapor content and temperature continually decrease. At this altitude, one encounters the *tropopause*, the boundary between the lower atmosphere or *troposphere* and the upper atmosphere or *stratosphere*. (The height of

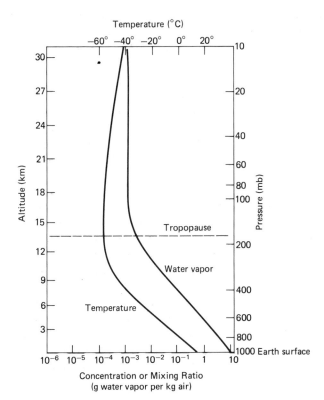

Figure 3.2 The decrease in concentration (or mixing ratio) of water vapor in the atmosphere with height (after Newell 1971) and the decrease in temperature (°C) with height (data from Miller et al. 1983). The approximate height of the tropopause (which varies with latitude) is shown by the dashed line. Pressure is in millibars (1 mb = 10^3 dynes/cm^2).

the tropopause varies from about 5 km at the poles to 15 km at the equator.) Air in the stratosphere contains a very small and nearly constant amount of water vapor (see Figure 3.2) because any air travelling to this height has already lost most of its moisture by condensation and sublimation in the troposphere. In this way the temperature minimum at the top of the troposphere serves as a *cold trap* that prevents loss of water from the earth to space (Newell 1971; see also Chapter 2).

The total amount of water vapor in the atmosphere over the whole globe is 13×10^{15} kg (which still represents only 0.001% of the total water on the earth; see Table 2.1). In a column of air overlying a square meter of the earth's surface from the ground to about 10 km (roughly the tropopause), this would amount to an average of 25 kg of water per square meter if the water were evenly spread over the land surface (or 2.5 cm of water if the water vapor fell as rain or snow). However, the water vapor content of air changes with latitude, as a result of temperature changes, from about 5 kg/m^2 at the North Pole through 20 kg/m^2 at 45°N, to 45 kg/m^2 over the equator. These values also change with the seasons. Over the continental United States the average water content varies from 9 kg/m^2 in the winter to about 27 kg/m^2 in the summer with the change again being due to temperature variation (Miller 1977).

As was pointed out in Chapter 2, the average *residence time* of water vapor in the atmosphere is only about ten days. This is the length of time the average water vapor molecule spends in the atmosphere between its evaporation from the earth's surface and its precipitation as rain or snow. Such rapid turnover is important for the removal of atmospheric pollutants that can be "washed out" by rain. It is not so rapid, however, that appreciable transport doesn't take place. Because of the speed of atmospheric winds, during a period of ten days a given mass of water vapor can be transported over great distances. For instance, the mean water vapor travel distance in temperate latitudes is around 1000 km (Peixoto and Kettani 1973).

Because of rapid transport, the amount of water vapor passing over a given landmass can be very large. For example, the average inflow of water vapor across the Gulf of Mexico coast of the United States, during the summer of 1949, was almost *ten times* higher than the flow of the Mississippi River during the same period (Benton and Estoque 1954). Much of the water vapor passing over a landmass, however, is not lost via precipitation. Benton and Estoque (1954) report that, on a mean annual basis, only 63% of the water vapor passing eastward from the Pacific Ocean across the entire continental United States and Canada is lost via precipitation. This can be compared to values for smaller regions of 40% for European Russia, 20% for the Mississippi Basin, and only 11% for Arizona (Sellers 1965). (Low removal percentages, as shown here for Arizona, pose a local problem for arid regions and efforts have been made to convert more water vapor to precipitation by means of such methods as cloud seeding. This is discussed below in the section on cloud formation.) Because of the large-scale transport of water vapor, most continental precipitation does not come from locally evaporated water.

Condensation

The transformation of water vapor in the atmosphere into rain involves two processes. First, the water vapor gas must *condense* to form water droplets (or *sublimate* to form ice crystals). This is cloud or fog formation. However, condensation does not necessarily lead to precipitation or the fall of water droplets or snowflakes. In order for cloud droplets to fall to the ground as rain, they must become large and heavy enough to reach the ground without evaporation. The average lifetime of a cloud is around 1 hour, while on the average water spends 11 hours in the atmosphere in the form of droplets before being removed as rain. Thus, cloud water evaporates and condenses several times before actually forming rain (Pruppacher 1973).

In order for condensation to occur, air must become supersaturated with water vapor, but this is not enough. Nuclei are needed to begin the condensation process. The nuclei can be any one of a number of small bodies suspended in air, chief of which are soil dust particles, combustion products, and sea salt. In air completely free of condensation nuclei, it would be possible to have relative humidities of as high as 800% without condensation taking place, whereas in actuality relative humidities (for condensation) never exceed 102% (Miller et al. 1983). This is because condensation nuclei promote the formation of water droplets and such nuclei are always present.

Different suspended nuclei bring about condensation of water with differing effectiveness. First of all, larger particles serve as better nuclei, because they have "flatter" surfaces which favor condensation (Neiburger, Edinger, and Bonner 1973; Pruppacher 1973). *Hygroscopic* particles, which are substances that readily absorb water and are very soluble in it, serve as the best nuclei. They are so efficient that they can initiate condensation at less then 100% relative humidity (Neiberger, Edinger, and Bonner 1973; Pruppacher 1973). Examples are NaCl from sea spray, and H_2SO_4 and HNO_3 from the burning of fossil fuels. Condensation may start at less than 100% because the hygroscopic particles are highly soluble in water and at high concentrations they lower the saturation vapor pressure of water (but because of dilution of the salts, continued condensation still requires relative humidities of more than 100%). Thus, hygroscopic nuclei require much lower degrees of supersaturation than their nonhygroscopic equivalents and it is for this reason that they are more efficient in forming water droplets.

Over the oceans, condensation nuclei are dominated by hygroscopic salt particles derived from sea spray. On land, the nuclei are both hygroscopic and nonhygroscopic with the latter being derived from soils and combustion products. The greater concentration of all kinds of particles over land, especially larger particles, results in competition for water and the formation of more but smaller water droplets in clouds. Over the oceans, the few but highly efficient hygroscopic particles produce relatively large droplets. Since larger cloud droplets favor rain formation (see below), rainout over the oceans (and coastal regions) is relatively easier than over land.   This aids in the efficient removal of marine aerosols and helps explain the high sea-salt content of marine and coastal rain (Junge 1963).

Sublimation

Instead of forming water droplets, water vapor is transformed to ice crystals via the process of *sublimation* whenever the temperature is sufficiently cold, as often occurs at the tops of clouds.   However, because ice crystals nucleate with much greater difficulty than water droplets, temperatures considerably lower than 0° C are necessary for sublimation to occur.   For ice to form, a nucleus is needed that promotes crystal growth by having interatomic spacings similar to those found in ice.   Only a few substances are suitable for this purpose, principally clay minerals from soils.   Hygroscopic particles and combustion products are much less important in sublimation than they are in condensation.   The degree of supercooling necessary to bring about crystallization is variable from cloud to cloud, the highest temperature at which ice crystallizes being about $-10°$ C.   Between $-10°$ and $-20°$ C clouds form that consist of both ice crystals and water droplets whereas below $-20°$ C only ice clouds are present (Neiburger, Edinger, and Bonner 1973).   An example of the latter are the high, wispy ice clouds known as cirrus clouds.

Rain (and Snow) Formation

The water droplets in most clouds average about 5 to 10 μm in diameter with the largest being about 20 μm.   Because of constant updrafts, these sizes are too small for the droplets to fall to the ground.   In order to have rain, there must be a process whereby the droplets can become big enough, on the average about 1000 μm in diameter, to fall as raindrops (Neiburger, Edinger, and Bonner 1973). Further condensation on existing droplets in clouds is not an efficient mechanism. Instead, the most commonly cited processes of rain formation are *collision-coalescence* and *ice crystal growth followed by melting*.   For snow formation the major process is ice crystal growth.

In collison-coalescence, droplets somewhat larger than the average begin to fall and, as they collide with smaller droplets, they grow in size by incorporating the small droplets.   As a result of growth they fall faster.   After a large number of collisions ($\sim$1 million), a rain-sized drop may be produced.   If the drop becomes big enough, it can split into two drops and these two can grow by further coalescence.   Continued splitting and growth by collision plus coalescence produces a chain reaction resulting ultimately in the formation of rain.   This process is especially effective over the oceans where, on the average, larger cloud droplets are found which can begin to fall and initiate the chain reaction.   It is also operative wherever air masses are too warm (greater than $-10°$ C) to allow ice crystal formation (Mason 1971) and it serves as a process for enlarging raindrops originally formed by ice crystal growth plus melting.

Ice crystal growth (or the Bergeron process) occurs in the upper or colder parts of clouds, which contain both water droplets and ice crystals at temperatures of about $-10°$ to $-20°$ C (Neiburger, Edinger, and Bonner 1973). At such temperatures the saturation vapor pressure of water is greater than that of ice. In other words, air containing a given amount of water vapor can be supersaturated over ice (so that sublimation will occur) while it is saturated or even undersaturated with respect to water (see Figure 3.1). As a result, the ice crystals will grow at the expense of coexisting water droplets. Eventually the ice crystals may become big enough that they will fall to the ground. If they melt on the way down, rain is produced; if they don't, snow results. If the ice grows very rapidly, it may fall as hail. (For further details on rain, snow, and hail formation see Miller, et al. 1983; Neiburger, Edinger, and Bonner 1973; and Pruppacher 1973.) Because of the smaller droplets found in clouds over the continents, much rain over land forms via the process of ice crystal growth followed by melting and enlargement by collision-coalescence.

In areas of deficient rainfall, humans have attempted to produce rain by artificial methods. Such cloud seeding proceeds along the lines of the two major processes of rain formation discussed above. Sometimes artificial ice nuclei, such as silver iodide or solid carbon dioxide (dry ice), are introduced at the cold tops of clouds in an attempt to induce the formation of ice crystals of a size large enough to fall. In other situations, hygroscopic nuclei (for example, $CaCl_2$) or large water droplets are added to clouds to initiate rain via the collision-coalescence process.

Air Motion in Cloud Formation

In considering the processes within the cloud that convert water vapor in the atmosphere to precipitation falling on the ground we should not neglect the fact that larger-scale air motions are important. Basically, we need to cool moist air masses since the amount of water vapor the air can hold will decrease with cooling (see Figure 3.1). Cooling can occur either directly or indirectly. Direct cooling comes about by the movement of warmer air over a colder land or water surface whereas indirect or *adiabatic cooling* occurs when an air mass is lifted vertically. Since air pressure decreases with height, an ascending air mass expands on uplift and this causes a drop in temperature. An example of adiabatic cooling is when an advancing cold front forces warm air upward in front of it.

Once the moist air has been cooled by one of these mechanisms, its relative humidity increases and, when supersaturation occurs, condensation can proceed around particles present in the air. Air motions are also involved in the formation of rain in clouds; they affect where and how fast the rain falls, and how much moist air is being brought in and converted to rain. Thus, although we are primarily concerned with the processes of condensation and rain formation within the cloud, because of their influence on the chemical composition of rain, we should not neglect the fact that cooling of moist air by air motion is essential to these processes.

Certain areas of the earth receive more rain than others because of these effects (see Chapter 2).

AEROSOLS

In the atmosphere, as we have noted, water always condenses on a particle or *nucleus* because the nucleus promotes condensation at a reasonably low relative humidity (a maximum of 102%). Therefore, the rain that forms from these cloud droplets of condensed water will have a chemical composition that reflects the composition of the particles on which it condenses and also the composition of other atmospheric particles that it contacts on the way down. For this reason we shall now consider particles in the atmosphere and their origin and composition.

Types of Aerosols

In addition to major and trace gases, the atmosphere contains *aerosols*, small particles of solid or liquid ranging in size from clusters of a few molecules to about 20 μm in radius. Particles larger than 20 μm do not remain in the atmosphere very long because they are heavy enough to settle out rapidly. There are two main types of particles or aerosols in the atmosphere: primary particles emitted directly into the atmosphere (such as dust or sea salt) and secondary particles formed from gaseous emissions that subsequently condense in the atmosphere. Gas-to-particle conversion results in the formation of fine particles (<0.5 μm), whereas directly emitted particles are dominantly coarse (>0.5 μm).

In chemical composition, aerosols may consist of one or more fractions (Rahn 1976): (1) water-soluble ions (such as sulfate, nitrate, ammonium, and several sea-salt-derived ions); (2) a mostly insoluble inorganic part (silicates, oxides, etc.); and (3) a carbonaceous part (soluble and insoluble organic matter). We shall be concerned mainly with the soluble ions that later appear in rain. In form, aerosols range from dry dust particles to sea-salt particles which are sometimes drops of salty water (at a high relative humidity). Most continental aerosols are a mixture of soluble and insoluble components (*mixed particles*) whereas most marine aerosols are soluble sea salt (Junge 1963).

Condensation occurs preferentially on the largest particles (radius of 0.1 to 20 μm) and most of the smaller particles are never used in condensation. The type of particles used for condensation is apparently controlled not only by the sizes available, but also by preferential use of certain chemical compositions, such as *hygroscopic* or soluble particles (see previous section). Precipitation tends to be quite efficient in removing particles from the atmosphere; that is, most particles in the atmosphere will end up in the rain, although not necessarily in the same relative concentrations as in the atmosphere bcause of their different solubilities and sizes (SMIC 1970; Junge 1972).

The main sources of particles or aerosols in the atmosphere are as follows (SMIC 1970; see also Table 3.1):

1. Sea salts from the ocean surface (dominantly coarse)
2. Soil and mineral dust from the continents (coarse)
3. Volcanic ash
4. Forest fire and slash burning products, dominantly anthropogenic; both ash and gases converted to particles
5. Biogenic aerosols (plant exudates), including volatile hydrocarbons which are converted to particles as well as direct particle emissions
6. Gaseous emissions, both natural and anthropogenic, which react in the atmosphere to form fine particles (excluding forest fires)
7. Solid particles from other anthropogenic activities, e.g., combustion and industry

TABLE 3.1 Estimates of Particles Smaller than 20 μm Radius Emitted into or Formed in the Atmosphere (in Tg/yr; Tg $= 10^{12}$ g)

	Natural	Anthropogenic	Percent of Total	Source of Data
Soil and rock dust:				
Long transport	500–900	?	35	See text
(Total production)	(3000–4000)			
Sea salt:				
Long transport[a]	540	—	27	See text
(Land deposition)	(180)			
(Total production)	(1800)			
Forest fires and slash burning				
debris	6–11	72–139	6	SMIC 1970
Volcanic debris	25–150	—	3	SMIC 1970
Particles from anthropogenic direct emissions: fuel, incinerators, and industry	—	10–90	2	SMIC 1970
Particles formed from gaseous emissions:				
Sulfate from H_2S	115	—	6	See text
Sulfate from SO_2	—	100	5	See text
Nitrate from NO_x	30	75	5	See text
Ammonium from NH_3	2	26	1	See text
Hydrocarbons: Biogenic aerosols	75–200	—	7	SMIC 1970
Fossil fuel	—	15–90	3	SMIC 1970
Approximate total	1293–1948	298–520		
	80%	20%	100	

[a] Three times land deposition (see text).

   Although the particle fluxes given in Table 3.1 are not well known and represent approximate estimates based, in the case of gases, on fluxes given later in this chapter, it is still apparent that the main particle sources are natural: soil dust, which contributes about one-third of the particles, and sea salt, which contributes about one quarter. Anthropogenic particles formed in the atmosphere from gaseous emissions (sulfate, nitrate, and hydrocarbons) amount to another 13%. Overall, roughly, about 80% of the particles are natural and 20% are anthropogenic.

   The dissolved ions Na^+, K^+, Ca^{++}, Mg^{++}, and Cl^- in rain come mainly from sea salt and soil dust whereas gaseous emissions (which form secondary particles) give rise to such ions as SO_4^{--}, NO_3^-, and NH_4^+. (Some ions such as SO_4^{--}, and Cl^- are formed by several of these processes). In this section, we shall be concerned mainly with aerosols that are directly emitted to the atmosphere, that is, with primary particles, and emphasis will be on sea salt and soil dust. The secondary particles derived from gaseous emissions will be discussed in more detail in the sections devoted to sulfur and nitrogen later in this chapter.

Sea-Salt Particles

Sea salt, produced by the oceans, makes a large contribution to atmospheric particles. The bursting of small air bubbles in the foam of breaking waves or "white caps" forms sea-salt particles, as shown in Figure 3.3 (Junge 1963). A bubble breaks when it reaches the ocean surface and the water rushes in to make a rapidly upward-moving jet that projects into the air. This jet breaks to form about ten droplets which are ejected $\sim$ 15 cm above the ocean surface and then carried upward by air currents. The droplets of seawater that are thrown into the air in this manner evaporate to produce small sea-salt particles of a radius ranging from 2 to 20 μm (Blanchard and Woodcock 1957). They are designated in Figure 3.3 as C. There are also numerous smaller particles produced from the bubble film itself, which range from 0.1 to 1 μm in radius (F in Figure 3.3). However, 90% of the sea-salt mass is carried by the larger particles.

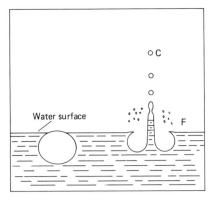

Figure 3.3 The formation of sea-salt particles from the bursting of bubbles. Large droplets originating from the jet are designated as C whereas smaller particles produced from the bubble are designated as F. (After C. E. Junge, *Air Chemistry and Radioactivity.* © 1963 by Academic Press, reprinted by permission of the publisher.)

Over the oceans, sea-salt particles make up a major part of the particles in the coarse size range (0.5–20 μm). By contrast, the smaller particles, known as the *background aerosol*, are derived from the continents. It is the sea salt, however, that is important in forming marine rain.

The average residence time in the atmosphere for sea salt is three days (Junge 1972); thus, it can be transported considerable distances to and over the continents. Since some sea-salt ions are common constituents of continental rains, it is important to ascertain how much sea salt is carried to the continents. Eriksson (1960) calculated that the yearly deposition of sea-salt Cl on the continents is 100 Tg/yr, (Tg = 10^{12} g) which represents 10% of the total marine sea-salt Cl production (1000 Tg Cl/yr). Since sea salt is 55% Cl by weight, about 1800 Tg of total sea salt would be generated over the oceans and 180 Tg carried to and deposited on land. Eriksson's calculation of sea-salt Cl deposition on the continents is problematic, however, in that both higher and lower estimates have been made by others. (See Chapter 5 for a discussion of estimates of atmospherically-derived chloride in rivers.) For example Petrenchuk (1980) also estimates sea-salt production of 1800 Tg/yr of which 20% falls on land, whereas Blanchard (1963) estimates that sea-salt production may be as high as 10,000 Tg/yr. For our purposes in this chapter as a rough guide we use the Eriksson estimate (see Table 3.1) and also assume that three times the land deposition of sea salt, or 540 Tg/yr, represents long-distance transport. (This accounts for transport over the ocean as well as the land; see SMIC 1970.)

Although most sea salt is removed over the oceans, the ~ 10% deposited over land as *cyclic sea salt* (Eriksson 1960) exerts an important influence on the composition of continental waters. Thus, there has been considerable interest in the chemical composition of marine aerosols, which are comprised to a large extent by sea salt. The composition of marine aerosols and marine rain (i.e., the ratios of the different ions), however, often differs from that of bulk seawater. The problem lies in deciding whether these differences are due to chemical fractionation in the change from seawater to atmospheric sea salt or are due to the presence in marine air of other particles besides sea salt that have another source and a different chemical composition.

Several authors (among them Duce and Hoffman 1976; Junge 1972; MacIntyre 1974; and Buat-Menard 1983) have discussed the question of possible chemical fractionation at the sea-air surface versus the presence of particles from outside (nonmarine) sources or changes in the ratios of sea-salt particles by interaction with gases in the atmosphere (such as HCl or SO_2 gas). The reason for expecting that chemical fractionation might occur is that certain substances are more concentrated in the top 10–20 cm of the ocean than they are at greater depth. However, recent work has shown that these substances are not major ions (see below), but are primarily *surface-active* organic substances that have one part of their molecule polar and water-soluble and another part nonpolar and non-water-soluble. The molecules are attracted to the interface between air and seawater because they can orient themselves with their water-soluble part in water and their nonsoluble part in the air.   This interface occurs both on the surface of air bubbles and on the surface of the sea itself.   In this manner, surface-active organic matter and other chemical substances associated with it (particulate and dissolved organic nitrogen and phosphate and trace metals such as iron, lead, and manganese) are concentrated near the ocean's surface.   Bubbles scavenge surface-active material from lower layers and transport it to the sea surface.   When these bubbles break, ejecting sea salt into the air, they essentially skim off a very thin and *enriched* microlayer at the ocean surface.   Thus, the substances enriched at the ocean surface should also be enriched, along with organic matter, in sea-salt particles (MacIntyre 1974).

The important plant nutrients, phosphate and organic nitrogen, are intimately associated with organic matter in seawater and they would be expected to be concentrated in marine aerosols.   *Organic phosphorus*, defined as the difference between total P and soluble reactive P, actually has been found to be enriched by a factor of 100–200 in marine aerosols (Graham, Piotrowicz, and Duce 1979). However, in the western North Atlantic, only 30% of this organic P is from sea salt, and the rest is of pollutive and soil dust origin (Graham and Duce 1982).   In addition to phosphorus and nitrogen, organic carbon in sea-salt aerosols is enriched by about a factor of 100 (Buat-Menard 1983).

The enrichment in sea-salt aerosols, relative to seawater, of the alkali and alkaline earth metals, whose ions are important in rainfall (Na, K, Mg, and Ca), appears to be slight or nonexistent (MacIntyre 1974; Duce and Hoffman 1976; Buat-Menard 1983).   A possible exception is the potassium enrichment found in very small (<0.5 μm) marine aerosols, but this enrichment may be due to long-range transport to the ocean of K-enriched particles, originating from terrestrial vegetation, rather than chemical fractionation at the time of sea-salt formation (Buat-Menard 1983).

Chloride and sulfate also probably do not undergo appreciable fractionation during the formation of sea salt.   Meinert and Winchester (1977), in studies of large sea-salt aerosols in the North Atlantic, found a two- to three-fold enrichment of sulfur and 25%–40% depletion of Cl relative to K and Ca.   They suggest that the process causing the seeming enrichment of S and depletion of Cl is not simple fractionation during the transfer from sea to air but rather reactions between the aerosols and SO_2 gas.   Duce and Hoffman (1976) also attribute observed changes in the ratio of Cl/Na to loss of Cl from the aerosol, as gaseous HCl, upon reaction with HNO_3 (or perhaps SO_2) from the atmosphere (see further discussion below in the section on gases in the atmosphere).

Soil and Mineral Dust—Continental Aerosols

Another major source of atmospheric particles is soil and rock debris.   Natural weathering breaks down rocks to produce soil and mineral dust which is then transported by the wind.   Typically the windblown dust consists of yellow-brown aggregates of quartz (SiO_2), mica, and clay minerals (hydrous cation aluminosilicates; for a further description of clay minerals see Chapter 4), the latter formed in part by weathering. The yellow-brown color of windblown dust is from iron oxides which are strongly enriched in the dust. Calcareous minerals—calcite, $CaCO_3$; dolomite, $Ca,Mg(CO_3)_2$; anhydrite, $CaSO_4$; and gypsum, $CaSO_4 \cdot 2H_2O$— are also found, particularly downwind from dry desert areas.

Since Al, Fe, and Si come almost exclusively from soil dust, their presence in aerosols is a strong indication of soil dust origin. Al is most commonly used as an aerosol tracer for soil dust, and the ratio of another soil element, such as Ca, to Al in an aerosol is compared to the known ratio in soils, rocks, or the earth's crust to determine enrichment of the element. In general, the element ratios in windblown soils are similar to bulk crustal rock (Rahn 1976). The major soluble ions in rain, which come commonly from soil dust include Ca, Mg, and K, while SO_4 and Cl are sometimes locally important. (For further discussion of soil dust, as it affects rainfall composition, consult later sections for each element.)

Although windblown mineral particles are derived from the continents, they are found in considerable concentration over the oceans as well. For example, over the North Atlantic there is a large contribution of particles in the 0.1 to 20 μm size range from Sahara dust storms (Junge 1972). Over the Pacific, dust from Asian continental sources, such as the Gobi Desert, is transported long distances. However, aerosol dust concentrations are very variable in both time and space. For example, most of the annual dust deposition on the ocean in a certain area may result from one dust storm (Buat-Menard 1983).

The flux of soil dust to the atmosphere is not well known. Estimates of soil dust that is transported long distances by the atmosphere and deposited in the oceans range from 500 to 900 Tg/yr (Goldberg 1971; Prospero 1979; Graham and Duce 1979; Petrenchuk 1980). The total dust production over the continents is considerably larger, 3000–4000 Tg/yr (Graham and Duce 1979) or more. Also, some part of this is anthropogenic, resulting from roads, construction, farming, and other human activities. (In some anthropogenic inputs, such as fertilizer dust, the chemical composition may be considerably different from that for average soil, rock, etc.)

Volcanic Particles

Volcanic debris accounts for a small part of total particle formation but the production of volcanic material tends to be episodic. One violent volcanic eruption will produce dust and ash that circle the earth in the lower atmosphere and ultimately in the stratosphere for several years. The eruption of Krakatoa in the East Indies in 1883 ejected an estimated 25,000 Tg of material into the atmosphere or around 300 times the estimated normal yearly production of volcanic material (Goldberg 1971).

The particles produced by volcanoes consist of finely divided lava or ash, consisting of silicate minerals, and sulfuric acid aerosols, which originate from the oxidation of SO_2 from the volcanic plume. Rain influenced by volcanic eruptions has high concentrations of SO_4, Cl, and F and is very acid (Harding and Miller 1982).

Biogenic Aerosols and Aerosols from Forest Burning

Vegetation can emit volatile organic matter (mainly hydrocarbons) directly to the atmosphere where fine organic aerosols are produced by gas-to-particle conversion (Went 1960). Sources include emanations from trees in coniferous forests as well as the decomposition of dead organic matter. Major biogeochemically important elements, such as K, Na, Ca, and Mg, may also be directly released by plants to the atmosphere in sub-micron-sized particles from plant waxes or airborne salt from leaves and/or as larger particles from leaf abrasion (see Curtin, King, and Mosier 1974; and Beauford, Barber, and Barringer 1977). Large biogenic particles have a relatively short residence time in the atmosphere and chiefly affect *local* aerosol concentrations (and rain concentrations). However, sub-micron-sized particles can be more widely transported.

Lawson and Winchester (1979) found evidence for the biogenic production of sulfur, potassium, and phosphorus in *coarse* aerosol particles (but not in fine particles) over tropical rain forests in the Amazon Basin in South America. Similarly Crozat (1979) found that during the rainy season, tropical forests in the Ivory Coast emit a fine aerosol rich in potassium. Crozat suggests that plant fluids containing potassium evaporate on plant leaves, leaving salts that may become airborne. The potassium enrichment found in fine particles in the marine atmosphere is attributed to this continental vegetation source (Buat-Menard 1983).

In the marine atmosphere, most of the total particulate organic carbon is in small particles (Buat-Menard 1983), which along with other evidence suggests that gas-to-particle conversion of volatile hydrocarbons released by vegetation may be the dominant source. (However, there is also a seawater source of surface-active organic carbon in large particles as discussed above.)

Forest fires, both those set by humans (the dominant cause) and natural, are another source of particles. Lawson and Winchester (1978) suggest that large-scale forest burning, which is used seasonally to clear the land in Brazil, may be a source of fine particulate potassium and sulfur, which are wood constituents. Similarly in West Africa, Crozat et al. (1978) found increased potassium concentrations in fine aerosols due to brush fires.

Gaseous Emissions

For the sake of completeness, fine secondary aerosols formed from gases emitted to the atmosphere are mentioned here. They include small aerosols of sulfuric acid formed from anthropogenic SO_2, aerosols of $(NH_4)_2SO_4$ and NH_4HSO_4 formed by the reaction of NH_3 with the sulfuric acid aerosols, and nitric acid, HNO_3, formed from emissions of nitrogenous gases. The process of secondary particle formation is important to the atmospheric chemical cycles of sulfur and nitrogen and is discussed in detail in the sections of this chapter devoted to these topics. Here for the fluxes of Table 3.1, we assume that 50% of anthropogenic SO_2 production (Rodhe and Isaksen 1980) and 75% of natural biogenic SO_2 production from H_2S (Granat, Rodhe, and Hallberg 1976) is changed to sulfate aerosols.

Other Anthropogenic Aerosols

Direct addition of primary aerosols also results from fuel combustion (e.g., coal fly ash and fuel oil soot). The aerosols often contain trace elements, such as lead and vanadium, which can be used as tracers of the anthropogenic origin of aerosols. Certain combinations of trace elements in fine primary anthropogenic aerosols, whose proportions vary from area to area, are used as signatures of different regional pollution sources and can be correlated with secondary sulfate aerosols over long transport distances (>200 km) (Rahn and Lowenthal 1984).

DRY DEPOSITION OF AEROSOLS

In addition to aerosol removal from the atmosphere in precipitation by condensation processes (rainout) and by impaction with falling raindrops (washout), there is also *dry deposition* from the atmosphere. The processes of dry aerosol removal include (1) *sedimentation*, which involves gravity settling of larger (>20μ), and thus heavier, particles; and (2) *dry impaction* of aerosol particles on trees and foliage. If dry deposition is ignored, estimates of the amount of a substance delivered on land from precipitation may underestimate the total amount being delivered.

The amount of substances removed by sedimentation or gravitational settling is sometimes measured separately and referred to as *dry fallout*. Certain types of precipitation collectors that are continuously open to the atmosphere will include contributions by dry fallout in their precipitation collections and this combination of dry fallout and rainfall is referred to as *bulk precipitation* (Whitehead and Feth 1964). Because only large particles are heavy enough to settle out of the air rapidly, dry fallout is greatly subject to local influences. For example, Whitehead and Feth found considerable Ca, Mg, and SO_4 in dry fallout in Menlo Park, California, which they attribute to the presence of local gypsum and cement industries combined with SO_2 air pollution from San Francisco. They also found NaCl in dry fallout when the air was coming from the nearby Pacific Ocean.

In other areas away from the ocean or industry, the elements most represented in dry fallout are soil elements: K, Na, Ca, and Mg from windblown soil dust. Galloway and Likens (1978) feel that including dry fallout with precipitation (as bulk precipitation) creates problems because the composition of dry fallout is so variable by element and by season.

Dry impaction of aerosols on plants and trees is another way aerosols are removed from the atmosphere outside of precipitation. Foliage, particularly evergreens, scavenges aerosols from the air as the air passes by. Material also settles out on trees and this material, combined with biological exudates, is washed off by subsequent precipitation. The combination of rainfall plus soluble exudates and captured aerosols that wash off in passing through trees is known as *throughfall* (see Chapter 4).

In an area 30–40 miles from the ocean in England, White and Turner (1971) measured the quantities of various elements impacted on trees and collectors. They found considerable amounts of Na, K, and Mg, elements that had come from sea salt. In a more inland area in the mountains of New Hampshire, Schlesinger and Reiners (1974) measured the difference in composition between rainfall captured in the open (including sedimented aerosols) and throughfall (from artificial plastic "trees"), which also included dry impacted aerosols. They found that Ca was 3.5 times greater and Mg 1.4 times greater in this throughfall than in rain, but that Na and K were not greatly affected.

Graustein (1981) estimated the dry impaction input of various elements to throughfall in the mountains of New Mexico and found that the dust trapped by spruce and fir trees accounts for a large fraction of the total dissolved throughfall flux of Na, Ca, and SO_4. The type of vegetation is very important in dust entrapment; aspen in the same area of New Mexico as the spruce-fir was found by Graustein to trap very little dust.

In the marine atmosphere, dry deposition is largely from the recycling of coarse sea-salt aerosols. For large particles over the oceans (mainly sea salt and soil dust), wet and dry deposition are about equal. However, wet deposition dominates for small particles, which represent most of the atmospheric net particle flux from the continents to the oceans (Buat-Menard 1983).

CHEMICAL COMPOSITION OF RAINWATER: GENERAL CHARACTERISTICS

The major dissolved element composition of a large number of rainfalls is presented in Table 3.2. As can be seen, rainwater can be characterized as being dilute (with average total dissolved salt contents of a few milligrams per liter) and weakly acidic (pH 4–6). Dilution is brought about by the way rain forms. Evaporation into the atmosphere involves extensive separation of water molecules from dissolved salts in surface waters. The resulting water vapor ultimately condenses to form rain, and the overall process can be viewed as purification by natural distillation. However, rainwater is not totally pure. Solid particles and gases in the atmosphere are dissolved by rainwater, which results in a wide range in chemical composition, as well as in variations of pH. This section briefly summarizes, in tabular form, compositional variation, and ensuing sections cover in detail the origin of each major element in rain. In general, there are two characteristics of rain data that will be considered: (1) concentrations of the various ions, and (2) relative amounts of ions (i.e., ion ratios).

TABLE 3.2 Composition of Precipitation—World (in mg/ℓ)

Area	Na$^+$	K$^+$	Mg^{++}	Ca^{++}	Cl$^-$	SO$_4^{--}$	NO$_3^-$	NH$_4^+$	pH	Cl$^-$/Na$^+$	Reference
Coastal Europe											
S.W. Sweden coast 1967–1969	1.96	0.27	0.36	0.84	3.48	4.9	2.0	0.91	4.65	1.8	Granat 1972
S. Norway coast (polluted) 1972	11.0	0.59	1.58	0.90	20.38	7.87	3.35	0.43	4.15	1.85	Likens et al. 1979
W. Ireland coast (unpolluted) 1967	21.3	0.94	2.59	1.52	36.42	6.29	0.06	0.02	5.8	1.71	Likens et al. 1979
World average coastal (<100 km from ocean)	3.45	0.17	0.45	0.29	6.0	1.45	—	—	—	1.74	Meybeck 1983
Inland Eurasia											
W. Sweden 1956 (unpolluted)	0.16	0.12	0.10	0.70	0.36	1.39	0	0	5.4	2.25	Likens et al. 1979
N. Sweden 1967–1969	0.30	0.20	0.12	0.64	0.39	2.0	0.31	0.12	—	1.3	Granat 1972
S. and Central Sweden 1973–1975	0.35	0.12	0.17	0.52	0.64	3.31	1.92	0.56	4.3	1.83	Granat 1978
S. Norway 1974–1975 (polluted)	0.21	0.12	0.15	0.16	0.39	2.5	1.61	0.39	4.32	1.85	Likens et al. 1979
Belgium 1967–1969	0.97	0.23	0.36	1.32	1.95	6.0	2.23	0.48	4.42	2.0	Granat 1972
France 1967–1969	0.92	0.16	0.39	0.68	2.13	2.8	1.9	0.29	4.8	2.3	Granat 1972
Switzerland 1977	0.18	0.27	0.11	0.82	0.82	4.0	3.1	0.003	4.47	4.5	Zobrist and Stumm 1980
N. Europe 1955–1956 (average)	2.05	0.35	0.39	1.42	3.47	2.19	0.27	0.41	5.47	1.7	Carroll 1962
USSR (average pptn from cloud fronts)	0.4	0.2	0.3	0.4	0.8	2.7	0.2	0.5	—	2.0	Petrenchuk 1980
USSR-European (pptn=57 cm)	2.4	0.7	0.5	2.0	1.8	5.7	0.8	0.6	5.9	0.75	Zverev and Rubeikin 1973
USSR-Asian (pptn=45 cm)	1.55	0.7	0.2	2.1	1.5	4.35	0.7	0.8	6.0	0.97	Zverev and Rubeikin 1973

TABLE 3.2 Composition of Precipitation—World (in mg/ℓ) (continued)

Area	Na$^+$	K$^+$	Mg^{++}	Ca^{++}	Cl$^-$	SO$_4^{--}$	NO$_3^-$	NH$_4^+$	pH	Cl$^-$/Na$^+$	Reference
USSR-European:											
North	1.6	0.5	0.4	0.7	2.5	4.4	0.6	0.7	5.4	1.56	Petrenchuk and Selezneva 1970
Northwest	1.2	0.7	1.4	1.2	1.4	7.4	0.7	0.9	5.2	1.17	Petrenchuk and Selezneva 1970
Miscellaneous Land											
Bankipur, India (monsoon; 100 cm rain)	0.47	0.23	0.23	1.4	0.92	0.63	—	—	—	1.96	Handa B.K., 1971
S.E. Australia (average)	2.46	0.37	0.50	1.20	4.43	Trace	—	—	—	1.8	Hutton and Leslie 1958
Katherine, N. Central Australia (99 cm rain)	0.16	0.04	0.02	0.05	0.42	0.3	0.27	0.04	4.8	2.63	Galloway et al. 1982
Tavapur, India (70 km from Bombay)	2.4	0.16	0.32	1.4	4.4	1.3	—	0.13	6.15	1.8	Sequeira 1976
Japan (average)	1.1	0.26	0.36	0.97	1.2	4.5	—	—	—	1.1	Sugawara 1967
Kampala, Uganda (10 km from L. Victoria)	1.7	1.7	—	0.05	0.9	1.8	1.7	0.63	7.9	0.53	Visser 1961
Greenland (ice and snow)	0.007	—	—	0.007	0.021	0.12	—	0.006	—	3.0	Busenberg and Langway 1979
Marine											
Samoa	1.25	—	—	—	—	(0.07)[a]	0.01	[<0.03]	5.53	—	Pszenny, MacIntyre, and Duce 1982; [Logan 1983]
Pacific Ocean (34°46' N 177°15' W)	24	1.0	—	4.3	43	8.0	—	—	—	1.86	Gambell and Fisher 1966
Hawaii (near ocean)	5.46	0.37	0.92	0.47	9.63	1.92 (0.57)[a]	0.2	0.1	4.8	1.76	Eriksson 1957

64

Location										Reference
N. Atlantic Ocean (120 mi off N. Carolina)	2.8	—	0.2	0.2	5.1	1.2 (0.61)[a]	0.2	0.1	—	1.82	Gambell and Fisher 1964
N.W. Atlantic: Westward source	2.41	0.2	0.24	0.19	4.58	1.50 (0.87)[a]	0.42	0.07	4.66	1.90	Galloway, Knap, and Church 1983
Eastward source	3.62	0.2	0.38	0.19	6.46	1.22 (0.29)[a]	0.26	0.045	5.07	1.78	Galloway, Knap, and Church 1983
Bermuda 1955–1956	7.23	0.36	—	2.91	12.41	2.12	0.10	0.10	—	1.72	Junge and Werby 1958
Bermuda 1980–1981	3.38	0.17	0.41	0.19	6.2	1.74 (0.88)[a]	0.34	0.04	4.8	1.83	Galloway et al. 1982
S. Atlantic (250 km off Brazil)	5.34	0.18	0.73	0.17	10.26	2.87 (1.53)[a]	—	—	—	1.92	Stallard and Edmond 1981
S. Atlantic (85 km off Brazil)	2.99	0.17	0.39	0.15	5.01	2.38 (1.63)[a]	—	—	—	1.67	Stallard and Edmond 1981
Amsterdam Is., Indian Ocean (112 cm rain)	4.07	0.14	0.46	0.15	7.38	1.47 (0.42)[a]	0.11	0.07	4.9	1.81	Galloway, Knap, and Church 1982
South America **Amazon R. Basin** (mean)	0.285	0.039	0.029	0.044	0.49	0.49	0.13	—	5.03	1.72	Stallard and Edmond 1981
Over Amazon R. (670 km inland)	0.50	0.020	0.036	0.028	0.87	0.64	0.19	0.002	4.71	1.74	Stallard and Edmond 1981
Over Amazon R. (1700 km inland)	0.23	0.039	0.024	0.056	0.30	0.55	0.25	0.007	5.32	1.31	Stallard and Edmond 1981
Over Amazon R. (1930 km inland)	0.21	0.035	0.034	0.060	0.41	0.70	0.18	0.00	4.97	1.95	Stallard and Edmond 1981
Over Amazon R. (2050 km inland)	0.12	0.094	0.012	0.056	0.24	0.56	—	—	5.04	2.01	Stallard and Edmond 1981

TABLE 3.2 Composition of Precipitation—World (in mg/ℓ) (continued)

Area	Na^+	K^+	Mg^{++}	Ca^{++}	Cl^-	SO_4^{--}	NO_3^-	NH_4^+	pH	Cl^-/Na^+	Reference
Over Amazon R. (2230 km inland)	0.23	0.012	0.012	0.008	0.39	0.28	0.056	—	5.31	1.70	Stallard and Edmond 1981
Peru (3000 km from Atlantic)	0.039	0.039	0.020	0.184	0.12	0.18	—	—	5.67	3.08	Stallard and Edmond 1981
Venezuela (near coast)	2.2	0.6	0.7	1.14	2.6	2.2	0.2	0.3	—	1.2	Lewis 1981
Venezuela (San Carlos rain forest; 400 cm rain)	0.04	0.03	0.01	0.01	0.09	0.14	0.16	0.04	4.81	2.3	Galloway, Knap, and Church 1983
U.S. Coastal											
Bodie Is., N.C. 1955–1956	7.16	0.1	1.3	1.02	15.8	3.41	0.59	—	5.4	2.21	Gambell and Fisher 1966
Cape Hatteras, N.C. 1955–1956	4.49	0.24	—	0.44	6.9	1.22	0.04	0.01	—	1.54	Junge and Werby 1958
Cape Hatteras, N.C. 1962–1963	4.36	0.1	0.59	0.41	8.2	1.97	0.23	—	5.4	1.88	Gambell and Fisher 1966
N.J. Pine Barrens 1970–1972	1.39	0.32	0.23	1.10	2.82	5.09	0.39	—	—	2.02	Means et al. 1981
Stevensville, Nfld., 1955–1956	5.16	0.32	—	0.78	8.85	2.16	0.29	0.05	—	1.72	Junge and Werby 1958
Menlo Park, Calif., 1957–1959	2.0	0.25	0.37	0.79	3.43	1.39	0.16	—	6.0	1.7	Whitehead and Feth 1964
Tatoosh Is., Wash. (remote)	14.30	0.59	—	0.73	22.58	3.40	0.38	0.02	—	1.57	Junge and Werby 1958

Location											Reference
Brownsville, Tex., 1955–1956	22.3	1.0	—	6.5	22.0	10.68	0.13	0.01	—	1.0	Junge and Werby 1958
San Diego, Calif., 1955–1956	2.17	0.21	—	0.67	3.31	3.35	1.5	0.05	—	1.5	Junge and Werby 1958
U.S. coastal (average)	3.68	0.24	—	0.58	4.83	2.45	—	—	—	1.31	Whitehead and Feth 1964 (data from Junge and Werby 1958)
U.S. Inland (and Canada)											
Hubbard Brook, N.H., 1963–1974 (average)	0.12	0.07	0.05	0.17	0.51	2.87	1.47	0.22	4.1	4.25	Likens et al. 1977
N.E. U.S. 1978–1979 (average): All	0.36	—	—	—	0.40	2.81	1.58	0.31	4.2	—	Pack 1980
Noncoastal	0.32	—	—	—	0.29	2.70					
N.E. U.S. 1965–1968 (average)	0.27	0.16	0.11	0.60	0.45	4.3	0.34	0.22	4.4	1.7	Pearson and Fisher 1971
Ithaca, N.Y., 1972–1973	0.15	0.09	0.08	0.83	0.47	4.96	2.88	0.32	4.05	3.13	Cogbill and Likens 1974
N. Carolina and Virginia 1962–1963 (average)	0.56	0.11	0.14	0.65	0.57	2.18	0.62	0.1	4.9	1.01	Gambell and Fisher 1966
Gatlinburg, Tenn., 1973	0.05	0.07	0.03	0.20	0.15	3.19	1.24	0.19	4.19	3.0	Cogbill and Likens 1974
Tallahassee, Fla., 1978–1979: N. air	—	0.16	—	0.37	—	1.69	—	—	4.4	—	Tanaka, Darzi, and Winchester 1980
S. air	—	0.10	—	0.38	—	0.65	—	—	5.3	—	
Average	—	0.12	—	0.38	—	1.09	—	—	—	—	

TABLE 3.2 Composition of Precipitation—World (in mg/ℓ) (continued)

Area	Na$^+$	K$^+$	Mg^{++}	Ca^{++}	Cl$^-$	SO$_4^{--}$	NO$_3^-$	NH$_4^+$	pH	Cl$^-$/Na$^+$	Reference
Salem, Ill., 1972–1973	0.39	0.24	0.05	0.38	0.62	3.11	1.36	0.59	4.39	1.6	Miller 1974
Huron, S.D., 1972–1973	0.91	0.28	0.27	1.39	1.21	3.92	2.47	0.68	6.0	1.3	Miller 1974
Glasgow, Mont., 1955–1956	0.40	0.26	—	1.72	0.17	2.62	0.71	0.24	—	0.43	Junge and Werby 1958
Grand Junction, Colo., 1955–1956	0.26	0.17	—	3.41	0.28	4.76	0.98	0.26	—	1.08	Junge and Werby 1958
Columbia, Mo., 1955–1956	0.33	0.31	—	2.82	0.15	3.6	0.6	0.17	—	0.45	Junge and Werby 1958
Tewaukon, N.D., 1978–1979	0.27	0.23	0.27	1.05	0.20	1.74	1.59	0.86	5.27	0.74	Munger 1982
Itasca, W. Minn., 1978–1979	0.20	0.17	0.23	0.69	0.15	1.53	1.24	0.60	5.0	0.48	Munger 1982
Hovland, E. Minn., 1978–1979	0.14	0.13	0.13	0.40	0.10	1.89	1.18	0.67	4.67	0.71	Munger 1982
Amarillo, Tex., 1955–1956	0.22	0.23	—	2.7	0.14	1.86	0.68	0.05	—	0.64	Junge and Werby 1958
Tom Green Co., Tex., 1972–1973	0.86	0.15	0.05	0.14	0.61	3.17	1.5	1.5	5.98	0.7	Miller 1974
Bishop, Calif., 1972–1973	0.84	0.42	0.08	0.67	0.64	2.26	1.03	0.47	6.1	0.8	Miller 1974
Ely, Nev., 1955–1956	0.69	0.22	—	3.28	0.3	2.84	1.44	0.35	—	0.43	Junge and Werby 1958
Albuquerque, N.M., 1955–1956	0.24	0.18	—	4.74	0.09	2.39	0.86	0.09	—	0.37	Junge and Werby 1958

Location											Reference
Tesuque Mtn., N.M., 1975–1976	0.07	0.12	0.08	0.70	0.33	3.29	1.12	—	5.0	4.7	Graustein 1981
Santa Fe, N.M., 1975–1976	0.06	0.08	0.15	3.62	0.33	2.95	0.99	—	6.7	5.5	Graustein 1981
U.S. inland (average)	0.40	0.20	0.10	1.4	0.41	3.0	1.20	0.30	—	0.9	(From above)
U.S. (average) 1955–1956	0.90	0.23	[0.15]	1.0	1.13	2.02	0.70	—	—	1.26	Garrels and Mac-kenzie 1971; [Lodge et al. 1968]
Poker Flat, Alas. (70 km N. of Fairbanks; 29 cm ppm)	0.02	0.02	0.002	0.002	0.09	0.35	0.12	0.02	5.0	3.9	Galloway et al. 1982
Experimental Lakes Area, Ont.	0.19	0.13	0.11	0.45	0.35	4.32	0.11	0.38	5.0	1.84	Schindler et al. 1976
Haney, B.C., 1972–1973	0.3	0.1	0.1	0.2	0.6	1.3	0.7	0.1	4.5	2.0	Feller and Kimmins 1979

a Numbers in parentheses refer to excess sulfate—see text

The dissolved chemical components of rainwater can be divided into two groups: (1) those derived primarily from particles in the air (Na^+, K^+, Ca^{++}, Mg^{++}, and Cl^-), and (2) those derived mainly from atmospheric gases (SO_4^{--}, NH_4^+, and NO_3^-). The particles and gases, in turn, have a variety of sources, and the element associations in rain which result from these sources are given in Table 3.3. In addition, Tables 3.4 and 3.5 list the sources and typical ranges in concentration for each of the major dissolved components of rain.

The composition of condensation that ultimately falls to the ground as rainwater is determined by the composition of nucleating aerosols and soluble trace gases that react with water during both the condensation process and the fall to the ground. In the former case the process is called *rainout*, referring to reactions occurring within the clouds, while in the latter case the process is called *washout*, referring to reactions occurring below the clouds. Elements in rain that result from rainout will show little change or a slight rise in concentration with time. By contrast, elements contributed to rain by washout exhibit a sharp drop in concentration with time because the air becomes essentially cleansed (Junge 1963). Brief showers are washout-dominated. A drop in the concentrations of ions in rain with time due to washout always occurs for terrestrially dominated species (Ca^{++}, K^+, NO_3^-), which are concentrated in the lower atmosphere near the ground. Species derived from marine aerosols (Cl^-, Na^+, Mg^{++}) show washout near the coast where they are concentrated in the lower atmosphere, but inland, as marine aerosols become dispersed through the atmosphere, washout does not occur for these species (Stallard 1980).

TABLE 3.3 Primary Associations in Rain

Origin	Associations
Marine inputs	$Cl - Na - Mg - SO_4$
Soil inputs	$Al - Fe - Si - Ca - (K, Mg, Na)$
Biological inputs	$NO_3 - NH_4 - SO_4 - K$
Burning of vegetation	$NO_3 - NH_4 - P - K - Ca - Na - Mg - SO_4$
Industrial pollution	$SO_4 - NO_3 - Cl$
Fertilizers	$K - PO_4 - NH_4 - NO_3$

Sources: Modified after Stallard 1980; and Lewis 1981.

Na^+, Mg^{++}, K^+, Ca^{++}, AND Cl^- IN RAIN

The primary sources of dissolved Na^+, Mg^{++}, K^+, Ca^{++}, and Cl^- in rain are *marine* (sea-salt aerosols), *terrestrial* (soil dust, biological emissions), and *anthropogenic* (industrial, burning of vegetation) (see Table 3.4). The relative importance of marine sources varies with distance from the coast and levels off at a fairly

TABLE 3.4   Sources of Individual Ions in Rainwater

Ion	Origin		
	Marine Input	Terrestrial Inputs	Pollution Inputs
Na$^+$	Sea salt	Soil dust	Burning vegetation
Mg^{++}	Sea salt	Soil dust	Burning vegetation
K$^+$	Sea salt	Biogenic aerosols Soil dust	Burning vegetation Fertilizer
Ca^{++}	Sea salt	Soil dust	Cement manufacture Fuel burning Burning vegetation
H$^+$	Gas reaction	Gas reaction	Fuel burning to form gases
Cl$^-$	Sea salt Gas release from sea salt	—	Industrial HCl
SO$_4^{-}$	Sea salt Marine gases (DMS)	H$_2$S from biological decay Volcanoes Soil dust (Biogenic aerosols)	Burning of fossil fuels to SO$_2$ Forest burning
NO$_3^-$	N$_2$ plus lightning	NO$_2$ from biological decay N$_2$ plus lightning	Gaseous auto emissions Combustion of fossil fuels Forest burning Nitrogen fertilizers
NH$_4^+$	—	NH$_3$ from bacterial decay	Ammonia fertilizers Decomposition of human and animal wastes Combustion
PO$_4^{-3}$	—	Soil dust Biogenic aerosols Absorbed on sea salt	Burning vegetation Fertilizer
HCO$_3^-$	CO$_2$ in air	CO$_2$ in air Soil dust	—
SiO$_2$, Al, Fe	—	Soil dust	Land clearing

Source: Junge 1963; Mason 1971; Miller 1974; Granat, Rodhe, and Hallberg 1976; and Stallard and Edmond 1981.

TABLE 3.5 Typical Concentrations of Major Ions in
Continental and Marine Rainfall (in mg/ℓ)

Ion	Continental Rain	Marine and Coastal Rain
Na^+	0.2–1	1–5
Mg^{++}	0.05–0.5	0.4–1.5
K^+	0.1–0.5[a]	0.2–0.6
Ca^{++}	0.2–4[a]	0.2–1.5
NH_4^+	0.1–0.5[b]	0.01–0.05
H^+	pH = 4–6	pH = 5–6
Cl^-	0.2–2	1–10
SO_4^{--}	1–3[a,b]	1–3
NO_3^-	0.4–1.3[b]	0.1–0.5

[a] In remote continental areas: $K^+ = 0.02 - 0.07$; $Ca^{++} = 0.02 - 0.20$; $SO_4^{--} = 0.2 - 0.8$.

[b] In polluted areas: $NH_4^+ = 1 - 2$; $SO_4^{--} = 3 - 8$; $NO_3^- = 1 - 3$.

constant low level inland. A "hierarchy of ions" can be established (after Means et al. 1981 and Stallard and Edmond 1981) based on the relative importance of marine sea-salt sources and continental (terrestrial or pollutive) sources:

$$Cl^- = Na^+ > Mg^{++} > K^+ > Ca^{++} > SO_4^{--} > NO_3^- = NH_4^+$$

mostly marine   mostly continental

Of the principal cations, Na^+ is the dominant cation in areas of marine-influenced rain while Ca^{++} is the dominant cation in inland rain. In terms of both cations and anions, for an area near the ocean, the rain is basically a NaCl solution but rapidly changes to a $CaSO_4$ or $Ca(HCO_3)_2$ solution inland. In addition, there is a much greater content of total dissolved salts in marine-influenced rain than is usual for inland rains, except in arid areas.

Ratios of major ions can be used to compare rainfall compositions with that of sea salt. Sea-salt weight ratios with respect to the concentration of Cl^-, taken from the data for seawater composition given in Chapter 8, are shown in Table 3.6. Using known seawater ratios, the contribution of sea-salt ions to precipitation (assuming no fractionation on aerosol formation from seawater) can be determined by assuming that all Cl^- and the proportionate amount of other ions are derived from sea salt. The amount of the concentration of an ion that is greater than sea-salt proportions is referred to as *excess ion* (e.g., excess Na, etc.).

Unfortunately there have been only a few measurements of marine rain made over the oceans, but rain falling on oceanic islands like Bermuda, Hawaii, and Amsterdam Island (Indian Ocean) has a composition much like marine rain. As shown in Table 3.6, the ratio of Na/Cl is close to sea-salt proportions for all marine and island rains. The ratios of Mg^{++}, Ca^{++}, and K^+ to Cl^- are almost identical

TABLE 3.6 Major Element Composition, in Terms of Weight Ratios to Cl$^-$, for Seawater and Marine and Island Rains

Ion	Seawater	South Atlantic Rain	N. Atlantic Rain (1981) West Source	N. Atlantic Rain (1981) East Source	Pacific Rain	Bermuda Rain	Hawaii Rain	Amsterdam Is. Rain
Cl$^-$	1.00	1.00	1.00	1.00	1.00	1.00	1.00	1.00
Na$^+$	0.56	0.52	0.53	0.56	0.55	0.55	0.57	0.55
SO$_4^-$	0.14	0.28	0.33	0.20	0.22	0.28	0.20	0.20
Mg^{++}	0.067	0.071	0.052	0.059	—	—	0.096	0.062
Ca^{++}	0.021	0.016	0.041	0.029	0.037	0.031	0.048	0.020
K$^+$	0.021	0.018	0.044	0.032	0.035	0.027	0.038	0.019

Sources: See Chapter 8 for seawater data and Table 3.2 for rainwater data.

to sea-salt proportions for South Atlantic rain (Stallard and Edmond 1981) and Amsterdam Island rain (Galloway et al. 1982). There are varying small amounts of contamination by continental cations (Ca^{++}, Mg^{++}, and K$^+$) in the other island and marine rains. However, the ratio of SO$_4$/Cl in all these rains is about twice the sea-salt ratio; this is presumably due either to transported sulfate pollution or to marine sulfate sources other than sea salt (see discussion below in section on sulfate in rain).

Marine coastal rainwater generally has close to sea-salt ratios to Cl$^-$ for Na$^+$, and Mg$^+$, but Ca^{++} and K$^+$ ratios may be greater than sea salt due to varying continental inputs (see Table 3.2). The definition of "coastal" rain varies considerably from less than 40 km from the coast in the northeastern United States (Pearson and Fisher 1971) to less than 100 km from the coast in Europe (Meybeck 1983).

Over the oceans the Cl content of rain is around 10–15 mg/ℓ, but over land the concentration drops rapidly within 10–20 km of the coast (Junge 1963). In the United States, Cl drops off to a fairly constant concentration of 0.15–0.20 mg/ℓ by 600 km inland. This is shown in Figure 3.4. A similar pattern is shown by Munger and Eisenreich (1983). The concentration of chloride in precipitation over the Amazon Basin also decreases sharply with increasing distance from the Atlantic Coast. The average value levels off at 1200 km inland to 0.35 mg/ℓ (Stallard and Edmond 1981), which is somewhat higher than the U.S. value, and declines gradually thereafter. A larger and deeper penetrating marine contribution might be expected here because the prevailing winds in the Amazon Basin are from off the Atlantic Ocean, while they are from the continental interior in the eastern United States.

The decrease in Cl$^-$ content inland is usually interpreted as rapid deposition of sea salt in precipitation near the coast. However, Junge (1963) has offered an

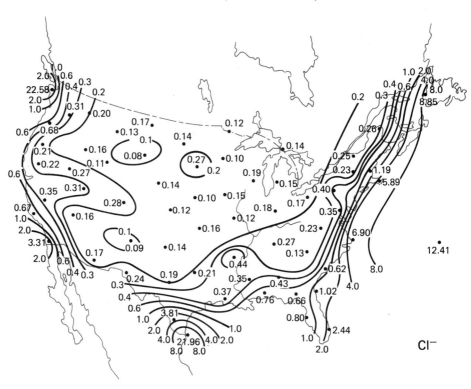

Figure 3.4 Average Cl⁻ concentration (mg/ℓ) in rain over the United States, July 1955–June 1956. [After C. E. Junge and R. T. Werby. "The Concentration of Chloride, Sodium, Potassium, Calcium and Sulfate in Rainwater over the United States," *J. of Meteorology*, 15 (October 1958) 418. © 1958 by the American Meteorological Society.]

additional explanation involving air mixing. Sea salt over the oceans is concentrated in the lower 500 m of air, but when the maritime air moves over the land, there is intense vertical mixing of the air up to 7000 m. Thus, even if the *amount* of sea salt in the air remains constant, the *concentration* of sea salt in the air is considerably reduced because the sea salt is mixed through a much greater volume of air.

There is often a change in the Cl/Na ratio from the seawater value on going inland. This could be due to (1) fractionation—differential removal of sea-salt components; (2) addition of terrestrial ions to the rain (particularly soil dust sodium, which would lower the ratio); or (3) anthropogenic Cl⁻ (which would increase the ratio).

In the Amazon Basin, where the prevailing winds are off the Atlantic Ocean and where pollution is minimal and soil dust is low, Stallard and Edmond (1981) found that the Cl/Na ratio (and the Cl/Mg ratio) of inland rain remains close to sea-salt proportions even 2000 km inland. This confirms that there are not *major* changes due to fractionation of the marine component going inland and that explanation 1 above is generally unimportant.

The marine influence in U.S. inland rain is considerably less. The prevailing winds are from the west and, although they bring marine air onshore along the West Coast, much of the marine aerosols are not carried inland because of rainout and washout over the Pacific coastal mountains. On the eastern coast of the United States, the prevailing winds are from over the continent, particularly in the summer, although in the winter storms move up the coast from the Gulf of Mexico bringing marine air onshore. The United States is also considerably dustier than the Amazon Basin. All these factors cause concentration ratios of rain to vary considerably from that of sea salt. Junge and Werby (1958) measured the Cl/Na ratio of U.S. precipitation and found the rain ratios to be close to the seawater ratio (1.8) at the coast but to decrease rapidly inland, levelling off at a value of 0.5 to 0.8 by 500 miles inland (Figure 3.5). They calculated the size of the sea-salt contribution

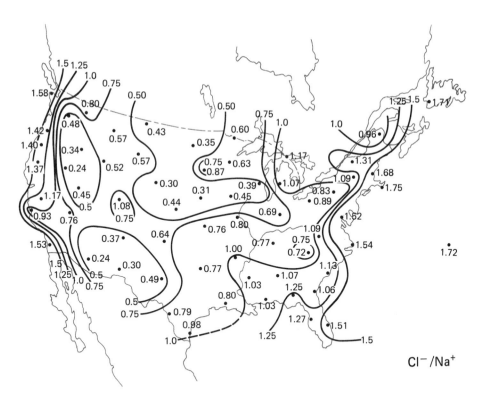

Figure 3.5 Average weight ratio of Cl$^-$/Na$^+$ in rain over the continental United States, July 1955–June 1956. Sea salt Cl$^-$/Na$^+$ = 1.8. [After C. E. Junge and R. T. Werby. "The Concentration of Chloride, Sodium, Potassium, Calcium and Sulfate in Rainwater over the United States," *J. of Meteorology*, 15 (October 1958), 419. © 1958 by the American Meteorological Society.]

of various ions to rain assuming that all the Cl^- originated as sea salt. On this basis sea salt contributes only from one-half to one-third of the total Na in interior U.S. rain (which averages 0.2 to 0.3 mg/ℓ); the remainder is thought to be from soil dust. Maps of total Na^+, and the "excess" not contributed by sea salt are shown in Figure 3.6. Note particularly the area of excess Na^+ in the Great Basin, an arid area of igneous rocks relatively enriched in Na^+ (over most sedimentary rocks), and one with many saline (Na-rich) dry lake beds (Gambell 1962), presumably good sources of Na-rich dust. (This high Na^+ area also appears on Munger and Eisenreich's [1983] map of Na^+ in rain.) Further evidence for a soil dust origin for Na comes from a decrease of Na in rain going from agriculturally developed (and presumably dusty) North Dakota to heavily forested Minnesota (Munger 1982) and in the high excess Na shown in Figure 3.6B for south Texas.

Gambell and Fisher (1966) measured Na^+, Cl^-, and Mg^{++} in rain from North Carolina and Virginia in an area ranging from the coast to 200 miles inland. The Cl/Na ratio was down to 1.5 within 20 miles of the coast and in the furthest inland location had dropped to 0.8 (vs. 1.8 in sea salt). Based on Cl/Na ratios, they felt that, for the area as a whole, about 60% of the Na^+ in rain was from sea salt and 40% from soil dust. Excess Na^+, excess Ca^{++}, and excess Mg^{++} (excess being above sea-salt ratios) all had similar monthly patterns and areal distributions, which again suggests a common soil dust origin for all three ions. Values of Cl/Na ratio in European rain, where the prevailing winds are off the Atlantic, range from 2.3 to 1.3, showing considerably more marine influence than U.S. continental values (see Table 3.2).

All Cl in rain is not derived from sea salt; some may come from Cl-containing gases. Sources of anthropogenic Cl, as HCl and other gases, in the atmosphere include automobiles, coal combustion, and burning of polyvinyl chloride in incinerators (Paciga and Jervis 1976). There has been an increase in pollutive Cl from these sources in the northeastern United States in the last 20 years (Cogbill and Likens 1974), which is reflected in Cl/Na ratios greater than the sea-salt ratio. Similarly in Europe and England, Cl/Na ratios greater than that for sea salt are found in industrial areas due to Cl pollution. Because of pollutive Cl, some workers tend to determine sea-salt contributions to precipitation on the basis of Na, not Cl. However, determining sea salt on the basis of Na is also problematical because of the variable soil dust contributions of Na (particularly inland) as pointed out earlier.

Figure 3.6   Dissolved Na^+ (in mg/ℓ) in continental United States rainfall, June 1955–June 1956. (A) Average Na^+ concentration. (B) Average excess Na^+ concentration, calculated as the difference between measured values shown in (A) and the values expected for seawater Cl^-/Na^+ ratios and the Cl^- concentrations shown in Figure 3.4. [After C. E. Junge and R. T. Werby. "The Concentration of Chloride, Sodium, Potassium, Calcium and Sulfate in Rainwater over the United States," *J. of Meteorology*, 15 (October 1958), 418, 420. © 1958 by the American Meteorological Society.]

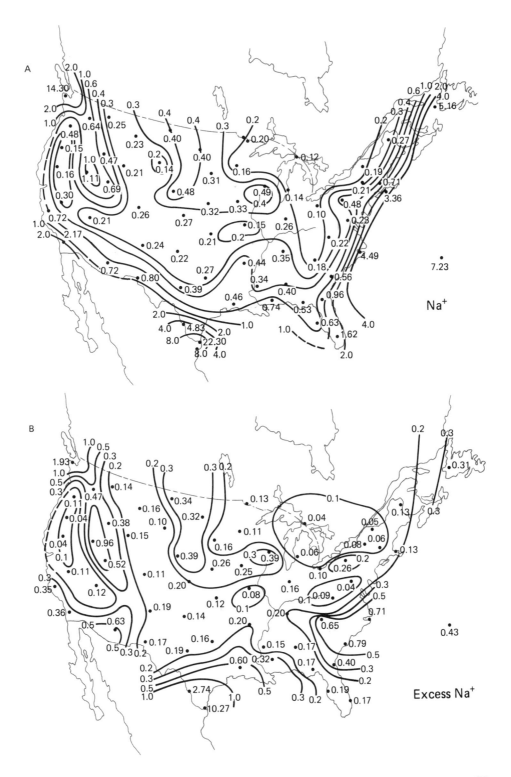

A

Na⁺

B

Excess Na⁺

The concentration of Mg in sea salt is less than that of Na by roughly a factor of 10; thus, sea salt is not as important a source of Mg except in coastal areas or in strongly marine-influenced areas such as the Amazon Basin where sea-salt Mg/Cl ratios are still found 2000 km inland (Stallard and Edmond 1981). Only one-third of the Mg delivered in North Carolina and Virginia rain (Gambell and Fisher 1966) is estimated to be from sea salt and the average concentration levels off at about 0.15 mg/l within about 40 miles of the coast. Soil dust is the dominant source of Mg in most interior U.S. rain (Munger and Eisenreich 1983). Most Mg values for interior U.S. rains (Lodge et al. 1968; see also Table 3.2) are around 0.1–0.2 mg/l except for the arid western and southwestern areas where they are considerably higher due to greater windblown soil dust. For example, in eastern North Dakota (Thornton and Eisenreich 1982), where there is considerable wind-blown agricultural dust, rain concentrations of Mg are 0.33 mg/l while even higher values are found in the western prairies of Canada (Munger and Eisenreich 1983).

The contribution of sea salt to Ca^{++} in continental rain is very small since the Ca/Cl ratio in sea salt is only 4% of the Na/Cl ratio. Instead Ca^{++} comes primarily from the dissolution of $CaCO_3$ in soil dust. Calcium carbonate dissolves in rain to form HCO_3^- and Ca^{++} by the reaction:

$$H^+ + CaCO_3 \rightarrow Ca^{++} + HCO_3^-$$

In this way neutralization of rain acidity comes about. In the case of strong acids like H_2SO_4 (when SO_2 is present in the air), this neutralization results in the production of CO_2 and acid anions such as SO_4^{--}:

$$H_2SO_4 + CaCO_3 \rightarrow CO_2 \uparrow + H_2O + SO_4^{--} + Ca^{++}$$

The average Ca^{++} concentration in U.S. rain is around 1.0 mg/l but there is considerable areal variation from 0.3 to 4.0 mg/l as can be seen in Figure 3.7A. In general, Ca^{++} is the dominant cation in inland U.S. rain. The maximum Ca^{++} concentrations are found in the arid zones of the Southwest where $CaCO_3$ is a major constituent of the topsoil due to continual evaporation of soil water at the ground surface, and where windblown soil dust is common. Large Ca^{++} (and other ion) concentrations in arid areas also are the result of small and infrequent amounts of precipitation.

Enrichment of Ca^{++} in rain also occurs over less arid areas such as the agricultural region of the midwestern United States in Kansas, Iowa, Nebraska, and Missouri. This can be attributed to cultivation of calcareous soils (Gambell

Figure 3.7 Concentrations (in mg/ℓ) in rainwater over the continental United States, June 1955–June 1956 of (A) dissolved Ca^{++}, and (B) dissolved K^+. [After C. E. Junge and R. T. Werby, "The Concentration of Chloride, Sodium, Potassium, Calcium and Sulfate in Rainwater over the United States," *J. of Meteorology*, 15 (October 1958), 421. © 1958 by the American Meteorological Society.]

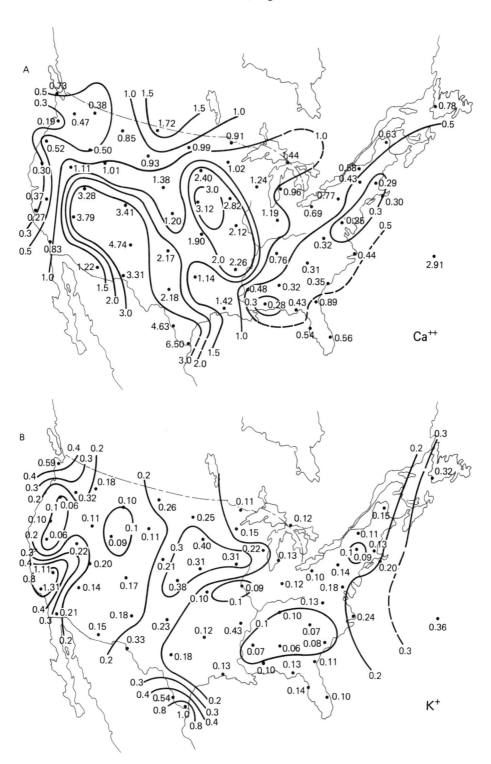

1962). Here, as in arid areas, most of the soil dust is transported only on the order of 200–400 miles. High concentrations of Ca^{++} tend to correlate with Mg in North American rain, presumably due to a similar soil dust source (Munger and Eisenreich 1983).

Gambell (1962) has used the Ca^{++}/Cl^- ratio in rain as an indication of natural continental aerosol versus marine aerosol sources. As can be seen in Figure 3.8, the Ca^{++}/Cl^- ratio in U.S. rain (in 1955–1956) is near that for sea salt ($\leqslant 1.0$) only for a very small fraction of the land surface (Gambell 1962). (The sea-salt value is 0.02.) For the rest of the country, the dominant source of calcium is from the land.

Ca^{++} can be produced by pollution from coal burning and from cement manufacture. Coal contains about 0.4% Ca and 3% sulfur but the elements are usually separated in the air after burning, since Ca appears in ash and would fall in rain near the source, while sulfur is released as SO_2 gas and is dispersed over a large area. Pearson and Fisher (1971) note that higher levels of both Ca^{++} and SO_4^{--} tend to occur in northeastern U.S. urban rain and that these ions also correlate in precipitation from industralized English areas. In Menlo Park, California, Whitehead and Feth (1964) attribute Ca^{++} in rain to industrial pollution from gypsum processing and cement manufacture.

Potassium in U.S. rain (see Figure 3.7B) tends to have small and fairly uniform concentrations (0.1 to 0.2 mg/l) over the whole country (Junge and Werby 1958),

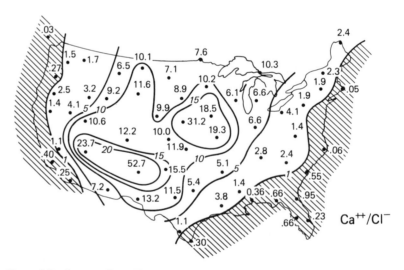

Figure 3.8 Average Ca^{++}/Cl^- ratios (g/g) in rain over the United States, July 1955–June 1956. Sea-salt $Ca^{++}/Cl^- = 0.02$. Note that Ca^{++}/Cl^- ratios $\leqslant 1.0$ are shaded. (After Gambell 1962, computed from data by Junge and Werby 1958.)

showing far less variation than Ca. European interior rain has similar concentrations, while Amazon interior rain has much lower K concentrations (0.03 to 0.07 mg/l), about one-fourth of U.S. and European concentrations (see Table 3.2). Based on the chloride content, sea salt contributes only about 10% of K in U.S. interior rain, whereas in Amazon rain, with its much lower K concentrations, it may account for about 25% of the K.

There are several possible nonmarine origins for K in continental rain, including (1) dissolution of soil dust; (2) K-containing fertilizers, which contribute K to soil dust; (3) pollen and seeds; (4) biogenic aerosols; and (5) forest burning, particularly in tropical areas. The relative importance of these sources varies strongly from area to area.

A dissolved soil dust origin for potassium in U.S. interior rains has been suggested by Junge (1963), who points out that the average Na/K ratio of such rains (1.7) is more like the average soil ratio (0.5) than like sea salt (27). Similarly, the Mg/K ratio of interior rain is more like soil than sea salt (see Table 3.7). Munger (1982) measured K$^+$, Ca^{++}, and P$_{total}$ in precipitation for three sites on a 600-km transect going from agricultural-prairie (eastern North Dakota) to mixed forest (eastern Minnesota). He found that K$^+$ follows Ca^{++} in having the highest average concentration in rain over the agricultural-prairie area and the highest seasonal concentration there in the fall (0.86 mg/l). Concentrations of K and Ca decrease sharply toward the forested area. This trend presumably indicates that K and Ca are dominated by agricultural soil dust from wind erosion. In the forested area, K concentrations are lower, vary less (from 0.08 to 0.2 mg/l), and are highest in spring precipitation paralleling concentrations of P$_{total}$, which are attributed to pollen and seeds in this area.

Saharan soil dust contributes a significant fraction (5%–10% usually, but up to 38%) of the water-soluble K in tropical North Atlantic aerosols, although the dominant K source is sea salt (Savoie and Prospero 1980). This shows again, that windblown dust can travel great distances, even across entire oceans, and hereby contribute to the composition of rainfall (Junge 1972).

Graustein (1981) has pointed out problems in identifying a source for K in rain if it comes from soil particles, because reasonable mineral candidates for soil sources of K (illite, feldspar, and mica) also contain SiO$_2$, but SiO$_2$ is virtually absent in most rainfall. This criticism can be met if the time of contact of rainwater with soil dust sources is short. Initial reaction in weathering of these minerals is K$^+$ exchange with H$^+$ ion, which would not involve addition of silica to solution. In addition, K may be enriched over usual soil concentrations and appear in a more soluble form in agricultural soil dust due to addition of K fertilizers.

Biogenic aerosols are another possible K source. As noted earlier in the aerosol section, plants exude waste products, on their leaf surfaces, which can then escape as particles into the atmosphere. This would produce a soluble salt (K$_2$SO$_4$) that could contribute K$^+$ and SO$_4^{--}$ to rain. Lawson and Winchester (1979)

TABLE 3.7 Composition of Various Sources of Aerosols, expressed as Weight Ratios to K[+], Compared with Rainfall over the Interior United States

Material	Na/K	Mg/K	Ca/K	S/K	Al/K	Reference
Rain:						
Interior U.S. Eastern interior	1.66	0.81	9.26	7.4	—	(Table 3.2)
U.S. Western interior	1.0	0.83	4.2	9.6	—	(Table 3.2)
U.S.	2.1	0.79	12.6	5.7	—	(Table 3.2)
Soil:						
U.S. average	0.52	0.4	1.04	—	2.5	Shacklette et al. 1971
E. U.S.	0.35	0.31	0.43	—	4.5	Shacklette et al. 1971
W. U.S.	0.6	0.46	1.06	—	3.2	Shacklette et al. 1971
U.S. from igneous rocks (less–more mature)	0.5–0.3	1.0–0.3	2.0–0.7	0.05	—	Bohn, McNeal, and O'Connor 1979
World average	0.46	0.36	1.0	0.05	5.1	Bowen 1966
Crustal rock	1.1	0.81	1.4	0.01	3.14	Mason 1966
Average platform sedimentary rock	0.34	1.1	4.7	0.26	3.0	Holland 1978 (Table 8.1)
Seawater	27.0	3.2	1.0	2.25	—	Bowen 1966
Angiosperms	0.09	0.23	1.29	0.24	0.04	Connor and Shacklette 1975
Trees (U.S.)	0.01	0.25	2.0	—	0.03	Connor and Shacklette 1975
Crops (U.S.)	0.001	0.12	0.14	—	0.003	Connor and Shacklette 1975
Soil ash of burned tropical vegetation (cane and grass)	0.42	0.25	0.8	0.2	—	Lewis 1981
Urban aerosol:						
Toronto	0.75	1.6	6.1	—	2.4	Paciga and Jervis 1976
Chicago	0.5	1.13	2.5	—	1.25	Gatz 1975

believe these plant exudates explain the presence of K (associated with P and S) in coarse aerosols of the Amazon Basin of South America, which is a humid, heavily forested area where dust production is minimal. Stallard and Edmond (1981) also believe that K in interior Amazon Basin rain is derived primarily from plant exudates and point out that K aerosol concentrations observed in the atmosphere would yield rain concentrations close to those observed. The concentrations of soluble K found by Crozat (1979) in biogenic aerosols of the Ivory Coast would give about 0.04 mg/l K if converted to rain. According to Crozat, the formation of biogenic aerosols should be favored by high heat and humidity, tropical conditions that do not generally exist in the United States. However, Graustein (1981) has evidence from his work in New Mexico, to suggest that K may be given off locally by trees to the atmosphere and thus appear in rain.

In tropical areas such as the Amazon Basin, Venezuela, and the Ivory Coast, forest burning is used during the dry season to clear the land, and this can contribute to K in rain. Lewis (1981) found that fine atmospheric particulates produced by forest burning were flushed out in considerable quantities at the onset of the rainy season in coastal Venezuela, resulting in high rain concentrations of K^+, Ca^{++}, SO_4^{--}, Na^+, and Mg^{++} followed by a rapid decline as the season progressed. In contrast to the tropics, there is little effect on rainfall composition by forest burning in North America, because of its much lower frequency there.

GASES AND RAIN

Gas-Rainwater Reactions

Because atmospheric water droplets and raindrops are small, they have large specific surface areas available for the exchange of gases between the droplets and the atmosphere. As a result, atmospheric gases become dissolved in rainwater both during the condensation process (rainout) and while the rain is on its way to the ground (washout). Washout occurs if the concentration of the gas below a cloud is greater than it is within the cloud or if the earlier reaction within the cloud is not complete.

The amount by which a gas dissolves in rainwater depends upon (1) its *partial pressure* or concentration in the atmosphere, and (2) its solubility in water. (Solubility, in turn, is a function of temperature.) Partial pressure is pressure exerted by each gaseous component in a mixture, so that together the sum of partial pressures equals total pressure. The ratio of partial pressure to total pressure is a measure of concentration and is equivalent to the volume fraction of a given gas in air. Volume fraction (expressed as a percent) for a number of atmospheric gases are listed in Table 3.8. Also shown in Table 3.8 are the concentrations in aqueous solution of some major gases for equilibrium with air at a total pressure

TABLE 3.8   Concentration of Gases in Air and in Solution

Gas	Volume Percent in Air	Equilibrium Concentration in Solution[a] (mg/ℓ)
Nitrogen (N_2)	78.084	13.5
Oxygen (O_2)	20.948	6.7
Argon (Ar)	0.934	—
Carbon dioxide (CO_2)	0.035	0.5
Neon (Ne)	0.0018	—
Helium (He)	0.0005	—
Methane (CH_4)	~0.0002	0.05
Sulfur dioxide (SO_2)	0–0.0001	—
Nitrogen dioxide (NO_2)	0–0.000002	—
Ammonia (NH_3)	0 to trace[b]	—
Krypton (Kr)	0.0001	—
Hydrogen (H_2)	0.00005	—
Ozone (O_3)	0–0.000001	—

Source: Air composition from Turekian 1972.

[a] For equilibrium with air at T = 25° C and a total pressure of 1 atmosphere.

[b] ~ 2×10^{-7} (Söderlund and Svensson 1976).

of 1 atmosphere (and T = 25° C).   The dissolved concentration divided by the volume fraction in air is a measure of the relative solubility for each gas.

Some gases not only dissolve in water, but also react with H_2O and with other gases, before and after dissolving, to form new species.   This includes carbon dioxide (CO_2), ammonia (NH_3), sulfur dioxide (SO_2), and nitrogen dioxide (NO_2). In the case of CO_2, SO_2, and NO_2, acids are produced: H_2CO_3 (carbonic acid), H_2SO_4 (sulfuric acid), and HNO_3 (nitric acid); whereas for NH_3, a base is formed: NH_4OH (ammonium hydroxide).   (The details of these reactions are discussed later in this chapter.)   Sulfuric and nitric are strong acids that completely dissociate:

$$H_2SO_4 \rightarrow 2H^+ + SO_4^{--}$$

$$HNO_3 \rightarrow H^+ + NO_3^-$$

whereas carbonic acid and ammonium hydroxide are weak and only partially dissociate:

$$H_2CO_3 \rightleftarrows H^+ + HCO_3^-$$

$$NH_4OH \rightleftarrows NH_4^+ + OH^-$$

Either way, the ions SO_4^{--}, NO_3^-, HCO_3^-, NH_4^+, and H^+ result and these are major components of rainwater.

Gases and Cl⁻ in Rain

Since the chloride ion (Cl^-) in rain comes primarily from sea salt, chloride is frequently used as a indicator of the amount of sea salt present, as noted earlier. However, chlorine can also exist as a gas (HCl) in the atmosphere produced by reactions with sea salt or arising from industrial sources. (Other gaseous forms of Cl may exist, but have not been well documented.) In these cases, the amount of chloride in rain may not be an accurate indication of sea-salt content.

Hydrogen chloride (HCl) gas can be formed by the reaction of hygroscopic sea-salt particles (NaCl) with polluted air containing either NO_2 or SO_2 gas. The NO_2 and SO_2 ultimately form HNO_3 and H_2SO_4, and the HCl is then formed by the reactions:

$$HNO_3 + Cl^- \rightarrow NO_3^- + HCl_{gas} \uparrow$$

$$H_2SO_4 + 2Cl^- \rightarrow SO_4^{--} + 2HCl_{gas} \uparrow$$

The HCl gas escapes to the atmosphere and the chloride content of the sea-salt aerosol is accordingly reduced (Eriksson 1952). However, since HCl gas is highly soluble in water, a good deal of it may be redissolved in precipitation as Cl^- if it is not removed from the area. Thus, while the Cl/Na ratio of sea-salt aerosol would be reduced, the rain Cl/Na ratio might not be.

Martens et al. (1973) found an average reduction of the sea-salt Cl/Na ratio of around 10% due to the reaction of the chloride in sea-salt aerosol with HNO_3 gas. This loss occurs as the sea salt is transported over land and is greater in smaller particles because of their greater surface area (per unit volume) for reaction. In North Atlantic sea salt aerosols, Meinert and Winchester (1977) found a 25%– 40% deficiency of Cl^- with respect to K^+ and Ca^{++} (which occur in sea-salt proportions to Na) accompanied by a S/K ratio two to three times that of sea salt. This suggests HCl loss by reaction with H_2SO_4.

As noted earlier, pollutive HCl is found in industrial areas and may raise the Cl/Na ratio of rain above the sea-salt ratio. This occurs in England, continental Europe, and the United States.

SULFATE IN RAIN: THE ATMOSPHERIC SULFUR CYCLE

As can be seen in Table 3.2, sulfate is the most abundant ion in almost all but purely marine rains. It is also the principal indicator of worldwide atmospheric pollution and the major culprit in the formation of acid rain. These observations demand that there be detailed discussion of both the natural and anthropogenic processes affecting sulfate concentrations in rain as well as discussion of general atmospheric sulfur pollution. This is done in the present section. Our ultimate goal will be to derive quantitative estimates of the major fluxes in the atmospheric cycle of sulfur.

Sulfate in rain has three major and roughly equal sources: (1) sea-salt aerosols; (2) sulfur dioxide from fossil fuel burning; and (3) biogenic reduced sulfur gases, such as H_2S and $(CH_3)_2S$, from natural metabolism and organic decay. Other less important sources include (4) volcanic emissions of sulfur dioxide, hydrogen sulfide, and particulate sulfates; (5) forest burning; (6) soil dust; and (7) plant aerosols.

Sea-Salt Sulfate

Since sulfate is a major seawater constituent, the amount of rain sulfate of sea-salt origin generally is determined from the known weight ratio of SO_4^{--}/Cl^- in seawater (which equals 0.14) and the measured concentration of Cl^- in rain. In polluted areas, because there is excess rain Cl^- of non-sea-salt (industrial) origin, and because some Cl^- may be lost as Cl_2 gas, the ratio of SO_4^{--}/Na in seawater and the measured concentration of Na in rain is often used instead. The use of the sea-salt ratio of SO_4/Cl (or SO_4/Na) assumes that there is no fractionation between seawater and sea-salt aerosol. This assumption has been challenged for sulfate by Garland (1981), who claims limited enrichments of approximately 10% in SO_4/Cl and SO_4/Na ratios during the formation of sea-salt aerosol from seawater, but his results can be explained without invoking fractionation (see Duce, MacIntyre, and Bonsang 1982). The amount of rain sulfate remaining, after sea-salt sulfate has been subtracted from total sulfate, is referred to as *excess sulfate*.

The amount of sea-salt sulfate cycled through the atmosphere has been estimated by Eriksson (1960) as approximately 44 Tg S/yr based on sea-spray production. (Note that 1 Tg $= 10^6$ metric tons $= 10^{12}$ g). Only 10% of this sea spray SO_4^{--} falls on the land (4 Tg S/yr) and the rest is simply redeposited on the ocean. As a result, the percentage of total land sulfur deposition due to sea spray is small ($<10\%$). Over the oceans sea spray is a major source of sulfur deposition ($\sim 50\%$), but even here about half of the sulfur comes from other sources. (Other estimates of sea-salt sulfur production range from 40 to 300 TgS/yr; see Andrae and Galbally 1984).

Anthropogenic SO$_2$

Much of the excess sulfate in rain in heavily populated areas arises from the oxidation of sulfur dioxide gas, which is given off to the atmosphere during the combustion of fossil fuels (coal and oil). The SO_2 forms during combustion from the oxidation of sulfur contaminants present as pyrite (FeS_2) in coal and organic sulfur compounds in both coal and oil. Fossil fuel combustion from stationary sources, particularly that from the burning of coal for electricity, accounted in 1975 for 78% of human-produced sulfur oxides in the United States, with industries such as refining and smelting accounting for most of the rest (see Table 3.9). Automobiles accounted for only 2% of the sulfur oxides produced, in contrast to their

TABLE 3.9 U.S. Emissions of Sulfur to the Atmosphere as Sulfur Oxides (SO_2 and SO_3) from Combustion in 1975

Source	Tg S/yr	Percent of Total
Power and heating:		
Electricity genera-tion—coal	8.2	55
Electricity genera-tion—oil	0.9	6
Heating	2.5	17
Subtotal	11.6	78
Industrial (smelting and refining)	3.0	20
Vehicles (e.g., auto emissions)	0.3	2
Total	14.9	100

Source: Based on data from Nader 1980.

much greater contribution to pollutive nitrogen oxides (~40%; see Table 3.16). Table 3.10 shows that a similar percent distribution of sources for sulfur oxides is found for the world as a whole.

Because of problems with oil supply and cost, the United States in the future will probably turn increasingly to coal for energy. Most coal has a higher sulfur content (mean 2% S by weight) than either oil (mean 0.3%—0.8% S) or natural gas (0.05% S), (Möller 1984). This is due, to a large extent, to pyrite in coal which does not occur in oil. In the case of oil, the pollutive sulfur added to the

TABLE 3.10 Global Emissions of Sulfur to the Atmosphere as Sulfur Oxides from Combustion

Source	Tg S/yr	Percent of Total
Coal burning	45.4	70
Oil burning	5.5	8
Industrial (smelting and oil refining)	13.6	21
Transportation	0.4	1
Total	64.9	100

Source: After J.P. Friend. "The Global Sulfur Cycle" in *Chemistry of the lower atmosphere*, ed. S.I. Rasool. © 1973 by Plenum Press. All rights reserved.

atmosphere has been reduced by U.S. laws requiring power plants to burn low-sulfur fuel oils ($<0.5\%$ S). The Environmental Protection Agency has ruled that 70% to 90% of gaseous sulfur must be removed from the emissions of all *new* coal-burning power plants; however, old plants are not similarly regulated (Likens et al. 1979). In 1975, U.S. SO_2 emissions were about 15 Tg SO_2-S/yr (Table 3.9), while in 1980 they had dropped to about 11.3 Tg SO_2-S/yr (Marshall 1983). Figure 3.9 shows U.S. SO_2 (and NO_x) emissions and European SO_2 emissions over the period 1940–1980. (Note the relative flatness of the U.S. SO_2 curve relative to that for Europe.)

The total world production of SO_2-S by humans during the period 1970–1975 amounted to about 65–70 Tg S/yr (Möller 1984; Friend 1973), and the 1985 production is estimated to be 90 Tg S/yr (Möller 1984). (In constructing our atmospheric sulfur cycle in Table 3.12 the value from 1975, 65–70 Tg S/yr, is used to correspond with the rain data from the 1960s and 1970s). Of the various estimates of fossil fuel that have been made, these values are a minimum; for instance, Cullis and Hirschler (1980) estimate 1976 production at 104 Tg S/yr. In any case, fossil fuel sulfur is roughly equal to natural sulfur sources (excluding sea salt), and thus is an important input to the world sulfur cycle.

On a global scale, pollution sources of human origin are not evenly distributed, but are concentrated in the Northern Hemisphere. In 1975, U.S. SO_2 emissions

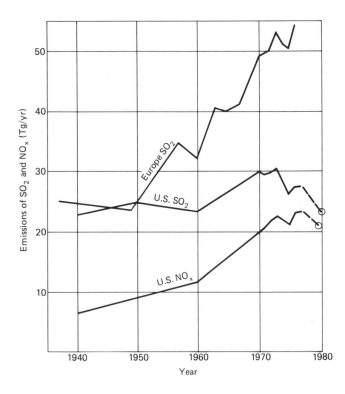

Figure 3.9 Rise in emissions of SO_2 and NO_x from 1940 to 1980. Top curve: European SO_2; middle curve: U.S. SO_2; bottom curve: U.S. NO_x. To compare SO_2 data here with data given elsewhere in the present chapter, divide by 2 to obtain Tg S/yr. 1980 U.S. SO_2 emissions from Marshall 1983; 1979 U.S. NO_x emissions from Logan 1983 [see also Table 3.16].(From G. E. Likens et al., "Acid Rain." Copyright © October 1979 by Scientific American, Inc. All rights reserved.)

were roughly 25% of the total worldwide emissions while Europe produced 40% (Figure 3.9). Thus, in northwestern Europe and eastern United States, pollution is a dominant sulfur source. A single smelter, that located at Sudbury, Ontario, produces sulfur emissions equal to the total global emissions of volcanic sulfur and around 1% of *all* sulfur emissions for the globe (Likens et al. 1979). Furthermore, Galloway and Whelpdale (1980) have calculated that anthropogenic sulfur emissions make up 90% of the total atmospheric sulfur input for eastern North America.

SO$_2$ gas has a residence time in the atmosphere of 1.5 days (corresponding on the average to a 500-km transport distance), before it is removed by dry deposition or is converted to sulfate. In addition, particulate sulfate often stays in the atmosphere for about 5 days and may be transported 1000 km or more before it is removed in precipitation (Rodhe and Isaksen 1980). Thus, the effects of anthropogenic sulfur oxide production tend to be felt over a large area, often as far as 1000 km from the sources of pollution. This is particularly true in Scandinavia where three-fourths of the pollutive sulfate in rain emanates from central Europe and the eastern British Isles (Rodhe 1972), and in the northeastern United States, which receives pollution from the industrialized portions of the Middle West. In addition, U.S. power plants have built taller smoke stacks in the last 20 years, which, while reducing local SO$_2$ concentrations, increase SO$_2$ and sulfate transport over wider areas.

Sulfur oxide pollution by humans has been going on since the Industrial Revolution, and the greater use of fossil fuels in the last 20–30 years has in many areas greatly increased the amount of SO$_2$ in the air and sulfate (and acidity) in rain (Likens et al. 1979). An example of this is shown in Figure 3.10. Here we compare excess sulfur deposition in 1960 (Figure 3.10A) for U.S. rain with total sulfate deposition in 1980 (Figure 3.10B). (Although the figures are not strictly comparable, the main difference between *total* sulfate-sulfur deposition and *excess* sulfate-sulfur deposition is at the coast; inland, the differences are negligible.) The bull's-eye pattern in the northeastern United States is similar in both years, and the greatest rate of sulfur deposition reached (12 kg S/ha/yr) is also similar, indicating a levelling off of sulfate concentrations (see also NRC 1983). However, the area covered in 1960 by high sulfur deposition (greater than 10 kg S/ha/yr in Figure 3.10A) has expanded considerably into the southeastern United States in 1980 (represented by greater than 30 mmoles SO$_4$/m^2/yr in Figure 3.10B, which is equivalent to 10 kg/S/ha/yr).

In Europe, while sulfur dioxide emissions increased greatly from 1960 to 1975 (Figure 3.9), most stations had constant or decreasing total sulfate deposition (wet and dry) suggesting considerable transport outside the area (Grant 1978).

Conversion of Sulfur Dioxide to Sulfate in Rain

There are two main processes, whose relative importance varies, involved in the conversion (oxidation) of SO$_2$ gas to sulfate (SO$_4^{--}$) which can be removed in rain or dry deposition. First, SO$_2$ gas reacts with neutral OH radicals in gaseous form in the atmosphere (via intermediate stages represented here by ". . . ") to form sulfuric acid aerosol (NRC 1983):

$$SO_2 + OH \rightarrow \ldots \rightarrow H_2SO_4$$

(Reactions of SO_2 with CH_3O_2 (Rodhe and Isaksen 1980) or with HO_2 (Galbally, Farquhar, and Ayers 1982) apparently also occur but are less important). In the second process, SO_2 dissolves in cloud droplets (or liquid aerosol particles) and is rapidly converted to sulfuric acid by reaction with H_2O_2 (Rodhe, Crutzen, and Vanderpol 1981):

$$SO_2 + H_2O_2 \rightarrow H_2SO_4$$
(in liquid cloud droplets)

Sulfuric acid (H_2SO_4) is highly soluble in water and is a strong acid which completely dissociates to hydrogen and sulfate ions:

$$H_2SO_4 \rightarrow 2H^+ + SO_4^{--}$$

In this way H_2SO_4 contributes to the formation of acid rain (see section below on acid rain).

Oxidation of SO_2 does not always result simply in the production of H_2SO_4. When ammonia is also present in polluted air, it reacts with sulfuric acid droplets to form ammonium sulfate particles, which raise the pH by consuming H^+ ions:

$$2NH_3 + 2H^+ + SO_4^{--} \rightarrow (NH_4)_2SO_4$$

$$NH_3 + 2H^+ + SO_4^{--} \rightarrow (NH_4)HSO_4$$

Other substances such as $CaCO_3$ can also react to neutralize sulfuric acid. (For further discussion of ammonium sulfate and acid netralization, consult sections on nitrogen in rain and acid rain respectively.)

Rain can also remove SO_2 directly from the air as it falls to the ground (washout). However, it has been found that even in heavily polluted air most of the sulfate in rainwater comes from the rainout of sulfate aerosols, not washout of SO_2 gas during rainfall (Gravenhorst et al. 1978).

Figure 3.10   (A) Excess sulfur deposition in precipitation over the United States in kg S/ha/yr (g $\times$ 10^{-1} S/m^2/yr) (Eriksson 1960). (B) Wet deposition of (total) sulfate in 1980 over the United States in mmoles SO_4/m^2/yr (NRC 1983). 1 mmole SO_4/m^2/yr = 0.32 kg S/ha/yr. Note that the 30 mmole SO_4/m^2/yr contour in (B) corresponds to the 10 kg S/ha/yr in (A).

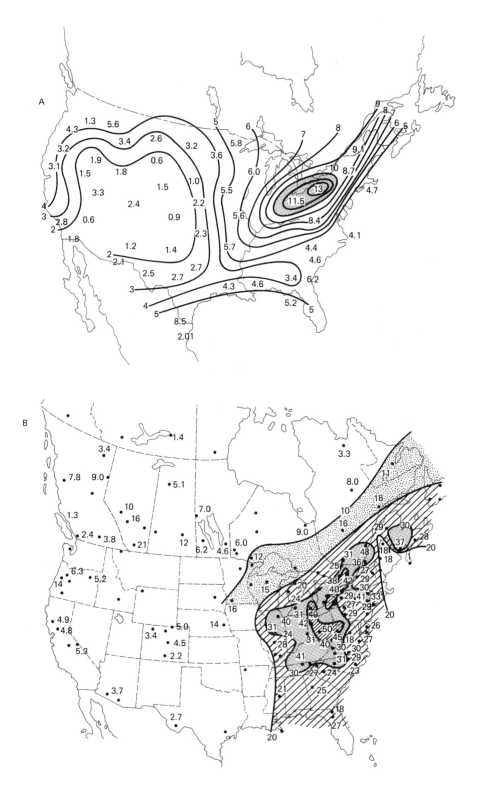

Sulfate from the Natural Biogenic Production of Reduced Sulfur Gases

The main *natural* source of sulfate in rain is from the oxidation in the atmosphere of biologically produced reduced-sulfur gases: hydrogen sulfide, H_2S; dimethyl sulfide, $(CH_3)_2S$, usually abbreviated as DMS; and other organic sulfur compounds. Oxidation probably proceeds via a large number of intermediate steps to the formation of sulfur dioxide gas (Rodhe and Isaksen 1980; Galbally, Farquhar, and Ayers 1982). For example,

$$H_2S + OH \rightarrow \ldots \rightarrow SO_2 \text{ (4 days)}$$

$$(CH_3)_2S + OH \rightarrow \ldots \rightarrow SO_2 \text{ (3 days)}$$

and the SO_2 is subsequently oxidized to H_2SO_4 as discussed above.

The amount of natural biogenic reduced sulfur (H_2S, DMS, etc.) added to the atmosphere over land and over the oceans is not yet well known. There has been considerable interest in biogenic sulfur release recently because of the necessity of providing a natural background with which to compare the importance of anthropogenic (pollutive) sulfur input. In addition, methods for the measurement of the release of reduced sulfur gases over the land have been perfected recently by a number of workers (Delmas et al. 1980; Adams et al. 1981; Aneja, Overton, and Aneja 1981). Many previous atmospheric sulfur budgets (Kellogg et al. 1972; Friend 1973; Granat, Rohde, and Hallberg 1976) based estimates of production of biogenic H_2S on the difference between removal of sulfur from the atmosphere in rain and dry deposition, and known sources of atmospheric sulfur. Because the production and deposition fluxes of sulfur are not very well known, these estimates of total (land and ocean) natural biogenic sulfur production have ranged from 35 Tg S/yr to 108 Tg S/yr (for example, see Table 3.13). In addition, these budgets have differed on how the production is divided between land and ocean. Here we shall not follow this route but, instead, shall try to independently estimate the natural biogenic flux itself.

Four types of potential source areas for natural biogenic reduced sulfur are (roughly in order of release rate) as follows: (1) marine coastal areas; (2) anoxic, high-organic inland soils; (3) aerobic (oxygenated) inland soils; and (4) marine open ocean. In the *marine coastal areas* fine-grained sediments in shallow near-shore marine tidal flats and estuaries contain *anoxic* (oxygen-free) high-sulfate interstitial water and are generally high in decomposable organic matter, some of which may be pollutive in origin. These are all ideal conditions for bacterial sulfate reduction, the process whereby dissolved sulfate is reduced to H_2S by bacteria (see Chapter 8). (Release of this H_2S in marine tidal flats is often high enough to produce a noticeable odor). In *anoxic, high-organic inland soils*, H_2S is produced on land in swampy areas, flooded soils, freshwater lakes, and high-organic soils—

all land areas with anoxic conditions. (Tropical rain forests are an outstanding example). With respect to *aerobic (oxygenated) inland soils*, rather surprisingly it has been found that H_2S and other natural sulfur compounds are released at a low level from aerated inland soils, probably as a result of plant metabolism. Because of the large areal extent of such soils, this can be an important source. In the *marine open ocean* environment, DMS (and other reduced organic sulfur compounds) are believed to be produced at a low rate by the metabolic activity of surface-dwelling marine organisms, including bacteria living on dead organic matter.

Although marine coastal areas have long been known to be a source of H_2S (Conway 1942), only recently have measurements been made of the H_2S (and DMS) fluxes there. Results show that coastal areas, even though they constitute only a small part of the land surface (~3%), are an important contributor of sulfur to the atmosphere. Measurements of average biogenic reduced sulfur fluxes from marine coastal areas (salt marshes and mud flats) range from 0.04 g H_2S-S/m²/yr in Denmark (Jaeschke 1978) to 1.2 g S/m²/yr (primarily H_2S and DMS) in the southeastern United States. (Adams et al. 1981). We, like Adams, feel that because bacterial activity is accelerated by increased temperature, a still higher flux should be used for a worldwide average. Using a worldwide area of marine coastal regions of about 0.4×10^{12} m² (Andrae and Galbally 1984) and an average flux of 2 g S/m²/yr, we obtain a very rough value of 1 Tg S/yr for sulfur given off to the atmosphere from this source. (Obviously, there is a need for more data on this subject.)

Water-logged, anoxic, highly organic soils are common in tropical forests, and high temperatures favor bacterial activity. Thus, these forests are potentially a good source of H_2S from the decomposition of organic matter. (H_2S is also emitted from the leaves of living plants; Winner et al. 1981). Delmas et al. (1980) measured an annual flux of H_2S from the humid Ivory Coast equatorial forest of 0.4–0.9 g S/m²/yr, more than sufficient to account for the entire rain deposition of sulfur. The release of H_2S from the Amazon Basin tropical forest has also been inferred as a source of rain sulfate concentrations by Stallard and Edmond (1981). Assuming an emission flux of 0.9 g S/m²/yr and an area of tropical forests of 17×10^{12} m² (Zehnder and Zinder 1980), we calculate that approximately 15 Tg of biogenic sulfur is produced annually. (This is less than Delmas and Servant's [1983] estimate of 25 Tg S/yr from tropical rain forests made using a box modelling scheme.) In any case, tropical rain forests seem likely to be the dominant natural sulfur source on land.

It has been generally assumed that aerated inland soils are not a good biogenic sulfur source. Recently, however, several workers have measured fluxes of H_2S and other reduced sulfur gases from dry mineral inland soils of one-fifth to one-tenth the rate from anaerobic organic soils. The release rate increases with increasing temperature. We have used the measured average flux of 0.06 g S/m²/yr from inland soils in the eastern and southeastern United States. (Adams et al.

1981), similar to 0.04 g $S/m^2/yr$ in France (Delmas et al. 1980), as typical of temperate soils. Since the area of temperate soils is 66×10^{12} m^2 (Zehnder and Zinder 1980), this would give an annual flux of biogenic sulfur of 4 Tg S/yr.

Thus, land production of natural biogenic reduced sulfur from tropical forests and inland soils is about 19 Tg S/yr, with another 1 Tg S from marine coastal areas, giving a total land production of 20 Tg S/yr. This is much less than the estimate of Adams et al. (1981) of 64 Tg S/yr which is based on a model using exponential increases in sulfur release with increased temperature toward the equator. In areas like the United States, the natural background biogenic sulfur production is swamped by anthropogenic SO_2 production as a contributor to sulfate in rain; by contrast, biogenic sulfur can be important in remote areas, particularly the tropics.

No direct measurements have been made of reduced sulfur release from the open ocean; consequently, it is only inferred from the composition of aerosols, concentration of sulfur gases, and from rain sulfate deposition over the ocean. Metabolic activity and biological decay of marine organisms, such as phytoplankton, produce volatile reduced biogenic sulfur compounds (primarily DMS), which are oxidized to SO_2 and sulfate aerosol. The resulting concentration of SO_2 gas and sulfate aerosol seem to be higher in zones of high biological productivity of surface seawater. Estimates have been made of oceanic biogenic sulfur production based on SO_2 concentrations in remote oceanic areas (27–72 Tg S/yr; Nguyen, Bonsang and Gaudry 1983) based on reduced sulfur gas concentrations (30 Tg S/yr; Graedel 1979), and excess sulfate and SO_2 concentrations (15 Tg S/yr; Kritz 1982). Excess aerosol sulfate concentrations and the sulfate deposition rate in remote marine atmospheres give a marine biogenic sulfur production of 27–28 Tg S/yr (Bonsang et al. 1980; Andrae 1982). Average concentrations of DMS in surface seawater give a large sulfur flux to the atmosphere: 39 Tg S/yr (Andrae and Raemdonck 1983). In our summary sulfur budget (see Table 3.12 and Figure 3.11), we have assumed a reduced sulfur flux from the oceans of about 28 Tg S/yr based on these estimates. This is more than half of the sulfur produced by sea spray (44 Tg S/yr).

Volcanic Sulfur

Sulfur in the atmosphere can also be produced by volcanic activity. Volcanic emissions are mainly SO_2 and sulfate with lesser amounts of H_2S. Berresheim and Jaeschke (1983) estimate a total production of 11.8 Tg S/yr from both eruptive and noneruptive volcanic activities. However, based on rain measurements near Kilauea volcano in Hawaii (Harding and Miller 1982), the effect of noneruptive volcanic emissions are very localized (within 10 km of source). Large volcanic eruptions, on the other hand, increase the number of sulfate particles in the so-called Junge layer at the bottom of the stratosphere for several years. Therefore, in our budget we have included only *eruptive* volcanic emissions of 2 Tg S/yr (0.5

Tg SO_2-S and 1.3 Tg SO_4-S; Berresheim and Jaeschke 1983) as being likely to be transported long distances in the atmosphere (see Table 3.12). Much of this sulfur should be deposited over the oceans since they occupy two-thirds of the earth's surface.

Forest Burning

Forest burning is used extensively in tropical areas to clear the land; wood is burned as fuel, and wildfires also consume a very small amount of biomass. The total biomass burned is estimated at about 7000 Tg of dry matter per yr (Seiler and Crutzen 1980; Logen et al. 1981). Assuming 0.2% S in land plants and 50% volatilization, Andrae and Galbally (1984) estimate a flux of 7 Tg S/yr from biomass burning. This is dominantly anthropogenic (96%; Logan et al. 1981).

Soil Dust Sulfate

Windblown dust can contribute sulfate to precipitation as $CaSO_4$. The top layers of soil in arid areas may contain $CaSO_4$ due to the evaporation of soil water and, because of the dryness, the dust is easily picked up by the wind. Most of the soil particles are probably large enough to have a short residence time in the atmosphere. However, on a regional or local scale in arid regions, dust may be an important source of sulfate in rain. Lodge et al. (1968), as well as Junge (1963) and Munger (1982), suggest that the high concentration of sulfate in rain in the arid areas of the western United States may be due to $CaSO_4$ dust. (An alternative explanation is that the $CaSO_4$ in the rain is actually derived from the reaction of windblown $CaCO_3$ with H_2SO_4 in droplets.) Russian sulfate measurements of rain in semiarid and arid areas are very high, presumably also due to dust. From the estimate of 700 Tg of windblown dust which is transported long distances (see Table 3.1) each year and an average S content of dust of 0.1%, we assume that 1 Tg S is added to the atmosphere each year in the form of dust.

Plant Aerosols

Biogenic sulfur-containing particles are known to be produced from decaying organic matter, from wax-covered plant surfaces in heavily forested areas, and from pea plants and pine trees. Lawson and Winchester (1979) found evidence for *biogenic production* of sulfur, potassium, and phosphorus in *coarse* aerosol particles over tropical rain forests in South America. (This is distinct from biogenic H_2S production, which would be in fine aerosols.) Since the coarse biogenic aerosol concentrations measured by Lawson and Winchester could produce at most 0.01 mg/ℓ SO_4 in Amazon rain, which is only 2% of the measured rain concentration of sulfate, plant aerosols are not very important even in this tropical area (Stallard and Edmond 1981). Similarly Delmas et al. (1980) in the Ivory Coast ruled out major local contributions of sulfate to rainwater by plant fragments.

Sulfate Deposition in Rain

In order to construct an atmospheric sulfur budget it is necessary to know how much sulfate is delivered to the earth's surface in rain. Sulfate concentrations in rain vary both with time and with location (see Table 3.2). (There are also analytical problems in accurately measuring sulfate at the low concentrations often found in rain.) The usual range in excess sulfate over land is from less than 1 to 10 mg SO_4/ℓ with higher values being due to pollution. The *unpolluted* remote land background concentration of excess sulfate in rain has been estimated as 0.5–0.6 mg/ℓ SO_4 (Kramer 1978; Granat et al. 1976) and about 0.5 mg/ℓ SO_4 has been measured in Amazon Basin rain (Stallard and Edmond 1981). These values are considerably less than most North American or European values, which reflect varying degrees of pollution.

Estimates of the total amount of excess sulfate deposited on the continents in rain range from 43 to 84 Tg S/yr (Robinson and Robbins 1975; Kellogg et al. 1972; Friend 1973; Granat, Rodhe, and Hallberg 1976), based on different estimates of the *average* excess sulfate concentration in rain. Granat, Rodhe, and Hallberg (1976) who estimate continental deposition of 43 Tg S/yr, feel that continental deposition values given in earlier studies are too high because of an overemphasis on sulfate concentrations in relatively polluted U.S. and European rain, which do not represent a true global average.

We have independently calculated rain sulfate deposition on land to be 53 Tg S/yr based on the average concentrations of excess sulfate in rain, average amount of precipitation, and different types of land area (see Table 3.11). Adding 4 Tg S/yr as due to cyclic sea salt gives 57 Tg S/yr for total land deposition in precipitation.

Estimates of excess sulfate deposition in rain over the oceans also vary a great deal, because the concentration of excess sulfate in ocean precipitation is even less well known than that over land. (Values range in Table 3.2 from 0.07 to 1.5 mg/ℓ.) Junge (1963), using 0.5 mg/ℓ excess SO_4 in rain, calculated excess ocean sulfur deposition in rain as 57 Tg S/yr, while Granat, Rodhe, and Hallberg's (1976) estimate of 23 Tg S/yr corresponds to an average rain excess concentration of 0.2 mg/ℓ SO_4. A large part of the excess sulfur in rain over the North Atlantic may be anthropogenic, representing transport of sulfate aerosol from the North American continent (Galloway, Knap, and Church 1983).

Calculating the amount of excess sulfate deposition in rain over the oceans based on excess sulfate in marine rain does not seem very practical for several reasons: (1) a small change in excess rain sulfate concentrations results in a large change in sulfate deposition because of the large area involved; (2) there are very few observations of marine rain sulfate (see Table 3.2) and they do not give very

TABLE 3.11   Rate of Excess Sulfate Deposition on Land by Precipitation (Predominantly Rainfall)

Continent	Mean Annual Precipitation[a] (cm)	Nondesert Area[b] (10^{12} m^2)	Excess SO_4^- (mg/ℓ)	Deposition Rate (Tg SO_4-S/yr)
Africa	69	21.8	1.0	5.0
Asia:	73			
Clean		24.8	1.5	9.1
Polluted		9.0	4.0	8.8
Europe: polluted	73	10.0	4.0	9.5
North America:				
Clean	67	18.0	1.5	6.0
Polluted[c]	100	5.2	3.0	5.2
South America	165	17.1	0.9	8.5
Antarctica	17	12.2	0.25	0.2
Australia	44	4.9	0.8	0.5
Total				52.8

NOTE: Excess sulfate defined as total sulfate minus sea-salt sulfate. Tg = 10^{12} g.

[a] From Lvovitch 1973.

[b] From Meybeck 1979.

[c] N. America polluted area = eastern United States and Canada (Galloway and Whelpdale 1980).

reliable excess sulfate estimates; (3) estimates of the average amount of oceanic precipitation also vary greatly. As a result we shall try to make a rough estimate of marine rain sulfate deposition based on balancing the sulfur cycle. (See section on atmospheric sulfur cycle below.)

Dry Deposition of Sulfur

Besides sulfur deposition in precipitation ("wet" deposition), there is also "dry" deposition of sulfur. SO_2 gas may be absorbed or dissolved directly by ocean water and on land by vegetation, soil, and other surfaces. Also, sulfate particles may settle out of the air or be trapped by vegetation (dry fallout). The size of these fluxes in the sulfur cycle are even less well known than those involving precipitation because they are harder to measure.

Dry deposition of SO_2 gas over land is dependent upon the characteristics of the surface, small-scale meteorological effects, and the atmospheric concentration of SO_2 (NRC 1983). Dry deposition is greater near the emission sources of the gas and decreases more rapidly with distance from the source than does wet deposition. For example, the rate of dry deposition of SO_2 is twice that of wet deposition in central Europe near pollution sources and about equal to wet deposition downwind in northern Scandinavia (Granat, Rodhe, and Hallberg 1976). In the northeastern United States, dry deposition is estimated to be about equal to wet deposition for pollutant sulfur (NRC 1983).

For the continents, we have calculated a total dry deposition of SO_2-S to be about 28 Tg S/yr. This is based on a gas deposition velocity of 0.8 cm/sec (Garland 1978) and a concentration of 0.5 μg SO_2-S/m^3 in clean areas (Robinson and Robbins 1975), plus a concentration of 2.5 μg/m^3 of SO_2-S in urban areas (Eliassen 1978). The land areas are the same as those used in Table 3.11 for precipitation. The same result (~30 Tg S per year) is obtained by letting dry deposition equal wet deposition for polluted land areas (in Table 3.11) and letting dry deposition equal one-fourth of wet deposition for nonpolluted land areas (the method used by Granat, Rodhe, and Hallberg 1976). Fine sulfate aerosols (0.1–1 μm) formed by the oxidation of SO_2 are, because of their slow settling rate, removed largely in precipitation and generally contribute little to the dry deposition flux (Garland 1978; Granat, Rodhe, and Hallberg 1976). (In forested areas, fine aerosols may be trapped by leaves, enchancing dry deposition, but this flux should be minor, on the order of 2 Tg S/yr.) Also, dry deposition of coarse sulfate-containing aerosols (sea salt and dust) is only of local importance over the land.

Seawater, with a pH of 8, is sufficiently basic to be a good absorber of SO_2. Several authors (Friend 1973; Varhelyi and Gravenhorst 1981, quoted in Church et al. 1982) consider dry deposition of SO_2 to constitute 20% of oceanic sulfur deposition (with the remainder in wet deposition). Mészáros (1982) calculated dry deposition of SO_2-S as 8 Tg S/yr based on SO_2 concentrations and deposition velocity over the oceans. This is the value used in our budget (Figure 3.11 and Table 3.12). Fine sulfate aerosols are removed mainly in rain over the oceans, not by dry deposition (Granat, Rodhe, and Hallberg 1976). Sea-salt aerosols are coarse and do undergo dry deposition but this flux is included in the overall return sea-salt flux (wet plus dry) of 40 Tg S/yr to the oceans.

Atmospheric Sulfur Cycle

In constructing an overall atmospheric sulfur cycle we have already derived and discussed most major fluxes. However, we have not considered the transport of sulfur between continental and marine air masses. The transfer of sulfur from the continental air mass to the oceanic air mass is assumed to be about 17 Tg S/yr. This number is not well known and estimates range from 8 to 17 Tg S/yr (Friend 1973; Kritz 1982; Ivanov 1981; Granat, Rodhe, and Hallberg 1976). Granat, Rodhe, and Hallberg's estimate was that one-fourth of anthropogenic emissions of SO_2-S (or about 17 Tg S/yr) are transported from land to sea, since much heavy industry is located upwind from the ocean (in the United States and Japan, for example). It has been estimated that one-third of U.S. and Canadian SO_2-S production (or 4 Tg S/yr) is exported (NRC 1983), as is one-third of western European production (~5 Tg S/yr; Granat, Rodhe and Hallberg 1976), although only part of the latter actually reaches the ocean. We have also assumed a reverse transport to land of marine sulfur from biogenic sources of 4 Tg S/yr, which is equal to, and in addition to, the sea-salt sulfur flux. This is based on measured ratios of SO_4/Cl in unpolluted marine rain and aerosols that are approximately

TABLE 3.12 Atmospheric Sulfur Cycle: (see also Figure 3.11)

	Flux (Tg S/yr)	Source
Air Over Continents		
A. *Input from Land*		
Anthropogenic SO_2	65–70	Möller 1984 (for 1970–1975)
Forest burning (mainly anthropogenic)	7	Andrae and Galbally 1984
Volcanic emissions (long transport)	2	(See text)
Dust SO_4 (long transport)	1	(See text)
Biogenic H_2S, DMS, etc., from land		
(tropical forests and soil)	19	(See text)
Biogenic H_2S, DMS, etc., from coastal areas	1	(See text)
Total input from land	95–100	
B. *Deposition over Land*		
Rainfall:		
As excess sulfur	53	(See text)
As sea salt	4	Eriksson 1960
Dry deposition as SO_2 (mainly, and minor sulfate)	30	(See text)
Total deposition over land	87	
C. *Input via Transport from Oceans*		
Sea salt	4	Eriksson 1960
Marine biogenic S	4	(See text)
Total transport from ocean	8	
D. *Transport to Marine Atmosphere*	17	(See text)
Balance: A + C − B − D = −1 to 4		
Air Over Oceans		
A. *Input from Ocean*		
Biogenic sulfur from open ocean	28	Bonsang et al. 1980; Andrae 1982
Sea spray	44	Eriksson 1960
Total input from oceans	72	
B. *Deposition over Ocean*		
Rainfall:		
As excess sulfate-sulfur	33	(For balance; see text)
As sea salt (includes dry deposition of sea salt)	40	Eriksson 1960
Dry deposition as SO_2	8	(See text)
Total deposition over oceans	81	
C. *Input via Transport from Land*	17	(See text)
D. *Transport to Continental Atmosphere*		
Sea salt	4	Eriksson 1960
Marine biogenic S	4	(See text)
Total transport to continental atmosphere	8	
Balance: A + C − B − D = 0[a]		

[a] Balanced by rainfall deposition of excess sulfur.

239004

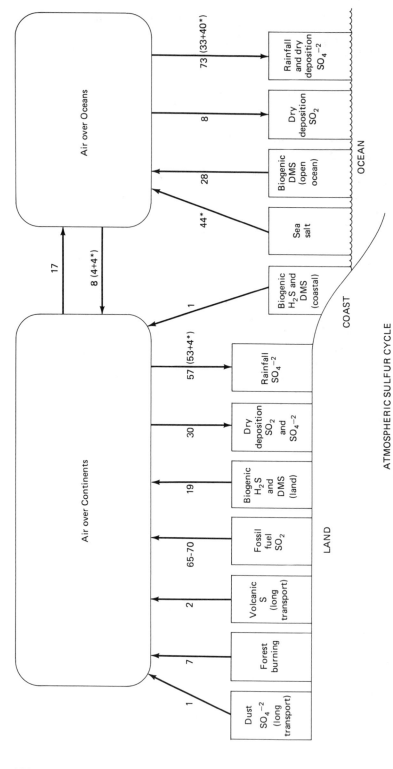

Figure 3.11 The atmospheric sulfur cycle. Values are fluxes in Tg S/yr (Tg = 10^{12} g). Values denoted by an asterisk refer to sea salt.

twice the sea-salt ratio (see Table 3.6; Bonsang et al. 1980; Stallard and Edmond 1981; Galloway et al. 1982).

After all these transport considerations, we have a net surplus of non-sea-salt sulfur of 41 Tg S/yr over the oceans which must be removed. This is done by apportioning 8 Tg SO_2-S to dry deposition of SO_2 (see previous discussion; this number is poorly known), so that the remaining 33 Tg SO_4-S/yr must be deposited in rain. Marine rainfall deposition of 33 Tg of excess sulfur per year would amount to 0.26 mg/ℓ of excess sulfate in rain (based on an ocean area of 3.6×10^{14} m² and 106 cm of rain per year.) This falls within the range of other previously mentioned estimates (0.2–0.5 m/ℓ SO_4). Redeposition of sulfate on the ocean as sea salt in rain (and dry deposition) is about 40 Tg S/yr (Eriksson 1960). Thus, total oceanic sulfur deposition in rain and dry deposition is about 81 Tg S/yr (see Table 3.12).

Our global atmospheric sulfur budget is summarized in Figure 3.11 and Table 3.12. As can be seen, we have obtained a balance within -1 to 4 Tg S/yr for air over the continents. Considering the uncertainties involved, this suggests that our estimated fluxes may be reasonable. In summary, for the world as a whole (land plus sea), the inputs to the atmosphere are 51 Tg S/yr (31%) for natural non-sea-salt emissions (primarily biogenic reduced sulfur); 44 Tg S/yr (26%) for sea salt; and 72–77 Tg S/yr (~43%) for pollutive sulfur (including forest burning, which is mainly anthropogenic). Excluding sea salt, natural emissions contribute about 40% of the input of excess atmospheric sulfur (see Table 3.13). This is lower than the 60%–65% of non-sea-salt natural emissions (see Table 3.13), estimated in some earlier calculations in order to balance atmospheric sulfur budgets. We feel that the earlier natural emission values are too high because they were used to balance abnormally high sulfur deposition rates. Highly polluted European and U.S. rains were used to represent world average rain sulfate concentrations and this included vast areas little affected by pollution.

TABLE 3.13 Estimates of Annual Global Emissions of Gaseous Sulfur into the Atmosphere by Natural and Anthropogenic Processes (in Tg S/yr)

Natural Emissions[a]	Anthropogenic Emissions	Natural as a Percent of Total	Reference
92	50	65	Kellogg et al. 1972
108	65	62	Friend 1973
98	64	60	Robinson and Robbins 1975
35	65	35	Granat, Rodhe, and Hallberg 1976
50–100	69	42–59	Möller 1984
100	100	50	Andrae and Galbally 1984
51	72–77[b]	~40	Present book

[a] Includes biological decay and volcanism, but does *not* include sea salt.

[b] Includes anthropogenic forest burning of 7 Tg S/yr.

The pollutive contribution to the total atmospheric sulfur input is large, about 40% (75% over the land alone) and illustrates, as is the case for nitrogen, the importance of human activities as they affect the composition of major reservoirs and fluxes at the earth's surface, in this case the atmosphere and rainwater. For sulfur, atmospheric pollution is a global as well as a local problem. As we shall see, pollution also plays a major role as it affects sulfur in continental waters and in the oceans.

NITROGEN IN RAIN: THE ATMOSPHERIC NITROGEN CYCLE

In this section we shall discuss the sources of nitrogen in rain. Nitrogen is both an essential nutrient for plants and animals (and, thus is strongly involved in biogeochemical cycles) and a major pollutant. It occurs in gaseous form in the atmosphere where the most abundant nitrogen gases are elemental nitrogen (N_2), nitrous oxide (N_2O), nitrogen dioxide (NO_2), nitric oxide (NO), and ammonia (NH_3). The gases NO and NO_2 are referred to collectively as NO_x. Because nitrogen occurs in so many different forms in the atmosphere, and has so many sources, we shall be concerned in this section not only with rainwater composition, but also with the rates by which nitrogen is transported to and from the atmosphere and converted from one form to another. In other words, we shall be concerned with the *atmospheric nitrogen cycle*, particularly the conversion of nitrogen to and from gases that react with rainwater, namely NO, NO_2, and NH_3.

N_2, Nitrogen Fixation, and Total Nitrogen Fluxes

Elemental nitrogen (N_2) is by far the most abundant atmospheric nitrogen gas, making up nearly 80% of the atmosphere by volume, but nitrogen in the form of N_2 is very unreactive because of the strong bonds between the nitrogen atoms. The conversion of N_2 into chemically reactive and biologically available compounds by the combination of nitrogen with hydrogen, carbon, and/or oxygen is called *nitrogen fixation*. This is a very important process since nitrogen is an essential nutrient for life.

The major nitrogen fixation processes are listed and quantitatively evaluated in Table 3.14. In biological fixation, N_2 is combined with H, C, and O by marine organisms, particularly by blue-green algae, and on land by plants such as legumes and lichens. This is the dominant natural nitrogen fixation process. However, some 30% of nitrogen fixation by plants results from human cultivation of legumes, such as peas and beans, which are excellent nitrogen fixers. Minor natural fixation of atmospheric nitrogen (to form NO_x and NO_3^- in rain) also occurs in the heat generated by lightning bolts. Humans fix nitrogen in internal combustion engines and power plants by heating N_2 to high temperatures and also by the production of nitrogen fertilizers. A final source of N fixation is in forest fires which can

TABLE 3.14 Summary of Major Nitrogen Fluxes involving N_2
(in Tg N/yr; Tg = 10^{12} g)

Process	Flux	Anthropogenic Flux	Reference
Nitrogen Fixation			
Land (biological)	139	44	Burns and Hardy 1975
Ocean (biological)	10–90		Simpson 1977
Atmosphere (lightning)	4		Hill, Rinker, and Wilson 1980
Fertilizers and industrial	40	40	CAST 1976
Fossil fuel combustion			
NO_3-N	21	21	Logan 1983
NH_4-N	<1	<1	See text
Forest fires	12	12	Logan 1983
Total	216–296	117	
Denitrification			
Land	107–161		Söderlund and Svensson 1976
Ocean	40–120		Simpson 1977

occur naturally but are predominantly set by humans. Since humans cause approximately 50% of total nitrogen fixation, by converting atmospheric nitrogen into reactive forms, one might expect to see changes in the atmospheric nitrogen cycle. However, only around one quarter of anthropogenic fixation is due to combustion that involves direct release to the atmosphere of fixed nitrogen. The rest of anthropogenic nitrogen fixation involves the biogeochemical cycles of plants and animals and its effect on atmospheric nitrogen is more complex and harder to determine.

N_2, once fixed, is released back to the atmosphere via the process of *denitrification*, or bacterial nitrate reduction, which is necessary to maintain its atmospheric concentration. Estimated denitrification fluxes are also shown in Table 3.14. These values, as pointed out in Chapter 5, are not well established and better values are needed before one can say whether denitrification does or does not balance fixation worldwide.

The forms of nitrogen in rain that are derived from fixed nitrogen gases include nitrate (NO_3^-) from NO and NO_2, and ammonium (NH_4^+) from NH_3. In addition, rain contains some dissolved organic nitrogen (DON). Together NO_3^-, NH_4^+, and DON are delivered to the surface of the land in an annual rain flux of about 41 Tg N/yr (Tg = 10^{12} g) with an additional 16 Tg N/yr from dry deposition (see Table 3.15). If we compare the rain and dry deposition N flux to land with other N sources on land (Table 3.14), we see that rain and dry deposition (57 Tg N/yr) are a somewhat greater N source than fertilizers and industry, (40 Tg N/yr) but only about 40% of biological N fixation on land (139 Tg N/yr). Thus, biological processes provide a large input to the overall N cycle.

TABLE 3.15 Fluxes of Nitrogen Forms in Rain and
Dry Deposition (in Tg N/yr)

Process	Flux			
	NO_3-N	NH_4-N	Dissolved Organic N	Total N
To land:				
Rain	17	15	9	41
Dry deposition	15	1	?	16
				57
To the oceans:				
Rain	9	7	?	16
Dry deposition	4	?	?	4
Total				20

Sources: See text.

Organic Nitrogen in Rain

Our knowledge of dissolved organic nitrogen (DON) is limited because it has been measured in rain at only a relatively few locations compared to the many measurements of nitrate and ammonium. The measured concentration of DON in rain (from a few values collected by Meybeck 1982) is on the order of 0.10 mg/l N, and from this we have estimated that about 9 Tg N/yr might be delivered to land, an amount which is about half the input of either NO_3-N or NH_4-N coming from rain (see Table 3.15). One possible origin suggested for dissolved organic nitrogen in rain is from sea salt transported to the land, since DON is concentrated in the surface layer of the ocean where sea salt is formed (see section on aerosols). Söderlund and Svensson (1976) estimate that sea salt could supply about 10–20 Tg N/yr. Other possible sources are anthropogenic: volatiles released from domestic animal wastes and organic volatiles released by forest burning.

Nitrate in Rain: Natural Sources

Nitrate (NO_3^-) in rain results both from natural processes and from human activities. Most nitrate in rain comes from nitrogen oxide gases (NO_x), directly from nitrogen dioxide (NO_2) and indirectly from nitric oxide (NO). Nitric oxide can be formed by the reaction of nitrogen and oxygen in the air at high temperatures, in excess of 2000° C. This process occurs in lightning, and during combustion, and can be summarized as

$$N_2 + O_2 \rightarrow 2NO \text{ (nitric oxide)}$$

Nitric oxide is quite reactive and combines easily in the atmosphere with ozone

(O_3), or with peroxides (HO_2 or organic peroxides) to form nitrogen dioxide (NO_2):

$$NO + O_3 \rightarrow NO_2 + O_2$$

or

$$NO + HO_2 \rightarrow NO_2 + OH$$

NO_2 in turn reacts with OH in the air (catalytically) to form nitric acid (HNO_3) (Logan 1983):

$$NO_2 + OH \rightarrow HNO_3$$

Nitric acid is then removed in rain (or surface deposition). Nitric acid is a strong acid and completely dissociates in rain water to form NO_3^- and H^+:

$$HNO_3 \rightarrow H^+ + NO_3^-$$

In this way nitric acid can lower the pH of rain.

The conversion of NO_x to HNO_3 in the air is fairly rapid, being less than one to two days in summer and about ten days in winter according to Logan (1983), and the HNO_3 is removed from the air in rain or surface deposition after only a few days.

The sources of NO_3^- in rain, or essentially the sources of NO_x gas, include the following:

1. Lightning
2. Photochemical oxidation in the stratosphere of N_2O gas to NO and NO_2
3. Chemical oxidation in the atmosphere of ammonia to NO_x
4. Soil production of NO by microbial processes
5. Fossil fuel combustion by humans
6. Forest burning

The first four of these processes are predominantly natural and the last two predominantly anthropogenic.

Atmospheric lightning can heat up the air enough to cause nitrogen and oxygen to combine to form nitric oxide (NO) and ultimately NO_3^- in rain by the previous reactions. However, the direct production of nitrate ions by lightning is generally too small to be detected in rain water (Hill and Rinker 1981; Gambell and Fisher 1964). Likewise, the total amount of atmospheric NO_2-N produced by lightning is not well known. Hill, Rinker, and Wilson (1980) estimate that the maximum global production rate of NO_2-N by lightning is about 4 Tg N/yr. (This is the number used in our summary budget, Table 3.18). Some other estimates are similar: 8 Tg N/yr (Logan 1983) and 3 Tg N/yr (Dawson 1980) but others are greater by a factor of 10 (Chameides 1979).

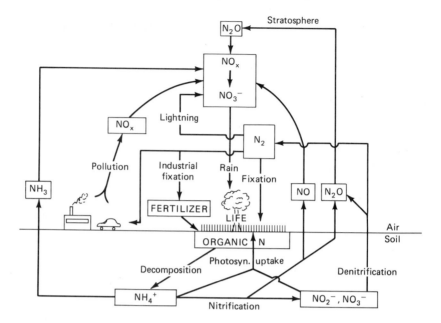

Figure 3.12   The atmosphere-soil nitrogen cycle, emphasizing sources of NO_x and NO_3^- in rain.

Before considering the three remaining *natural* sources of NO_x listed above it is helpful to briefly review the major features of the biological nitrogen cycle (see Figure 3.12) as it affects atmospheric nitrogen chemistry. (For further details on the nitrogen cycle see Chapters 5 and 8.) Organic nitrogen formed by N_2 fixation and/or photosynthetic uptake of NO_3^-, NO_2^-, or NH_4^+, upon the organism's death undergoes bacterial decomposition in soils and sediments resulting in the liberation of ammonia to solution. The ammonia may escape to the atmosphere or may remain in solution as NH_4^+ where it can be oxidized to dissolved NO_2^- and NO_3^- (*nitrification*). The produced nitrate may, in turn, be taken up by plants or undergo *denitrification* (nitrate reduction) by soil bacteria to N_2 and N_2O gases, which also escape to the atmosphere. This is the usual path in the nitrogen cycle. However, sometimes the soil conversion of NH_4^+ to NO_3^- (nitrification) by microorganisms is incomplete and part of the NH_4^+ is converted to NO (and N_2O) gas. As a result of these soil processes, the gases N_2, NH_3, NO, and N_2O are all released to the atmosphere.

Nitrous oxide gas (N_2O) released to the atmosphere by denitrification, or during ammonia oxidation, is relatively inert chemically. However, it is oxidized to NO in the stratosphere by exposure to the sun's radiation. NO reacts with ozone (O_3) to form NO_2, which can then find its way back into the troposphere and form NO_3^- in rain. However, the proportion of nitrate in rain originating from the breakdown of N_2O is only about 1% (0.3 Tg N/yr; Crutzen 1976). The importance of these reactions is not nitrate formation in rain, but the destruction of ozone during the process of converting the NO (produced from N_2O) to NO_2. This has been the cause for considerable concern recently since stratospheric ozone blocks harmful solar ultraviolet radiation from reaching the earth's surface.

As mentioned above, NH_3 also escapes to the atmosphere during the soil N cycle. The oxidation of atmospheric NH_3 has been suggested as an NO_x source (McConnell 1973). Logan (1983) estimates the size of this source as 1–10 Tg NO_x-N/yr. Calculations on the extent of this process in the atmosphere over rural West Germany (based on high measured NH_3 concentrations) by Lenhard and Gravenhorst (1980) would give a global destruction rate of NH_3-N of 0.1 Tg NH_3-N/yr. For this reason we have used < 1 Tg NO_x-N as the amount of NO_x resulting from the oxidation of NH_3.

It has been suggested (Junge 1963; Galbally 1975; Söderlund and Svensson 1976) that nitrate in rain might come partly from direct soil losses of NO to the atmosphere where it is oxidized to NO_2, producing NO_3^- in rain. One possible mechanism for the formation and release of NO gas is as a byproduct (along with N_2O) of the bacterial oxidation (nitrification) of NH_4^+ in soils (see Figure 3.12). The production of NO by soil-nitrifying bacteria has been observed in the laboratory by Lipschultz et al. (1981). (Chemical denitrification has also been suggested as a source of soil NO but this process is not documented and needs verification.) Galbally and Roy (1978) measured release of NO from grazed and ungrazed pastures (world area = 24×10^6 km^2; Table 5.4) of $(0.5 - 1.1) \times 10^5$ g NO-N/km^2/yr; this amounts to 1–3 Tg NO-N/yr. Johannson and Granat (1984) found that NO release from arable soils (world area = 14×10^6 km^2; Table 5.4) was $(0.2 - 0.6) \times 10^5$ g NO-N/km^2/yr which would give another 1 Tg NO-N/yr. Thus, the soil release of NO-N from utilized crop and pasture land might be on the order of 2–4 Tg N/yr and total soil release about 5 Tg N/yr. If the estimates of soil production of NO_x-N are correct, it amounts to some 12% of total NO_x-N production and is a major natural NO_x source. (However, some part of soil production may be indirectly anthropogenic due to increased fertilizer use.)

There should be seasonal variations in the nitrate content of rain, if it comes from soil bacterial processes, with more nitrate being produced in warm weather when bacterial decay is accelerated. Junge and Werby (1958) did find a maximum of rain nitrate in warm weather in unpolluted north central U.S. areas which they felt pointed to a soil source there. Freyer (1978) studied nitrogen isotope ratios in an agricultural area in Germany and found higher nitrate concentrations in summer rain, accompanied by lower $^{15}N/^{14}N$ ratios. Lower ^{15}N content suggests a soil-gas origin for the nitrate (as opposed to a fossil fuel origin).

Regardless of seasonality, release of nitrogen-containing gases to the atmosphere may be important as a local source of nitrate in rain. This is suggested in Figure 3.13 by maxima in the upper Midwest and along the U.S.-Canadian border where agricultural activity is intense. Perhaps such activity results in extra inputs of biologically derived nitrogenous gases to the atmosphere.

Overall, the natural processes discussed above, predominantly soil production of NO_x and lightning, account for about 25% of the total NO_x-N production. The

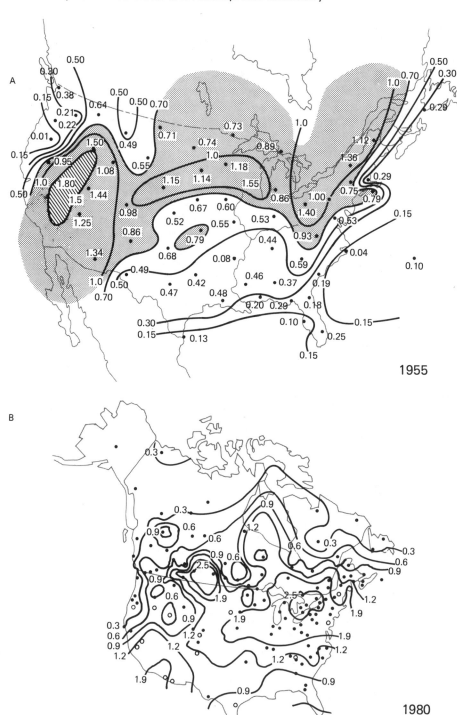

other 75% of NO_x-N comes from anthropogenic sources; fossil fuel combustion and forest burning.

Nitrate in Rain: Anthropogenic Sources

Humans undoubtedly provide the largest source of nitrate in rain ($\sim$ 50%) by the production of the nitrogen oxides NO and NO_2 from fossil fuel combustion, mainly in automobile engines and power plants. The NO_x gases are formed by the high-temperature reaction between atmospheric N_2 and O_2 (see above). Table 3.16 gives the relative importance of various combustion sources (and industrial sources) in the 1979 world production of nitrogen oxides. Worldwide, 21 Tg NO_x-N were produced in 1979, almost entirely from fossil fuel combustion. North America, Europe, and the rest of the world each contribute about one-third of the total.

In the United States and Canada, power and heating combustion sources generate about 55% of the NO_x emissions, and motor vehicles generate the remainder. (This can be compared to SO_2 from combustion, which is produced dominantly by power plants, with only 2% from vehicles.) The major sources of fossil fuel combustion NO_x are in urban areas, and thus there tends to be a much greater concentration of NO_x in urban air than in rural areas (Logan 1983). This is reflected in U.S. rain NO_3^-, which exhibits an area of high concentrations over the Great Lakes both in 1955–1956 and 1980 (Figure 3.13A and B). (This area correlates with high sulfate concentrations supporting an anthropogenic source for both nitrate and sulfate.)

High nitrate concentrations, due to excessive inputs of anthropogenic NO_x, are spread out over large areas (see Figure 3.13B) because modern power plants have been located in more remote, less populated areas and have taller smoke-stacks. In this way nitrogen oxide pollutants are dispersed over wider and less urban areas. In addition, appreciable dispersal is possible because the residence time of NO_x in the atmosphere before conversion to HNO_3 varies from one or two days to ten days (Logan 1983) and because HNO_3 itself may remain in the atmosphere a few days before removal in rain.

The U.S. emissions of NO_x in 1980 were about double what they were in 1955, although between 1970 and 1980 the emissions did not increase nearly as fast as they did prior to 1970 (see Figure 3.14). (As of 1983, the use of emission control

Figure 3.13   Dissolved nitrate (mg/ℓ NO_3^-) in rain over the continental United States and its change with time. (A) Average concentrations, July–September 1955. (After C. E. Junge. "The Distribution of Ammonium and Nitrate in Rainwater over the United States," *Transitions American Geophysical Union*, 39, 244. © 1958 by the American Geophysical Union.) (B) Average concentrations in 1980 for North America. Open circles indicate stations for which less than 20 weeks data were available. (Data from National Atmospheric Deposition Program.) (After J. A. Logan. "Nitrogen Oxides in the Troposphere: global and regional budgets," *J. of Geophysical Research*, 88 (C15), 10795. © 1983 by the American Geophysical Union.)

TABLE 3.16 Production of Nitrogen Oxides from Fossil Fuel Combustion and Industry in 1979 (in Tg NO$_x$-N/yr; Tg = 10^{12} g)

Source	U.S. and Canada	Europe[a]	Rest of World	Total
Power and heating combustion:				
Coal	1.6	2.4	2.4	6.4
Oil	0.8	1.3	1.1	3.2
Gas	1.1	0.9	0.3	2.3
Subtotal power + heating	3.5	4.6	3.8	11.9
Transportation[b]	2.5	2.8	2.5	8.0
Industrial sources[c]	0.2			1.2
Total	6.2			21.1

Source: After J. A. Logan. "Nitrogen oxides in the troposphere: global and regional budgets," *J. of Geophysical Research*, 88 (C15), 10792. © 1983 by the American Geophysical Union.

NOTE: NO$_x$ is the sum of NO$_2$ and NO.

[a] Europe includes 80% of USSR fuel.

[b] Total includes 0.2 Tg from air traffic.

[c] Petroleum refining and manufacture of nitric acid and cement in the United States.

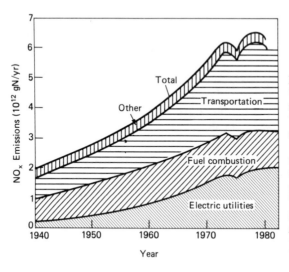

Figure 3.14 Emissions of NO$_x$ in the United States from combustion of fossil fuels and industrial processes in units of 10^{12} g N/yr (10^{12} g N = Tg N). The sources shown separately for 1940 to 1980 are power generation by electric utilities, industrial fuel combustion, transportation combustion and industrial processes ("other"). [J. A. Logan. "Nitrogen Oxides in the Troposphere: global and regional budgets," *J. of Geophysical Research*, 88 (C15), 10790. © 1983 by the American Geophysical Union.] (After U.S. EPA 1982).

devices for NO$_x$ on new vehicles had not yet had very noticeable effect on NO$_x$ totals [Logan 1983].) It appears that there have been considerable increases over the past 20–30 years in the nitrate concentration of rain over the United States, reflecting increased NO$_x$ emissions. For example, at Hubbard Brook, New Hampshire (Likens et al. 1977), the nitrate concentration of rain in 1970 was double that in 1964 or 1955 which were similar to each other. However from 1970 to 1981 there was a levelling off or slight decrease in nitrate concentrations. These changes appear to reflect NO_x emission trends in the northeastern United States (NRC 1983). Increases in nitrate concentrations resulted in a greater contribution of nitric acid to acidity in rain (see section below on acid rain for further discussion). Overall increases in nitrate in rain can also be demonstrated for the eastern United States as a whole. In 1978–1979, Pack (1980) found the average NO_3^- concentration for the eastern United States was 1.6–1.7 mg/ℓ NO_3^- versus an average for 1955–1956 on the order of ~0.7 mg/ℓ NO_3^-. Comparing Figure 3.13A (1955–1956) with Figure 3.13B (1980) for the eastern United States, we see that the pattern is similar but that the area of highest concentration over the Great Lakes, the 1.0 mg/ℓ NO_3 contour in 1955–1956, corresponds to 2.5 mg/l in 1980.

Reactions involving pollutant nitrogen oxides are particularly obvious in photochemical smog, where automobile exhausts accumulate as a result of a temperature inversion. These photochemical smog reactions are characteristic of California but also have become a problem in the U.S. Southwest and other areas. Automobile emissions of NO_x are presumably responsible for the area of high nitrate concentrations in California rain in 1955–1956 (Júnge 1963) and in the southwestern United States in 1980 (see Figure 3.13A and B).

Another important source of rain nitrate (28%) is from forest and grass fires, most of which are set by humans for land clearing in tropical areas such as South America and Africa. In addition, some wood is used as a fuel. Wild fires set by lightning amount to only 4% of the biomass burned and thus 96% of biomass burning is *anthropogenic* (Logan et al. 1981). Lewis (1981) found that combustion of vegetation on a large scale in Venezuela during the dry season caused an accumulation of nitrogen oxides in the air which were removed at the onset of the rainy season as nitric acid, in some cases acid rain. The amount of NO_x-N released to the atmosphere from biomass burning is estimated as 12 Tg N/yr by Logan (1983). This is derived from estimates of the amount of biomass fuel burned annually (7000 Tg/yr), the nitrogen content of the biomass burned, and the assumption that only 25% of the nitrogen is converted to NO. (Other estimates are larger where it is assumed that all N is converted to NO). Thus, as we noted earlier, the combination of forest burning and fossil fuel combustion accounts for the 75% of NO_x-N in the atmosphere, which is anthropogenic in origin.

Nitrate Deposition in Rain and the Nitrate-Nitrogen Cycle

Table 3.2 gives the concentration of nitrate in rain from various areas. (Nitrate concentrations in precipitation are less reliable and less numerous than those for other major ions because of analytical problems.) In estimating worldwide flux rates, however, there is a problem due to limited measurements of nitrate from remote continental areas that might be presumed to be unpolluted by humans. The minimum nitrate concentrations we find for inland continental rain (0.1–0.2

mg/ℓ NO_3) are from Poker Flat, Alaska, and Venezuela (Galloway et al. 1982) and from the Amazon Basin (Stallard and Edmond 1981). The more usual un-polluted continental values probably average around 0.5–0.6 mg/ℓ NO_3 (see Table 3.2). Europe and North America (particularly the U.S.) generally have much higher nitrate concentrations in rain reflecting the high combustion of fossil fuel in particular (around two-thirds of the world total, see Table 3.16) and probably also intense use of fertilizers in agriculture. There are many more measurements from these areas as well.

Using the areas of the continents, average annual precipitation, and an estimate of the average nitrate concentration in rain, we have estimated nitrate deposition over the continents (Table 3.17) as 17 Tg NO_3-N/yr. This is near the low end of other recent estimates: 13–30 Tg NO_3-N/yr (Söderlund and Svensson 1976) and 8–30 Tg NO_3-N/yr (Logan 1983).

The few measurements of marine and marine-influenced rain that have been made (Table 3.2) range from 0.1–0.2 mg/ℓ NO_3^-, except in the North Atlantic where pollution is likely to be important. Low oceanic values support a continental origin for most rain nitrate (Junge 1963). Over the oceans, assuming an average concentration of nitrate in precipitation of 0.1 mg/ℓ and an average precipitation of 107 cm (Table 2.2), 9 Tg NO_3-N/yr would be delivered by rain to the oceans (see Table 3.17).

In addition to NO_3-N deposition in rain, there is also dry deposition of NO_3-N, primarily through uptake of NO_x gas by plants and water and from deposition of particulates; however, this flux is poorly known. Using deposition rates given

TABLE 3.17 Rate of Nitrate Deposition on Land and Ocean by Precipitation

	Mean Annual Precipitation[a] (cm)	Nondesert Area[b] (10^6 km^2)	[NO_3^-] (mg/ℓ)	Deposition Rate (Tg N/yr)
Land: Continent				
Africa	69	21.8	0.7	2.1
Asia:				
Clean		24.8	0.6	2.5
Polluted		9.0	1.5	2.2
Europe: polluted	73	10.0	2.0	3.3
N. America:				
Clean	67	18.0	0.6	1.6
Polluted[c]	100	5.2	1.5	1.8
S. America	165	17.1	0.5	3.2
Antarctica	17	12.2	0.1	0.05
Australia	44	4.9	0.5	0.2
Total on land				17.0
Ocean	107	361.0	0.1	8.7

[a] From Lvovitch 1973.

[b] From Meybeck 1979.

[c] N. American polluted area = eastern United States and Canada (Galloway and Whelpdale 1980).

by Fowler (1978) for unpolluted regions (0.1 g NO_2-N/m^2/yr) and populated regions (0.24 g NO_2-N/m^2/yr) combined with land areas (from Table 3.17), we obtain dry deposition on land of 15 Tg NO_3-N/yr. Dry deposition over the oceans amounts to about 4 Tg NO_3-N/yr or about 0.011 g N/m^2/yr based on concentrations of HNO_3 and NO_3^- and deposition velocities given in Logan (1983). Thus, the total dry deposition would be about 19 Tg NO_3-N/yr (Table 3.18). This agrees well with Logan (1983) who estimated 12–22 Tg N/yr from dry deposition. In summary, total N deposition in rain and dry deposition (Table 3.18) amounts to 45 Tg NO_x-N/yr.

Now that we have derived values for N deposition and for fluxes of NO_x to the atmosphere (discussed in the previous section) it is possible to construct an overall budget for NO_3^- in the atmosphere. This is given in Table 3.18. The total production of NO_x-N from various sources amounts to about 43 Tg NO_3-N/yr. (This estimate is very similar to that of Logan [1983].) Because of the relatively short residence time of NO_x in the atmosphere (1–10 days), the atmospheric nitrate cycle should balance (Söderlund and Svensson 1976). Since estimated production and deposition of NO_x-N are roughly equal (43 vs. 45 Tg N/yr), the estimate of 5 Tg N/yr for soil production of NO_x (the most uncertain source flux) does not seem too unreasonable. However, there are uncertainties in all of the NO_3-N fluxes, particularly the size of the dry deposition, forest burning, lightning, and soil production fluxes of NO_x-N. (If greater dry deposition of NO_x occurs, then greater production of NO_x would be needed to balance the cycle.)

Since there are no known oceanic sources of NO_x (except lightning), there must be a net flow of NO_x in the atmosphere from over the land to over the oceans. Looking at Table 3.18, it would seem that fossil fuel combustion, which has been increasing over the last few decades, contributes roughly half of the NO_x-N production and, when combined with forest burning, anthropogenic sources amount to about 75% of NO_x-N production. By contrast, natural sources, predominantly soil production, and lightning, contribute only ~25% and part of the soil production may be anthropogenic. Thus, again we see how the activities of humans have made a major modification in the geochemical cycle of a major element.

Ammonium in Rain: Atmospheric Ammonium-Nitrogen Cycle

Ammonium (NH_4^+) is the other major nitrogen-containing ion found in rain. Ammonium ion results from the reaction of ammonia gas (NH_3) with water:

$$NH_{3\ gas} + H_2O \rightarrow NH_4^+ + OH^-$$

This reaction tends, because of the production of hydroxyl ions (OH^-), to raise the pH of rainwater and partly counteract the acid effects of CO_2, NO_x, and SO_2. Ammonia is the only atmospheric gas that can do this. Junge (1963) has calculated that, in the absence of other gases, the average concentration of NH_3 in the air of 3 μg/m^3 would result in a high pH for rainwater, about 8.5. However, since most

TABLE 3.18 Fluxes of NO_x-N and NO_3-N: The Atmospheric Nitrate Cycle (in Tg N/yr; Tg = 10^6 metric tons = 10^{12} g)

Process	Tg N/yr	Percent of Total	Anthropogenic as a Percent of Total	Reference
Production Source				
Lightning	4	9		Hill, Rinker, and Wilson 1980
Conversion of N_2O to NO_2 in stratosphere	0.3	1		Crutzen 1976
Conversion of ammonia to NO_x	<1	<1		(See text)
Soil production of NO_x	~5	12	?	(See text)
Combustion of fossil fuel	21	49	49	Logan 1983
Forest burning	12	28	27	Logan 1983
Total production	43	100	76	
Deposition				
Land:				
Rainfall	17			Calculated (see text)
Dry deposition	15			Calculated (see text)
Total land	32			
Ocean:				
Rainfall	9			Calculated (see text)
Dry deposition	~4			Calculated (see text)
Total ocean	13			
Total deposition	45			

rainwater has a pH of 4 to 6, ammonia obviously does not control its pH; the acidic gases CO_2, NO_x, and SO_2 are more important.   In fact, since most rainwater is acid, NH_3 should be very soluble in rain.   (The above reaction also shows why alkaline soils tend to give off NH_3 and acid soils to take it up; see below.) NH_3 gas remains in the atmosphere around one to four days before being converted to NH_4^+ (Söderlund and Svensson 1976), and once NH_4^+ is formed, it is removed in rain in about a week.

The measurement of NH_4^+ in rain is often subject to serious errors.   For instance, contamination by bird droppings may be large if precautions are not taken to keep birds away from rain collectors.   Thus, one should exert caution in interpreting NH_4^+ data (particularly old data) unless one has some knowledge of how the rain was collected.

There are four sources of ammonia in the air: (1) bacterial decomposition of domestic animal and human excreta; (2) bacterial decomposition of natural nitrogenous organic matter in soils; (3) release from fertilizers (and fertilizer manufacture); and (4) burning of coal (which contains organic nitrogen compounds).   Note that all of these sources are continental in nature.   This helps to account for the observations of higher concentrations of NH_4^+ in rain over the continents (0.03–1.0 mg/ℓ NH_4^+) than over the oceans (0.01–0.04 mg/ℓ NH_4^+).   (See Table 3.2.) In addition, the above sources are seasonal, the first three being at a maximum in the spring and summer and the last in winter.   The increase in NH_4^+ concentrations in rain (and NH_3 concentrations in the air) in warmer weather has been noted by several authors (Junge 1963; Freyer 1978; and Lenhard and Gravenhorst 1980).

Decomposition of domestic animal excrement is considered an important source of atmospheric ammonia.   In fact, urea from animal urine is the principal NH_3 source in air over the United Kingdom (Healy et al. 1970) and in some rural areas.   Estimates of release rates of NH_3 from animal excrement have been calculated in West Germany for a rural area (Lenhard and Gravenhorst 1980).   Multiplying the average release rate from these studies (0.33 g NH_3-N/m²/yr, based on 10% NH_3 volatilization from excrement) by the world agricultural area (44 × 10¹² m²; Burns and Hardy 1975) gives a production of about 15 Tg NH_3-N per year. This is the value used in our summary ammonia budget (Table 3.20).   (Söderlund and Svensson [1976] get a somewhat larger flux: 20–35 Tg NH_3-N/yr from animal plus human wastes.)

Release of NH_3 from soils as a result of bacterial decomposition of natural organic matter is another possible source of ammonia in rain, particularly in remote areas.   Ammonia release from the soil is favored by alkaline conditions, (Junge 1963), and low soil moisture.   Because soil production of ammonia is dependent upon bacterial decomposition, it is greater in warm weather.

Studies by Denmead, Nulsen, and Thurtell (1976) in an ungrazed grass-clover pasture indicate a large production of NH_3 at the soil surface (from the microbial breakdown of soil organic matter), but with almost complete reabsorption of NH_3 by the overlying plant cover, resulting in a nearly closed NH_3 cycle.   (NH_3 is absorbed by leaf stomatal uptake and by dissolution in water on the leaves.)   Plants apparently attempt to maintain a constant atmospheric NH_3 concentration of 2–6 $\times$ 10^{-9} atm. (or ~1 $\mu g/m^3$), referred to as the compensation point (Farquhar et al. 1980), due to their internal chemistry. If the surrounding atmospheric concentration of NH_3 is greater than the compensation point, they take up NH_3; if it is less they release NH_3. Thus, it is only the *net loss* of NH_3 above the plant canopy that results in ammonia in the atmosphere and in rain.

Since clover is a legume, the net flux from an ungrazed grass-clover pasture (1.76 g $N/m^2/yr$; Denmead, Freney, and Simpson·1976) might be typical of legumes in general (areally 2.5 $\times$ 10^{12} m^2; Burns and Hardy 1975) resulting in a worldwide release for legumes of 4 Tg NH_3-N/yr. Legumes fix N_2 at a rate about ten times that of other crops and grasses. Corn crops apparently take up NH_3 in wet soil and give off NH_3 from dry soil (at an atmospheric NH_3 concentration near the compensation point; Denmead, Nulsen, and Thurtell 1978). Thus, it is difficult to estimate whether, overall, most crops would lose or gain NH_3. Georgii and Lenhard (1978) measured a low loss rate of NH_3 (0.03 g $N/m^2/yr$) from uncultivated, fertilized soil. Assuming this value to be typical of forest and grasslands (71 $\times$ 10^{12} m^2 areally; Burns and Hardy 1975), we obtain a flux of NH_3 of about 2 Tg N/yr, which we add to legume soil losses of 4 Tg N/yr, for a total soil loss of about 6 Tg NH_3-N/yr from ungrazed land.

Ammonia can be released to the atmosphere also by the addition of ammonium fertilizers and other fixed nitrogen fertilizers to the soil. Soil nitrogen, if not in the form of NH_3, is reduced to it by microorganisms. The release of fertilizer ammonia is influenced by the same factors as is the release of soil NH_3; that is, it is favored by dry, warm, high pH soils. Estimates of the percentage of applied fertilizer nitrogen which is volatized as ammonia range from 1% to 5% (Lenhard and Gravenhorst 1980). Since world fertilizer use is about 40 Tg N/yr (CAST 1976), about 1–2 Tg NH_3-N/yr is probably released to the atmosphere from fertilizers. The use of nitrogen fertilizers is increasing at a rate of 6% per year and this could indirectly result in greater atmospheric NH_3 production.

Ammonia is a very minor product of combustion either from fuel use or from forest burning. Using the estimate of NH_3-N released by the burning of hard coal (11.8 g NH_3-N per ton of coal) and brown coal (7.4 g NH_3-N per ton of coal) from Freyer (1978) and worldwide coal consumption of 2100 Tg hard coal and 800 Tg brown coal (Söderlund and Svensson 1976), only 0.03 Tg NH_3-N/yr are produced. This is considerably less than earlier estimates (Robinson and Robbins 1975; Söderlund and Svensson 1976). Low NH_3 release from fuel combustion tends to be confirmed by studies of NH_4^+ concentrations in rain and NH_3 concentrations in the atmosphere which show little areal correlation with other fuel combustion products such as SO_2 (Healy et al. 1970; Junge 1963). (NH_3-N produced by fuel combustion is small compared to NO_x-N production by combustion; i.e., most nitrogen in fuels ends up in rain as nitrate.) In addition, some NH_3 may be released by forest burning in the tropics, but most of the combusted nitrogen is probably given off, instead, as NO_x (Logan 1983; Crutzen et al. 1979).

Junge (1963) found that ammonium (NH_4^+) in U.S. precipitation in 1955–1956 (Figure 3.15) was generally 0.1–0.2 mg/ℓ. By 1972–1973 Miller (1974) found greatly increased concentrations and the relative increases in NH_4^+ were greater than those for either sulfate or nitrate. The average rain concentration of NH_4^+ in the eastern United States in 1978–1979 was 0.31 mg/ℓ according to Pack (1980) with the highest values in Ohio, Indiana, and Illinois. Munger and Eisenreich (1983) found in the north central United States those areas which had concentrations of about 0.3 mg NH_4/ℓ in 1955–1956 had concentrations of about 0.55 mg NH_4/ℓ in the 1970s. These increases of NH_4^+ in rain seem to correlate with a tripling of the U.S. NH_3 production (mainly used in fertilizer) from 1962 to 1975 (NRC 1979).

Using the mean annual precipitation and the average NH_4^+ concentration in rain over each of the various continents, we have calculated the annual amount of nitrogen delivered as NH_4^+ in rainfall over all continents as 15 Tg NH_4-N/yr (Table

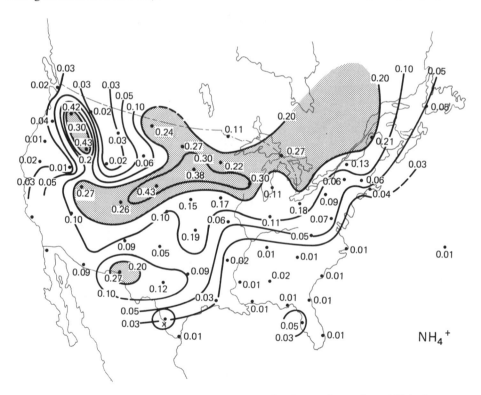

Figure 3.15 Ammonium (NH_4^+) concentration in rain over the continental United States (values in mg NH_4^+/ℓ) for July–September 1955. (After C. E. Junge. "The Distribution of Ammonium and Nitrate in Rainwater over the United States," *Transactions American Geophysical Union*, 39, 242. © 1958 by the American Geophysical Union.)

3.19). This is somewhat less than other estimates (Söderlund and Svensson 1976; Meybeck 1982). The removal of NH_4-N in precipitation over the oceans is low (about 7 Tg NH_4-N/yr), assuming an average concentration of 0.025 mg/ℓ NH_4^+ (from Table 3.2). This is because most ammonia sources appear to be continental, and thus NH_4^+ in oceanic rain is mostly derived from the continents. Total (continents plus oceans) rain removal of NH_4-N is estimated in Table 3.19 as 22 Tg N/yr.

Because we have used *net* soil release of ammonia, which is the difference between soil volatilization and reabsorption of ammonia, we have not included a dry deposition flux of ammonia gas (i.e., direct uptake of NH_3 by vegetation and soil). However, NH_3 reacts extensively with H_2SO_4 to form aerosol $(NH_4)_2SO_4$ (see following discussion), which is primarily removed in rain. In forested rural Massachusetts, where the concentration of particulate NH_4^+ is fairly high due to transport from other areas, dry deposition of particulate NH_4^+-N on leaf surfaces is a significant flux in contrast to dry deposition of NH_3, which is unimportant (Tjepkema, Cartica, and Hemond 1981). We have included a dry deposition flux of particulate NH_4^+ of 1 Tg NH_4-N/yr based on the polluted area in Table 3.11 and a NH_4^+-N flux of about 0.3 kg N/ha/yr. The total of rain and dry deposition of NH_4^+-N thus amounts to 23 Tg N/yr.

Our ammonia nitrogen budget, including both inputs and outputs, is summarized in Table 3.20. Although there are considerable uncertainties involved in all ammonia fluxes, we find that the atmospheric ammonia cycle is in balance as

TABLE 3.19   Rate of Ammonium Deposition on Land and Ocean by Precipitation (Predominantly Rainfall)

	Mean Annual Precipitation[a] (cm)	Nondesert Area[b] (10^6 km²)	$[NH_4^+]$ (mg/ℓ)	Deposition Rate (Tg N/yr)
Land: Continent				
Africa	69	21.8	0.2	2.3
Asia:	73			
Clean		24.8	0.15	2.0
Polluted		9.0	0.3	2.1
Europe: polluted	73	10.0	0.65	3.7
N. America:				
Clean	67	18.0	0.1	0.9
Polluted[c]	100	5.2	0.3	1.2
S. America	165	17.1	0.1	2.3
Antarctica	17	12.2	0.01	0.02
Australia	44	4.9	0.05	0.1
Total on land				14.6
Ocean	107	361.0	0.025	7.5

[a] From Lvovitch 1973.

[b] Nondesert area from Meybeck 1979.

[c] N. American polluted area = eastern United States and Canada (Galloway and Whelpdale 1980).

TABLE 3.20   Fluxes of Ammonia: The Atmospheric Ammonia Cycle

Process	Flux (Tg NH_4-N/yr)	Percent of Total	Anthropogenic as Percent of Total
Input Sources			
Domestic animal and human waste decomposition	15	65	65
Net soil loss from organic matter decomposition (excluding fertilizer)	6	26	17
Fertilizer release	2	9	9
Fossil fuel combustion	0.03	≤ 1	≤ 1
Total input	23	100	91
Removal			
Conversion in atmosphere to NO_x	<1		
Land—rainfall	15		
Ocean—rainfall	7		
Dry deposition—particulate NH_4	1		
Total removal	23		

Source: See calculations in text.

it should be because of the short turnover time of NH_3 in the atmosphere (1–4 days). Production and removal are both estimated at 23 Tg N/yr. As with NO_x-N, since production is mostly over land, there is net transport to the oceanic atmosphere. Table 3.20 again illustrates the importance of anthropogenic influences on the cycle of a major rainwater constituent. Animal wastes (the largest NH_3 source), fertilizer release, and fossil fuel combustion are almost entirely due to humans. We have assumed that, of the net soil NH_3 release (excluding fertilizers) of 6 Tg N/yr, at least the 4 Tg N/yr from legumes, which are extensively cultivated, are also anthropogenic. Thus, about 90% of total NH_3 production is due to humans.

(NH$_4$)$_2$SO$_4$ Aerosol Formation: Interaction of N and S Cycles

As noted earlier, NH_3 gas in the atmosphere reacts very readily with aqueous H_2SO_4 aerosols to form $(NH_4)_2SO_4$ aerosols. This reaction partially neutralizes the acid H_2SO_4 and converts gaseous NH_3 to NH_4^+ aerosol, which can be transported and removed in rain or dry deposition. It also provides a link between the atmospheric cycles of N and S, (Galbally, Farquhar, and Ayers 1982). In the atmospheric NH_3-N cycle, as pointed out by Galbally, Farquhar, and Ayers, there are essentially two competing processes regulating the atmospheric concentration of NH_3: (1) the tendency of plants to take up or give off atmospheric NH_3 gas, maintaining a concentration of NH_3 around 1 $\mu g/m^3$, due to their internal chemistry (see earlier discussion); and (2) the reaction of NH_3 with H_2SO_4, which converts gaseous NH_3 to NH_4^+ in aerosols and prevents NH_3 from being recycled by plants. Thus, conversion of gaseous NH_3 to NH_4^+ in sulfate aerosol essentially removes ammonia gas from the biogenic N cycle, and reduces the amount of atmospheric N available for plants. Also, NH_4^+ in aerosol form can be transported long distances (up to 5000 km) before removal in rain and dry deposition. Thus, the net effect of the formation of $(NH_4)_2SO_4$ aerosols is to spread NH_4^+-N around and probably increase NH_4^+ concentrations in rain in areas far from sources of NH_3 gas.

The other important effect of the formation of $(NH_4)_2SO_4$ aerosols is the neutralization of H_2SO_4 acidity in rain. For example, the average concentration of NH_4^+ in eastern U.S. rain is 0.3 mg/ℓ (Pack 1980), and this should represent neutralization of 0.8 mg/ℓ SO_4^{--} in rain. However, since the average eastern U.S. rain concentration of SO_4^{--} is 2.7 mg/ℓ SO_4^{--}, it is not possible for available NH_3 to neutralize all the H_2SO_4 in eastern U.S. rain, and consequently acid rain of average pH = 4.2 results.

ACID RAIN

Besides bringing about changes in the concentrations of NO_3^-, SO_4^{--}, and Cl^- in rain, the gases NO, NO_2, SO_2, and HCl also produce hydrogen ions, H^+. The result is called *acid rain*. Acid rain is rain that is more acid than it would be in the absence of these gases, and much of it is pollutive in origin. In this section we shall discuss the factors that affect the acidity of rain and the processes by which acid rain is formed. Throughout the discussion we shall express acidity in terms of pH. The pH of a solution is defined as the negative logarithm (base 10) of the hydrogen ion concentration[1]; in other words

$$pH = -\log[H^+] = \log(1/[H^+])$$

Here the brackets refer to concentration in moles per liter. Solutions having a pH greater than 7 are referred to as being basic (or alkaline); conversely, those with a pH less than 7 are referred to as being acidic. Here we shall be interested in how the pH of rain can fall below 7, often (when there is appreciable pollution) *far* below 7. Before discussing pollution, however, it is instructive to inquire into what the pH of rainwater would be in the absence of pollution, that is, the pH of natural rainwater.

The pH of Natural Rainwater

Pure water containing no dissolved substances should have a pH of 7, in which case it is referred to as being neutral (neither acidic nor basic). Natural rainwater, however, is not pure water. First of all, as a result of the solution of atmospheric

---

[1]Strictly speaking, activity should be substituted for concentration but discussion of activity is beyond the scope of the present book. For a discussion of activity consult, for example, Stumm and Morgan 1981.

carbon dioxide (to equilibrium), rainwater becomes moderately acidic with a pH of 5.7. This comes about from the reaction of CO_2 with H_2O, which results in the formation of carbonic acid (H_2CO_3), which in turn partly dissociates to produce hydrogen and bicarbonate ions:

$$CO_{2_{gas}} \rightleftarrows CO_{2_{soln.}}$$

$$CO_{2_{soln.}} + H_2O \rightleftarrows H_2CO_3$$

$$H_2CO_3 \rightleftarrows H^+ + HCO_3^-$$

(Further dissociation of HCO_3^- to CO_3^{--} and H^+ is negligible at the pH of rain.) From the latter two reactions, it can be seen that CO_2 reacting with H_2O results in the formation of H^+ and HCO_3^- in equal amounts.

By using equilibrium expressions for the above reactions and the fact that the concentrations of H^+ and HCO_3^- are equal, one can readily calculate (see Garrels and Christ 1965) that at equilibrium:

$$[H^+] = 2.1 \times 10^{-6} \text{ moles per liter} = 10^{-5.67}$$

This is equivalent to a pH of about 5.7. Since the concentration of carbon dioxide in the atmosphere is everywhere about the same, one would, therefore, expect that natural rainwater, if no other reactions were involved, would exhibit a pH close to 5.7.; that is, it would be moderately acidic. (Note that small deviations, on the scale of tenths of a pH unit, from this value can result from the uptake of CO_2 at different temperatures and pressures, and that, due to fossil fuel burning, atmospheric CO_2 has been increasing with time.)

Under a variety of circumstances, however, the pH of natural (unpolluted) rainwater can be higher or lower than 5.7. Higher pH is a less common situation on a worldwide basis and comes about mainly in arid regions (where air pollution is absent) as the result of the dissolution of windblown dust, which contains high concentrations of $CaCO_3$ (Kramer 1978). The reaction is

$$CaCO_3 + H^+ \rightarrow Ca^{++} + HCO_3^-$$

This reaction not only results in the neutralization of acidity via the consumption of hydrogen ions but also in the production of Ca^{++} and HCO_3^- ions. Thus, Ca^{++} in rain in excess of sea-salt concentrations often indicates that the rain has reacted with $CaCO_3$ dust. The pH of rain in a number of areas of western North America is greater than 5.7 (see Figure 3.18C) presumably because of this reaction or a similar reaction involving FeOOH dust ("brown dust"; Kramer 1978).

Neutralization of natural acidity in unpolluted rain over land can also take place by reaction with ammonia gas (NH_3). In regions where ammonia is emitted to the atmosphere from biological decay, agricultural activity, and so forth, there may be enough to bring about a slight rise in pH via the reaction

$$NH_3 + H^+ \rightarrow NH_4^+$$

Charlson and Rodhe (1982) calculate that, in the absence of sulfate aerosol, NH_3, at the lowest concentrations formed in continental areas ($0.13 \ \mu g/m^3$), could raise the pH of CO_2-containing rain to 6.2.   Since NH_3 gas or fine aerosol NH_4^+ can travel further than coarse soil dust containing $CaCO_3$, the neutralization effect of NH_3 may be important in areas further from the source.   The relative importance of NH_3 and soil dust in neutralizing acidity varies from area to area but on the average about one-third of the acid neutralization is due to NH_3 (Munger and Eisenreich 1983).

Over the oceans, sea-salt aerosol, which is alkaline (from bicarbonate and borate), may neutralize acids in marine precipitation.   This occurs when acid aerosols are carried out over the ocean from land.   However, the effect of raising the pH of marine rain is minimal.   Only when the concentration of sea-salt aerosol in rain is large (Na > 3.0 mg/ℓ) will the original pH be raised more than about 0.05 pH units (Galloway, Knap, and Church 1983; Pszenny, MacIntyre, and Duce 1982).

Natural rainwater can also have a pH less than 5.7, in which case it falls in the category of acid rain (see below).   This is due to the presence of naturally occurring H_2SO_4 from the oxidation of biogenic reduced sulfur gases (Charlson and Rodhe 1982).   In order for "naturally acid" rain to occur there must be a lack of neutralizing $CaCO_3$ soil dust and NH_3.   Charlson and Rodhe(1982) suggest that theoretically an average pH of about 5 might occur in pristine areas (which lack neutralizing substances) because of emissions of natural sulfur gases.   This is based on an annual global release of 65 Tg of non-sea-spray natural sulfur (see section on sulfur cycle) spread evenly over the globe.   Locally, because of variations in natural sulfur emissions, pH values in natural rain might range from 4.5 to 5.6 with the lower pH values being favored by high SO_4 aerosol concentrations and low liquid water content of clouds.   Low natural pH values might occur over the oceans because of considerable emissions of reduced sulfur gases there (provided sea-salt neutralization is minimal) and the absence of continentally derived bases (NH_3 and $CaCO_3$ dust).

Weak organic acids, such as acetic or formic acid also have been suggested as an additional source of natural acidity in some local areas (Galloway et al. 1982).   These acids could come either from natural volatilization from vegetation or from the sea surface.   (These acids might also arise from agricultural burning, but in this latter case they would be considered as pollutive in origin.)

Acid Rain from Pollution

Acid rain is defined, more precisely, as that having a pH less than 5.7 due to reactions with acidic gases other than CO_2.   The acidic gases are SO_2, NO_2, NO, and (to a lesser extent) HCl, and they result in the formation in the atmosphere and in rain clouds of sulfuric, nitric, and hydrochloric acids respectively.   (See also sections on sulfate and nitrate in rain.) Overall reactions are

$$SO_2 + OH \rightarrow \ldots \rightarrow H_2SO_4 \text{ (sulfuric acid)}$$

$$SO_2 + H_2O_2 \rightarrow H_2SO_4 \text{ (sulfuric acid)}$$

$$NO_2 + OH \rightarrow HNO_3 \text{ (nitric acid)}$$

$$HCl_{gas} \rightarrow HCl \text{ (hydrochloric acid)}$$

followed by the dissociation of these acids in rain water to form H^+ ions:

$$H_2SO_4 \rightarrow 2H^+ + SO_4^{--}$$

$$HNO_3 \rightarrow H^+ + NO_3^-$$

$$HCl \rightarrow H^+ + Cl^-$$

Thus, as more and more of the precursor gases are added to the atmosphere by human activities, more and more hydrogen ions are produced and the pH of rainwater drops. This helps explain why the pH of rainfall at many locations has been decreasing over time.

Acid rain was first noted in northwest Europe in the early 1950s. Barrett and Brodin (1955) found that precipitation in southern Sweden had a pH between 4 and 5 and that the pH was lowest in the winter when the air flow is from the south, bringing pollution from central and western Europe. In 1968, Odén (1968) found that the acidity of northern European rain had increased since 1956; rain in some parts of Scandinavia was 200 times more acid than in 1956. A region of high acidity (pH 4 to 4.5) that was centered in the Benelux countries in the late 1950s had spread by the late 1960s to Germany, northern France, the eastern British Isles, and southern Scandinavia. By 1974 most of northwestern Europe was receiving acid precipitation (pH < 4.6) (Likens 1976). This is all shown in Figure 3.16. (It should be noted that maps such as Figure 3.16 can represent only general trends both because there is considerable local geographic variation of rain pH in any one area, and because monthly values of pH vary from yearly averages with the winter months being higher and the summer months lower; see Kallend et al. 1983.)

A large part of the gases that produce acid Scandinavian rain (SO_2 and NO_2) are due to industrial activity in England and central Europe located large distances away. Because anthropogenic SO_2 and the resulting sulfate remain in the air for several days and are consequently transported more than 1000 km, it is possible for SO_2 and sulfate in air masses from England and the Ruhr Valley to reach Sweden before being rained out (Bolin 1971). In Norway (median pH = 4.6), Forland (1973) found the greatest acidity came from rain derived from air masses that had passed over major industrial areas to the south in central Europe and

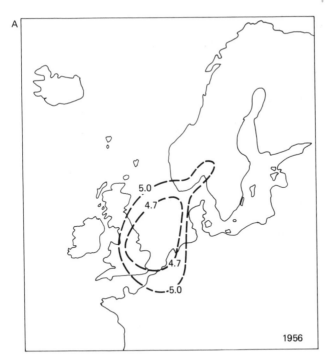

1956

1974

Figure 3.16 Acid rainfall in northwestern Europe and its change with time. Dashed lines represent contours of constant pH. (A) Situation in 1956. (B) Situation in 1974. (Modified from G. E. Likens et al., "Acid Rain." Copyright © October 1979 by Scientific American, Inc. All rights reserved.)

England.   By contrast, air from the Norwegian Sea to the north produced rain that was much less acidic (pH 5.1–6.6).

The acidity of European precipitation is due mainly to sulfuric acid (H_2SO_4) from SO_2 gas, but nitric acid (HNO_3) from NO_x is also significant.   Söderlund (1977) estimated the relative contribution of nitrate versus sulfate to acidity in northern European precipitation to be 1:3.   From 1956 to 1974, the period in which the spread of acidity shown in Figure 3.16 occurred, the emissions of SO_2 in Europe steadily increased by up to 50% (see Figure 3.9).   However, during the same period, sulfur deposition in European rain (Granat 1978) and acidity of European rain (Kallend et al. 1983) increased only up to about 1965 and remained constant thereafter.   In addition, the increases in rain acidity were mainly confined to Scandinavian stations with the lower-pH central European stations remaining constant.   In comparison with sulfate, nitrate increased rapidly and continually in European precipitation due to a doubling of NO_x emissions over the period 1956–1975 (Rodhe, Crutzen, and Vanderpol 1981).

Several explanations have been offered for the fact that European sulfur deposition and rain acidity did not continue to increase with increasing sulfur emissions.   First, because pH is a logarithmic measure, it becomes harder to continue to lower pH once the pH has become low.   Another explanation is meteorological; that is, changes in the frequency of winds from sulfur emission areas, which tend to produce acid precipitation, may have occurred.   Rodhe, Crutzen, and Vanderpol (1981) have offered an interesting third explanation.   They suggest that increases in concentrations of NO_x in the atmosphere may have retarded the rate of SO_2 transformation to H_2SO_4.   Since both SO_2 and NO_x oxidation depend upon the concentration of OH (see equations above), and because HNO_3 forms faster from NO_x oxidation than H_2SO_4 forms from SO_2 oxidation, a higher level of NO_x near emission areas may use up OH and delay the transfer of SO_2 to H_2SO_4.   This could, in turn, cause delays in increases of sulfate in precipitation until stations further from SO_2 emission areas are reached.   In support of this idea, the largest concommitant increase in both acidity and sulfur deposition did seem to occur at Scandinavian stations that were considerable distances from the main SO_2 emission sources in England and central Europe.

In addition to the effects of air transport direction and time on the spread of acid rain, windward mountain slopes tend to receive a larger amount of acid rain than lower-lying areas because more precipitation occurs in the mountains.   For example, the mountainous southern Norwegian coast receives large quantities of acid rain (pH = 4.2) from air masses that have travelled from the British Isles and central Europe over low-lying land and the North Sea.   By the time these air masses have moved inland several hundred kilometers past the mountains, a considerable amount of the pollutants have been removed, and the pH rises to 4.6 or 4.7 (Likens et al. 1979).

The distribution of pH values for rainwater in the Northern Hemisphere is plotted in Figure 3.17.   Note that areas with the most acidic rain, besides Europe,

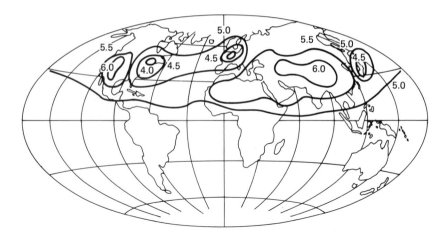

Figure 3.17   Estimated distribution of the pH of rainwater in the Northern Hemisphere. (After G. Gravenhorst et al. "Sulfur Dioxide Absorbed in Rainwater" in *Effects of Acid Precipitation on Terrestrial Ecosystems*, ed. T. C. Hutchinson and M. Havas. © 1978 by Plenum Press. All rights reserved.)

include the northeastern United States and Japan. Most of the United States east of the Mississippi River now has acid rain with pH values less than 4.6. This situation has presumably existed at least since the mid 1950s but the intensity and areal distribution of the acid rain has increased since then (Likens and Borman 1974). Figure 3.18A shows the pH in 1955–1956 in the eastern United States as calculated by Cogbill and Likens (1974) from Junge and Werby's (1958) data. Figure 3.18B is Likens' map for 1972–1973 and Figure 3.18C shows pH values for the entire United States in 1980 (NRC 1983). The area covered by acid rain had become greater from 1955–1956 to 1972–1973 and the rain in the New York–New England area had become more acid. Data for 1980 (NRC 1983) show a larger area affected by acid rain, a pH less than 4.2 in the central "bull's eye," and the entire eastern United States and southeastern Canada receiving rain with a pH < 5.0. The most rapid increase in the acidity of rain in the last 30 years has been in the southeastern United States, which has also had an increase in industrialization and urbanization during this period (Likens et al. 1979). (SO_2 emissions in the southeastern United States increased 33% from 1965 to 1978; NRC 1983.)

The acidity of U.S. rain is due primarily to H_2SO_4, but the relative contribution of HNO_3 is increasing faster than that of H_2SO_4. At Hubbard Brook, New Hampshire, which has a long precipitation record, the pH showed no significant

Figure 3.18   Contours of average pH of annual precipitation over the eastern United States (A) for 1955–1956 (From G. E. Likens, 1976, from C. V. Cogbill and G. E. Likens, 1974. "Acid Precipitation in the Northeastern United States," *Water Resources Research*, 10 (6), 1135. © 1974 by the American Geophysical Union.) (B) for 1972–1973 (From G. E. Likens. "Acid Precipitation." Reprinted with permission from the *Chemical and Engineering News*, 54 (48), 29–44. Copyright © 1976 by the American Chemical Society.) (C) For 1980 (NRC 1983).

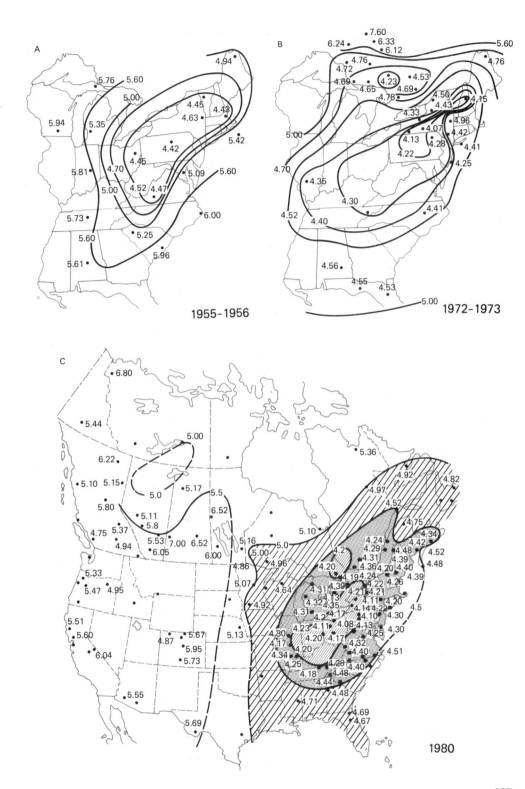

A

5.76 5.60
5.00
5.94
5.35   4.45   4.43
            4.63
5.81   4.42
    4.70   5.42
    4.45
5.00 4.52 4.47   5.09   5.60
5.73   6.00
5.60
5.25
5.61   5.96
1955-1956

B

                    • 7.60
            6.24 • 6.33
                  • 6.12
                  4.76
            4.72   4.53
            4.69
                4.65 4.69   4.50
                4.78   4.43   4.15
                        4.33   4.96
5.00   4.07 4.42
                    4.13   4.28   4.41
4.70   4.22   4.25
    • 4.35
            4.30
4.52 4.40   4.41
    4.56
        4.55
            4.53
                    5.00
1972-1973

C

• 6.80
• 5.44
• 6.22   • 5.36
• 5.10 5.15   5.00
            • 5.17   5.5
5.80   • 5.11   5.10
        5.8   6.52
4.75 5.37 5.53 7.00 6.52
    4.94   • 6.05 6.00
5.33   4.86 4.96
    5.47 4.95   4.64
5.51   4.92
    5.60   4.87 5.67 5.13
        • 5.95
        • 5.73
• 6.04
• 5.55
            5.69

1980

127

change from 1964 to 1981 (Likens, Borman and Eaton 1980; NRC 1983), but this may represent the situation, mentioned above, of the difficulty in lowering pH when it is already low.   The importance of HNO_3 relative to H_2SO_4 at Hubbard Brook increased from 1964 to 1979 by about 50%, particularly in the winter months (Galloway and Likens 1981).   The sulfate levels in rain there decreased 33% over this period (NRC 1983) while nitrate increased irregularly from 1965 to 1970 and levelled off or slightly decreased from 1970 to 1981.   The changes in concentrations in sulfate and nitrate in Hubbard Brook rain seem to parallel changes in emissions in the northeastern United States where SO_2 emissions declined 38% from 1965 to 1978 and NO_x emissions increased 26% from 1960 to 1970 and then decreased 4% by 1978.   The NRC (1983) concludes that in general in the eastern United States, long-term average emissions of SO_2 and NO_x correlate well with the deposition of sulfate and nitrate (which, as H_2SO_4 and HNO_3, represent the main acid sources).   This is in contrast to the European data mentioned earlier.

Studies at areas such as Whiteface Mountain and Ithaca, New York, and south central Ontario (summarized in NRC 1983) show that the source of most of the acidity in the precipitation (and most of the H_2SO_4 and HNO_3) is from air masses that have passed through the Ohio Valley and other Midwest areas to the southwest of the receptor areas.   Moreover, since most of the precipitation also comes from these air masses (because of air transport patterns), pollution from distant industrial areas has a dominant effect.   In general, it is difficult in the eastern United States to distinguish between distant and local sources because pollutants are well mixed over large areas (up to 1000 km on a straight line).

Acid precipitation is also found in the western United States (see Figure 3.18C).   For example, Lewis and Grant (1980), in a study of precipitation in a rural area in the Colorado Rockies, found increasingly acid precipitation, with the pH dropping from 5.43 in 1975 to 4.63 in 1978.   Although sulfuric acid was the major cause of acidity, the *increase* in acidity correlated with an increase in nitric acid in the rain.   Since there were no sizable local sources of nitrogen oxides, and the prevailing air movements would not tend to carry nearby Denver pollution into the area, they felt that these changes reflected a more widespread change in western U.S. precipitation chemistry due to increases in the release of nitrogen oxides from various pollutive western sources.

There is general agreement that the acid rain falling in most of the United States is a serious problem.   However, most of the U.S. government attempts to control SO_2 and NO_2 in atmospheric pollution are based on *local* air quality standards while acid rain seems to be related to *regional* air pollution problems, which are not regulated.   For example, while increasing smokestack height reduces *local* SO_2 pollution, it increases SO_2 pollution downwind and over a larger area.   Thus, the reduction of acid rain requires more regional air pollution standards that take into consideration regional air movements.

Acid rain definitely traceable to fossil fuel combustion is not confined to Europe and North America.   For example, in Machin, a remote small city in the eastern Tibetan Plateau in China, pH 2.25 rain has been measured (Harte 1983).

Nitric acid is the dominant acid in the rain apparently from the burning of coal with a high N/S ratio.

Problems in Distinguishing Naturally Acid Rain from that Due to Pollution

In areas like the United States and Europe where there are widespread anthropogenic acids in the atmosphere, it is difficult to see whether the "natural" pH of rain would in fact be about 5.7 as predicted from equilibrium with atmospheric CO_2, or whether rain might actually have a lower pH due to the removal of naturally occurring acids from the atmosphere, as has been suggested by several workers (Kerr 1981a; Charlson and Rodhe 1982; Galloway et al. 1982). Acid rain (pH < 5.7) has been found recently in a number of *remote* areas. Galloway et al. (1982) propose five possible causes of acid rain in such areas: (1) local fossil fuel combustion; (2) very long range transport of sulfate aerosol from distant anthropogenic sources; (3) agricultural and natural burning of vegetation; (4) natural emissions of reduced sulfur compounds both from the ocean and from the land (as suggested by Charlson and Rodhe 1982); and (5) natural emissions of organic acids from vegetation.

Acid rain can occur in remote areas from local fossil fuel combustion as we have already mentioned for China. The long-range transport of sulfate aerosols from distant anthropogenic sources has been implicated as a cause of acid rain in a number of localities. For example, in Bermuda, in the North Atlantic Ocean, the overall average pH of rain is 4.7–4.8, presumably due to sulfuric acid from long-range transport of sulfate aerosol from the North American continent 1100 km away (Galloway et al. 1982; Jickells et al. 1982; Church et al. 1982). Also, rain from storms originating in the North Atlantic east of Bermuda, where a pollutive source is less obvious, is less acid (pH 4.9) than that from storms originating in the eastern United States (pH 4.4) (Jickells et al. 1982). (The value of 4.9, nevertheless, indicates that considerable sulfate aerosol is transported to the east in the Atlantic even beyond Bermuda; see Church et al. 1982).

A persistent Arctic haze occurs in the winter and spring and has been attributed to very long range transport, over 5000 km or five to ten days travel time, of sulfate aerosols from industrial areas in Europe, Siberia, and perhaps North America (Kerr 1981b). This occurs when circulation patterns favor transport and where there is little precipitation to remove pollutants, which allows long-distance, near-surface transport. Galloway et al. (1982) found an average pH of 5.0 in rain at Poker Flat, Alaska, primarily due to sulfuric acid at least partly from the sulfate aerosol that produces Arctic haze although there may also be local sulfate pollution from Fairbanks, Alaska, 70 km away.

Acid rain (primarily from sulfuric acid) has also been found in Hawaii (Miller and Yoshinaga 1981; Kerr 1981a), where the pH decreases with increasing altitude from 5.2 at sea level to 4.3 at 2500 m. It is suggested that sulfuric acid from Asian sources thousands of kilometers eastward may be transported across the Pacific in the mid-troposphere at high altitudes (> 2000 m) where it is not removed by rainout. Convective rainstorms in Hawaii may reach up into the mid-troposphere to remove the sulfuric acid. Samoa, in the South Pacific receives rain only very slightly acid (pH 5.5) (Pszenny, MacIntyre, and Duce 1982). Apparently the Northern Hemisphere transport of sulfate aerosol which affects Hawaii is not carried southward across the Pacific equator (Kerr 1981a).

Amsterdam Island in the central Indian Ocean (Southern Hemisphere) receives sulfuric acid rain of pH 4.9, (Galloway et al. 1982), whose origin is unclear. If its origin is pollutive, the nearest downwind source is the African continent 5000 km to the west but industry is important only in South Africa. The origin of the sulfate may instead be the oxidation of natural biogenic reduced sulfur gases from the ocean itself since Bonsang et al. (1980) attribute Indian Ocean sulfate aerosol to this origin.

Agricultural burning apparently produces acid rain in several remote continental areas. In Katherine, Australia (mean pH = 4.7), the effect is seasonal with grassland burning during the dry season followed by high acidity in rain at the beginning of the wet season (Galloway et al. 1982). Similar dry-season combustion followed by the removal of built-up nitric acid at the beginning of the wet season, results in very acid rain (pH = 4.0) at Lake Valencia, Venezuela (Lewis 1981). At San Carlos, Venezuela (mean pH = 4.8), a tropical rain forest area with no wet season and where "slash and burn" agriculture is important, as well as at Katherine, Australia, agricultural burning results in a high contribution of organic acids (formic and acetic) to the acidity of rain (Galloway et al. 1982). In addition, the amount of nitric acid relative to sulfuric acid is higher from agricultural burning than from other sources.

Natural emissions of reduced sulfur compounds have been implicated in sulfuric acid rain (mean pH 5.05) in the Amazon River Basin, Brazil (Stallard and Edmond 1981). The pH, 4.7–5.7, is considerably higher than that of rain in the northeastern United States where the pH range is 3.4–5.4 with most values being less than 4.0 (see Figure 3.19). This area of the Amazon Basin has low dust, no pollution sources (including no agricultural burning), and might be expected to have high emissions of reduced biogenic sulfur since it is a tropical rain forest (see section on sulfur). This seems to be the best-documented occurrence of *naturally acid* continental rain (presumably due to biogenic sulfur emissions). Haines et al. (1983) also found acid rain (pH 4.7) of unknown source in the remote Amazon rain forest in Venezuela.

Galloway et al. (1982) suggest that after inputs from human activities are removed, the mean pH of natural rain is $\geqslant 5$. This seems to be true for the limited observations available (e.g., see Figure 3.19), although there are very few proven observations of truly "natural" acid rain. It is certainly true that acid rain (pH $<$ 5.7) of various origins is not uncommon in remote areas. However, as it appears that "natural" acid rain is less acidic than rain that has been attributed to obvious pollutive influences, there is, thus, definite evidence that pollution is a major source of acid rain!

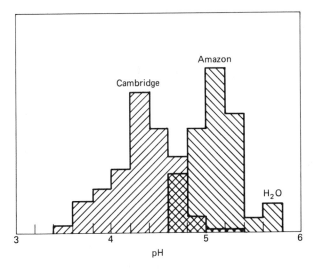

Figure 3.19 Histograms comparing the pH of Amazon rain (Brazil) with that from Cambridge, Massachusetts. Their mean pH is 5.05 and 4.19, respectively. The value for water equilibrated with atmospheric CO_2 is labelled as H_2O. (After R. F. Stallard and J. M. Edmond. "Chemistry of the Amazon, precipitation chemistry and the marine contribution to the dissolved load at the time of peak discharge," *J. of Geophysical Research*, 86 (C10), 9852. © 1981 by the American Geophysical Union.)

Effects of Acid Rain

There are various serious effects of acid rain, ranging from increased destruction of structures, such as corrosion of metal and weathering of buildings, to the effect of acid rain on lakes, soils, and vegetation.   There is evidence that acid rain causes increased leaching of nutrients from foliage and may disrupt leaf physiology and plant growth.   Increased leaching of cations, particularly Ca^{++} and Al^{+++} from the soil by acid rain has been noted (Overein 1972; Cronan and Schonfield 1979). In addition, lakes and rivers have become acid (pH < 5) in such areas as the Adirondacks and Scandinavia (Wright and Gessing 1976) and the loss of fish populations in such acid lakes has focused increasing attention on the acid rain problem (Schofield 1976).   Henriksen (1979) working on Norwegian lakes concludes that a calcium bicarbonate lake of the type common in North America and northern Europe, with pH 6.5, 2.0 mg/ℓ Ca, and 1.1 meq/ℓ HCO_3^-, will become acid with pH < 5 when the long-term average of precipitation is less than 4.6.   Thus, rain of average pH 4.6 is the approximate boundary level for damage to aquatic ecosystems.   When this occurs, the adverse effects of acid precipitation will be felt over areas much wider than those affected today.   (For a further discussion of the effects of acid precipitation on lakes and soils see the section in Chapter 6 on acid lakes.)

REFERENCES

ADAMS, D. F., S. O. FARWELL, E. ROBINSON, M. E. PACK, and W. L. BAMESBERGER. 1981. Biogenic sulfur source strengths, *Environ. Sci. Technol.* 15 (12): 1493–1498.

ANDRAE, M. O. 1982. Marine aerosol chemistry at Cape Grim, Tasmania, and Townsville, Queensland, *J. Geophys. Res.* 87 (C11): 8875–8885.

ANDRAE, M. O., and I. E. GALBALLY. 1984. Sulfur and nitrogen emissions in remote areas. Paper delivered at NATO Advanced Research Inst., Bermuda, October, 1984.

ANDRAE, M. O., and H. RAEMDONCK. 1983. Dimethyl sulfide in the surface ocean and marine atmosphere: A global view, *Science* 221: 744–747.

ANEJA, V. P., J. H. OVERTON, and A. P. ANEJA. 1981. Emission survey of biogenic sulfur flux from terrestrial surfaces, *Air Pollution Control Assn. Jour.* 31 (3): 256–258.

BARRETT, E., and G. BRODIN. 1955. The acidity of Scandinavian precipitation, *Tellus* 7: 251–257.

BENTON, G. S., and M. A. ESTOQUE. 1954. Water vapor transport over North American continent, *J. Meteorology* 11: 462–477.

BEAUFORD, W., J. BARBER, and A. R. BARRINGER. 1977. Release of particles containing metals from vegetation into the atmosphere, *Science* 195: 571–573.

BERRESHEIM, H., and W. JAESCHKE. 1983. The contribution of volcanoes to the global atmospheric sulfur budget, *J. Geophys. Res.* 88 (C6): 3732–3740.

BLANCHARD, D. C. 1963. Electrification of the atmosphere by particles from bubbles in the sea. In *Progress in oceanography*, vol. 1: 71–202. Oxford: Pergamon Press.

BLANCHARD, D. C., and A. H. WOODCOCK. 1957. Bubble formation and modification in the sea and its meteorological significance, *Tellus* 9: 145–158.

BOHN, H. L., B. L. MCNEAL, and G. A. O'CONNOR. 1979. *Soil chemistry*. New York: John Wiley. 329 pp.

BOLIN, B., ED. 1971. Air pollution across national boundaries: The impact on the environment. In *Report of the Swedish Preparatory Committee for the U.N. Conference on the Human Environment*. Stockholm: Royal Ministry of Agriculture. 96 pp.

BONSANG, B., B. C. NGUYEN, A. GANDRY, and G. LAMBERT. 1980. Sulfate enrichment in marine aerosols owing to biogenic gaseous sulfur compounds, *J. Geophys. Res.* 85: 7410–7416.

BOWEN, H. J. M. 1966. *Trace elements in biochemistry*. New York: Academic Press. 241 pp.

BUAT-MENARD, P. 1983. Particle geochemistry in the atmosphere and ocean. In *Air-sea Exchange of gases and particles*, ed. P. N. Liss and W. G. N. Slinn, pp. 455–532. Boston: D. Reidel Co.

BURNS, R. C., and R. W. F. HARDY. 1975. *Nitrogen fixation in bacteria and higher plants*. New York: Springer-Verlag. 189 pp.

BUSENBERG, E., and C. C. LANGWAY, JR. 1979. Levels of ammonium sulfate, chloride, calcium, and sodium in ice and snow from southern Greenland, *J. Geophys. Res.* 84 (C4): 1705–1709.

BYERS, H. R. 1965. *Elements of cloud physics*. Chicago: Univ. of Chicago Press. 191 pp.

CARROLL, D. 1962. *Rainwater as a chemical agent of geologic processes—A review.* U.S.G.S. Water Supply Paper no. 1535-G. 18 pp.

CAST (COUNCIL FOR AGRICULTURAL SCIENCE AND TECHNOLOGY). 1976. *Effect of increased nitrogen fixation on stratospheric ozone.* Report no. 53.

CHAMEIDES, W. L. 1979. Effect of variable energy input on nitrogen fixation in instantaneous linear discharges, *Nature* 277: 123–125.

CHARLSON, R. J., and H. RODHE. 1982. Factors controlling the acidity of natural rainwater, *Nature* 295: 683–685.

CHURCH, T. M., J. N. GALLOWAY, T. D. JICKELLS, and A. H. KNAP. 1982. The chemistry of western Atlantic precipitation at the mid-Atlantic coast and on Bermuda, *J. Geophys. Res.* 87 (C13): 11013–11018.

COGBILL, C. V., and G. E. LIKENS. 1974. Acid precipitation in the northeastern United States, *Water Resources Res.* 10 (6): 1133–1137.

CONNOR, J. J., and H. T. SHACKLETTE. 1975. Background geochemistry of some rocks, soils, plants and vegetables in the conterminous United States. U.S.G.S. Prof. Paper no. 574-F. 168 pp.

CONWAY, E. J. 1942. Mean geochemical data in relation to oceanic evolution, *Proc. Roy. Irish Acad.* 48B: 119–159.

CRONAN, C. S., and C. C. SCHOFIELD. 1979. Aluminum leaching in response to acid precipitation: Effects on high-elevation watersheds in the northeast, *Science* 204: 304–306.

CROZAT, G. 1979. Sur l'émission d'un aerosol riche en potassium par la forêt tropical, *Tellus* 31: 52–57.

CROZAT, G., J. L. DOMERQUE, J. BAUDET, and V. BONGI. 1978. Influence des feux de brousse sur la composition chimique des aerosols atmospheriques en Afrique de l'ouest, *Atmos. Environ.* 12 (9): 1917.

CRUTZEN, P. J. 1976. Upper limits on atmospheric ozone reductions following increased application of fixed nitrogen to the soil, *Geophys. Res. Letters* 3 (3): 169–172.

CRUTZEN, P. J., L. E. HEIDT, J. P. KRASNER, W. H. POLLOCK, and W. SAILER. 1979. Biomass burning as a source of atmospheric gases CO_2, H_2, N_2O, NO, CH_3Cl, and COS, *Nature* 282: 253–256.

CULLIS, C. F., and M. M. HIRSCHLER. 1980. Atmospheric sulfur: Natural and man-made sources, *Atmos. Environ.* 14: 1263–1278.

CURTIN, G. C., H. O. KING, and E. L. MOSIER. 1974. Movement of elements into the atmosphere from coniferous trees in subalpine forest of Colorado and Idaho, *J. Geochem. Explor.* 3: 245–263.

DAWSON, G. A. 1977. Atmospheric ammonia from undisturbed land, *J. Geophys. Res.* 82 (21): 3125–3133.

DELMAS, R., J. BAUDET, J. SERVANT, and Y. BAZIARD. 1980. Emissions and concentrations of hydrogen sulfide in the air of the tropical forest of the Ivory Coast and temperate regions of France, *J. Geophys. Res.* 85: 4468–4474.

DELMAS, R., and J. SERVANT. 1983. Atmospheric balance of sulphur above an equatorial forest, *Tellus* 35B: 110–120.

DENMEAD, O. T., J. R. FRENEY, and J. R. SIMPSON. 1976. A closed ammonia cycle within a plant canopy, *Soil Biol. Biochem.* 8: 161–164.

DENMEAD, O. T., R. NULSEN, and G. W. THURTELL. 1978. Ammonia exchange over a corn crop, *Soil Sci. Soc. Amer. J.* 42: 840–842.

DUCE, R. A., and E. J. HOFFMAN. 1976. Chemical fractionation at the air/sea interfaces. In *Annual Review of earth and planetary sciences*, pp. 187–228. Palo Alto, Calif.: Ann. Revues, Inc.

DUCE, R. A., F. MacINTYRE, and B. BONSANG. 1982. discussion of "Enrichment of sulfate in maritime aerosols" (Garland 1981), *Atmos. Environ.* 16 (8): 2025–2026.

ELIASSEN, A. 1978. The OECD study of long range transport modelling, *Atmos. Environ.* 12: 479–487.

ERIKSSON, E. 1952. Composition of atmospheric precipitation, I, Nitrogen compounds, *Tellus* 4: 215–232.

———. 1957. The chemical composition of Hawaiian rainfall, *Tellus* 9:509–520.

———. 1960. Yearly circulation of chloride and sulfur in nature, meteorological, geochemical and pedological implications, Part 2, *Tellus* 12: 63–109.

FARQUHAR, G. D., P. M. FIRTH, R. WETSELAAR and B. WEIR. 1980. On the gaseous exchange of ammonia between leaves and the environment: Determination of the ammonia compensation point, *Plant Physiol.* 66: 710–714.

FELLER, M. C., and J. P. KIMMINS. 1979. Chemical characteristics of small streams near Haney in southwestern British Columbia, *Water Resources Res.* 15 (2): 247–258.

FORLAND, E. J. 1973. A study of the acidity in the precipitation of southwestern Norway, *Tellus* 25: 291–299.

FOWLER, D. 1978. Wet and dry deposition of sulphur and nitrogen compounds from the atmosphere. In *Effects of acid precipitation on terrestrial ecosystems*, ed. T. C. Hutchinson and M. Havas, pp. 9–27. New York: Plenum Press.

FREYER, H. D. 1978. Seasonal trends of NH_4^+ + NO_3^- nitrogen isotope composition in rain collected in Jülich, Germany, *Tellus* 30: 83–92.

FRIEND, J. P. 1973. The global sulfur cycle. In *Chemistry of the lower atmosphere*, ed. S. I. Rasool, pp. 177–201. New York: Plenum Press.

GALBALLY, I. E. 1975. Emission of oxides of nitrogen (NO_x) and ammonia from the earth's surface, *Tellus* 27 (1): 67–70.

GALBALLY, I. E., and C. R. ROY. 1978. Loss of fixed nitrogen from soils by nitric oxide exhalation, *Nature* 275: 734–735.

GALBALLY, I. E., G. D. FARQUHAR, and G. P. AYERS. 1982. Interactions in the atmosphere of the biogeochemical cycles of carbon, nitrogen and sulfur. In *Cycling of carbon, nitrogen, sulfur and phosphorus in terrestrial and aquatic ecosystems*, ed. J. R. Freney and I. E. Galbally, pp. 1–9. Berlin: Springer-Verlag.

GALLOWAY, J. N., A. H. KNAP, and T. M. CHURCH. 1983. The composition of Western Atlantic precipitation using shipboard collectors, *J. Geophys. Res.* 88 (C15): 10859–10864.

GALLOWAY, J. N., and G. E. LIKENS. 1978. The collection of precipitation for chemical analysis, *Tellus* 30: 71–82.

———. 1981. Acid precipitation: The importance of nitric acid, *Atmos. Environ.* 15:1081–1085.

GALLOWAY, J. N., G. E. LIKENS, W. C. KEENE, and J. M. MILLER. 1982. The composition of precipitation in remote areas of the world, *J. Geophys. Res.* 87 (11): 8771–8786.

GALLOWAY, J. N., and D. M. WHELPDALE. 1980. An atmospheric sulfur budget for eastern North America, *Atmos. Environ.* 14: 409–417.

GAMBELL, A. W., JR. 1962. Indirect evidence of the importance of water-soluble continentally derived aerosols, *Tellus* 14: 91–95.

GAMBELL, A. W., and D. W. FISHER. 1964. Occurrence of sulfate and nitrate in rainfall, *J. Geophys. Res.* 69 (20): 4203–4210.

————. 1966. Chemical composition of rainfall, western North Carolina and southeastern Virginia. U.S.G.S. Water Supply Paper no. 1535-K. 41 pp.

GARLAND, J. A. 1978. Dry and wet removal of sulfur from the atmosphere, *Atmos. Environ.* 12: 349–362.

————. 1981. Enrichment of sulfate in maritime aerosols, *Atmos. Environ.* 15: 787–791.

GARRELS, R. M., and C. L. CHRIST. 1965. *Solutions, minerals, and equilibria.* New York: Harper. 450 pp.

GARRELS, R. M., and F. T. MACKENZIE. 1971. *Evolution of sedimentary rocks.* New York: W. W. Norton. 397 pp.

GATZ, D. F. 1975. Relative contributions of different sources of urban aerosols: Application of a new estimation method to multiple sites in Chicago, *Atmos. Environ.* 9: 1–18.

GEORGII, H. W., and U. LENHARD. 1978. Contribution to the atmospheric NH₃ budget, *Pure and Applied Geophysics* 116: 385–392.

GOLDBERG, E. 1971. Atmospheric dust, the sedimentary cycle and man. In *Comments on earth science: Geophysics*, vol. 1: 117–132.

GRAEDEL, T. E. 1979. Reduced sulfur emission from the open oceans, *Geophys. Res. Letters* 6: 329–331.

GRAHAM, W. F., and R. F. DUCE. 1979. Atmospheric pathways of the phosphorus cycle, *Geochim. Cosmochim. Acta* 43: 1195–1208.

————. 1982. The atmospheric transport of phosphorus to the western North Atlantic, *Atmos. Environ.* 16: 1089–1097.

GRAHAM, W. F., S. R. PIOTROWICZ, and R. F. DUCE. 1979. The sea as a source of atmospheric phosphorus, *Marine Chem.* 7: 325–342.

GRANAT, L. 1972. On the relation between pH and the chemical composition in atmospheric precipitation, *Tellus* 24: 550–560.

————. 1978. Sulfate in precipitation as observed by the European atmospheric chemistry network, *Atmos. Environ.* 12: 413–424.

GRANAT, L., H. RODHE, and R. O. HALLBERG. 1976. The global sulfur cycle. In *Nitrogen, phosphorus, and sulfur global cycles*, ed. B. H. Svensson and R. Söderlund, pp. 89–134. SCOPE Report no. 7. Ecol. Bull. (Stockholm) 22.

GRAUSTEIN, W. C. 1981. The effects of forest vegetation on solute acquisition and chemical weathering: a study of the Tesuque watersheds near Santa Fe, New Mexico. Ph.D. dissertation, Yale University, New Haven, Conn. 645 pp.

GRAVENHORST, G., S. BEILKE, M. BETZ, and H. GEORGII. 1978. Sulfur dioxide absorbed in rain water. In *Effects of acid precipitation on terrestrial ecosystems*, ed. T. C. Hutchinson and M. Havas, pp. 41–55. New York: Plenum Press.

HAINES, B., C. JORDAN, H. CLARK, and K. E. CLARK. 1983. Acid rain in an Amazon rain forest, *Tellus* 35B: 77–80.

HANDA, B. K. 1971. Chemical composition of monsoon rain water over Bankipur, *Indian J. Meteorol. Geoph.* 22: 603.

HARDING, D., and J. M. MILLER. 1982. The influence on rain chemistry of the Hawaiian volcano Kilauea, *J. Geophys. Res.* 87 (C2): 1225–1230.

HARTE, J. 1983. An investigation of acid precipitation in Quinghai Province, China, *Atmos. Environ.* 17 (2): 403–408.

HEALY, T. V., A. C. MCKAY, A. PILBEAM, and D. SCARGILL. 1970. Ammonia and ammonium sulfate in the troposphere over the United Kingdom, *J. Geophys. Res.* 75: 2317–2321.

HENRIKSEN, A. 1979. A simple approach for identifying and measuring acidification of freshwater, *Nature* 278: 542–545.

HILL, R. D., and R. G. RINKER. 1981. Production of nitrate ions and other trace species by lightning, *J. Geophys. Res.* 86 (C4): 3203–3209.

HILL, R. D., R. G. RINKER, and W. D. WILSON. 1980. Atmospheric nitrogen fixation by lightning, *J. Atmos. Sci.* 37: 179–192.

HOLLAND, H. D. 1978. *The chemistry of the atmosphere and oceans.* New York: John Wiley. 351 pp.

HUTTON, J. T. and T. I. LESLIE. 1958. Accession of nonnitrogenous ions dissolved in rainwater to soils in Victoria, *Australian J. Agr. Research*, 9: 492–507.

IVANOV, M. V. 1981. The global biogeochemical sulfur cycle. In *Some perspectives of the major biogeochemical cycles*, ed. G. E. Likens, pp. 61–78. SCOPE 4th Gen. Ass. Stockholm. New York: John Wiley & Sons.

JAESCHKE, W. 1978. New methods for analysis of SO_2 and H_2S in remote areas and their applicability to the atmosphere, *Atmos. Environ.* 12: 715–722.

JICKELLS, T., A. KNAP, T. CHURCH, J. GALLOWAY, and J. MILLER. 1982. Acid rain on Bermuda, *Nature* 297: 55–57.

JOHANSSON, C., and L. GRANAT. 1984. Emission of nitric oxide from arable land, *Tellus* 36B: 25–37.

JUNGE, C. E. 1958. The distribution of ammonium and nitrate in rain water over the United States, *Trans. Amer. Geophys. Union* 39: 241–248.

———. 1963. *Air chemistry and radioactivity.* New York: Academic Press. 382 pp.

———. 1972. Our knowledge of the physico-chemistry of aerosols in the undisturbed marine environment, *J. Geophys. Res.* 77: 5183–5200.

JUNGE, C. E., and R. T. WERBY. 1958. The concentration of chloride, sodium, potassium, calcium and sulfate in rainwater over the United States, *J. Meteorology* 15: 417–425.

KALLEND, A. S., A. R. MARSH, J. H. PICKLES, and M. V. PROCTOR. 1983. Acidity of rain in Europe, *Atmos. Environ.* 17: 127–137.

KERR, R. A. 1981a. Is all acid rain polluted? *Science* 212 (29): 1014.

———.1981b. Pollution of the Arctic atmosphere confirmed, *Science* 212 (29): 1013–1014.

KELLOGG, W. W., R. D. CADLE, E. R. ALLEN, A. L. LAZRUS, and E. A. MARTELL. 1972. The sulfur cycle, *Science* 175: 587.

KRAMER, J. R. 1978. Acid precipitation. In *Sulfur in the environment, Part 1: The Atmospheric Cycle*, ed. J. O. Nriagu, pp. 325–370. New York: John Wiley.

KRITZ, M. 1982. Exchange of sulfur between the free troposphere, marine boundary layer, and the sea surface, *J. Geophys. Res.* 87 (C11): 8795–8803.

LAWSON, D. R., and J. W. WINCHESTER. 1978. Sulfur and trace element relationships in aerosols from the South American continent, *Geophys. Res. Letters* 5: 195–198.

———. 1979. Sulfur, potassium, and phosphorus associations in aerosols from South American tropical rain forests, *J. Geophys. Res.* 84 (C7): 3723–3727.

LENHARD, U., and G. GRAVENHORST. 1980. Evaluation of ammonia fluxes into the free atmosphere over Western Germany, *Tellus* 32: 48–55.

LEWIS, W. M., JR. 1981. Precipitation chemistry and nutrient loading by precipitation in a tropical watershed, *Water Resources Res.* 17 (1): 161–181.

LEWIS, W. M., JR., and M. C. GRANT. 1980. Acid precipitation in the western United States, *Science* 207: 176–177.

LIKENS, G. E. 1976. Acid precipitation, *Chemical and Engineering News* 54 (48): 29–44.

LIKENS, G. E., and F. H. BORMANN. 1974. Acid rain, a serious regional environmental problem, *Science* 184: 1176–1179.

LIKENS, G. E., F. H. BORMANN, and J. S. EATON. 1980. Variations in precipitation and streamwater chemistry at the Hubbard Brook Experimental Forest during 1964–1977. In *Effects of Acid Precipitation on Terrestrial Ecosystems*, ed. T. C. Hutchinson and M. Havas, pp. 443–464. New York: Plenum Press.

LIKENS, G. E., F. H. BORMANN, R. S. PIERCE, J. S. EATON, and N. M. JOHNSON. 1977. *Biochemistry of a forested ecosystem*. New York: Springer-Verlag. 146 pp.

LIKENS, G. E., R. F., WRIGHT, J. N. GALLOWAY, and T. J. BUTLER. 1979. Acid rain, *Sci. Amer.* 241 (4): 43–51.

LIPSCHULTZ, F., O. C. ZAFIRIOU, S. C. WOFSY, M. B. McELROY, F. W. VALOIS, and S. W. WATSON. 1981. Production of NO and N_2O by soil nitrifying bacteria, *Nature* 294: 641–643.

LODGE, J. P., JR., J. B. PAKE, W. BASBASILL, G. S. SWANSON, K. C. HILL, A. L. LORANGE, and A. L. LAZRUS. 1968. *Chemistry of U.S. precipitation*. Report of Natl. Precipitation Sampling Network, Natl. Center for Atmospheric Res., Boulder, Colo.

LOGAN, J. A. 1983. Nitrogen oxides in the troposphere: Global and regional budgets, *J. Geophys. Res.* 88 (C15): 10785–10807.

LOGAN, J. A., M. J. PRATHER, S. C. WOFSY, and M. B. McELROY. 1981. Tropospheric chemistry: A global perspective, *J. Geophys. Res.* 86 (C8): 7210–7254.

LVOVITCH, M. I. 1973. The global water balance, *Trans. Amer. Geophys. Union* 227 (3): 60–70.

MacINTYRE, F. M. 1974. The top millimeter of the ocean, *Sci. Amer.* 230: 62–77.

MARSHALL, E. 1983. Acid rain, a year later, *Science* 221: 241–242.

MARTENS, C. S., J. J. WESILOWSKI, R. C. HARRISS, and R. KAIFER. 1973. Chlorine loss from Puerto Rican and San Francisco Bay area marine aerosol, *J. Geophys. Res.* 78: 8778–8792.

MASON, B. 1966. *Principles of geochemistry*, 3rd ed. New York: John Wiley & Sons.

MASON, B. J. 1971. *Physics of clouds*, 2nd ed. New York: Oxford Univ. Press. 671 pp.

McCONNELL, J. C. 1973. Atmospheric ammonia, *J. Geophys. Res.* 78: 7812–7821.

MEANS, J. L., R. F. YURETICH, D. A. CRERAR, D. J. J. KINSMAN, and M. P. BORCSIK. 1981. *Hydrogeochemistry of the New Jersey Pine Barrens*. Dept. of Environmental Protection, N.J., Geol. Survey Bull. no. 76, Trenton, N.J. 107 pp.

MEINERT, D. L., and J. W. WINCHESTER. 1977. Chemical relationships in the North Atlantic marine aerosol. *J. Geophys. Res.* 82 (12): 1778–1782.

MÉSZÁROS, E. 1982. On the atmospheric input of sulfur into the ocean, *Tellus*, 34: 277–282.

MEYBECK, M. 1979. Concentration des eaus fluviales en éléments majeurs et apports en solution aux océans, *Rev. Géol. Dyn. Géogr. Phys.* 21: 215–246.

———. 1982. Carbon, nitrogen and phosphorus transport by world rivers, *Amer. J. Sci.* 282: 401–450.

———. 1983. Atmospheric inputs and river transport of dissolved substances. In *Dissolved loads of rivers and surface water quantity/quality relationships*, pp. 173–192. Proceedings of the Hamburg Symposium, August 1983. IAHS Publ. no. 141.

MILLER, A. C., J. C. THOMPSON, R. E. PETERSON, and D. R. HARAGAN. 1983. *Elements of meteorology,* 4th ed. Columbus, Ohio: Chas. E. Merrill.

MILLER, D. H. 1977. *Water at the surface of the earth.* New York: Academic Press. 557 pp.

MILLER, J. M. 1974. A statistical evaluation of the U.S. precipitation chemistry network. In *Precipitation scavenging*, ed. R. G. Semonin and R. W. Beadle, pp. 639–661. ERDA Sympos. ser. 41.

MILLER, J. M., and A. M. YOSHINAGA. 1981. The pH of Hawaiian precipitation, a preliminary report, *Geophys. Res. Letters* 8: 779–782.

MÖLLER, D. 1984. Estimation of the global man-made sulfur emission, *Atmos. Environ.* 18 (1): 19–27.

MUNGER, J. W. 1982. Chemistry of atmospheric precipitation in the north-central United States: Influence of sulfate, nitrate, ammonia and calcareous soil particulates, *Atmos. Environ.*, 16 (7): 1633–1645.

MUNGER, J. W., and S. J. EISENREICH. Continental-scale variations in precipitation chemistry, *Environ. Sci. Technol.* 17 (1):32A–42A.

NADER, J. S. 1980. Primary sulfate emissions from stationary industrial sources. In *Atmospheric sulfur deposition—Environmental and health effects,* ed. D. S. Shriner, C. R. Richmond, and S. Lindberg, pp. 121–130. Ann Arbor, Mich. Ann Arbor Science.

NEIBURGER, M., J. G. EDINGER, and W. D. BONNER. 1973. *Understanding our atmospheric environment.* San Francisco: W. H. Freeman. 293 pp.

NEWELL, R. E. 1971. The global circulation of atmospheric pollutants, *Sci. Amer.* 224 (1): 32–42.

NGUYEN, B. C., B. BONSANG, and A. GAUDRY. 1983. The role of the ocean in the global atmospheric sulfur cycle, *J. Geophys. Res.* 88 (C15): 10903–10914.

NRC (National Research Council). 1979. *Ammonia.* Washington, D.C.: National Academy of Sciences.

NRC (NATIONAL RESEARCH COUNCIL). 1983. *Acid deposition: Atmospheric processes in eastern North America.* Washington, D.C.: National Academy Press. 291 pp.

ODÉN, S. 1968. The acidification of air and precipitation and its consequences on the natural environment (In Swedish), *Statens Naturvetenskapliga Forskningsräd*, Stockholm. Bull. 1: 1–86.

OVEREIN, L. N. 1972. Sulfur pollution pattern observed: Leaching of calcium in forest soil determined, *Ambio* 1: 145–147.

PACIGA, J. J., and R. E. JERVIS. 1976. Multielement size characterization of urban aerosols, *Environ. Sci. Technol.* 10 (12): 1124–1128.

PACK, D. H. 1980. Precipitation chemistry patterns: A two-network data set, *Science* 208: 1143–1145.

PEARSON, F. J., JR., and D. W. FISHER. 1971. *Chemical composition of atmospheric precipitation in the northeastern United States.* U.S.G.S. Water Supply Paper no. 1535-P. 23 pp.

PEIXOTO, J. P., and M. A. KETTANI. 1973. The control of the water cycle, *Sci. Amer.* 228 (4): 46–61.

PETRENCHUK, O. P. 1980. On the budget of sea salts and sulfur in the atmosphere, *J. Geophys. Res.* 85 (C12): 7439–7444.

PETRENCHUK, O. P., and E. S. SELEZNEVA. 1970. Chemical composition of precipitation in regions of the Soviet Union, *J. Geophys. Res.* 75 (18): 3629–3634.

PROSPERO, J. M. 1979. Eolian transport to the world ocean. In *The oceanic lithosphere: The sea,* v. 7, ed. C. Emiliani, pp. 801–874. New York: John Wiley.

PRUPPACHER, H. R. 1973. The role of natural and anthropogenic pollutants in clouds and precipitation formation. In *Chemistry of the Lower Atmosphere,* ed. S. I. Rasool, pp. 1–62. New York: Plenum Press.

PSZENNY, A. A. P., F. MacINTYRE, and R. A. DUCE. 1982. Sea salt and the acidity of marine rain on the windward coast of Samoa, *Geophys. Res. Letters* 9: 751–754.

RAHN, K. A. 1976. *The chemical composition of the atmospheric aerosol.* Tech. Report, Grad. School of Oceanography, Univ. of Rhode Island, Kingston. 265 pp.

RAHN, K. A., and D. H. LOWENTHAL. 1984. Elemental tracers of distant regional pollutive aerosols, *Science* 223: 132–139.

ROBINSON, E., and R. C. ROBBINS. 1975. Gaseous atmospheric pollutants from urban and natural sources. In *The changing global environment,* ed. S. F. Singer, pp. 111–123. Dordvecht, Holland: D. Reidel Co.

RODHE, H. 1972. A study of the sulfur budget in the atmosphere over northern Europe, *Tellus* 24: 128–138.

RODHE, H., and I. ISAKSEN. 1980. Global distribution of sulfur compounds in the troposphere estimated in a height/latitude transport model, *J. Geophys. Res.* 85 (C12): 7401–7409.

RODHE, H., P. CRUTZEN, and A. VANDERPOL. 1981. Formation of sulfuric and nitric acid during long-range transport, *Tellus* 33: 132–141.

SAVOIE, D. L., and J. M. PROSPERO. 1980. Water-soluble K, Ca and Mg in the aerosols over the tropical North Atlantic, *J. Geophys. Res.* 85 (C1): 385–392.

SCHINDLER, D. W., R. W. NEWBURY, K. G. BEATTY, and P. CAMPBELL. 1976. Natural water and chemical budgets for a small Precambrian lake basin in central Canada, *J. Fish. Res. Bd. Can.* 33: 2526–2543.

SCHLESINGER, W. H., and W. A. REINERS. 1974. Deposition of water and cations on artificial foliar collectors in fir krummholz of New England Mountains, *Ecology* 55: 378–386.

SCHOFIELD, C. L. 1976. Acid precipitation: Effects on fish, *Ambio* 5 (5–6): 228–230.

SEILER, W., and P. J. CRUTZEN. 1980. Estimates of gross and net fluxes of carbon between the biosphere and the atmosphere from biomass burning, *Climatic Change* 2: 207–247.

SELLERS, W. D. 1965. *Physical climatology,* 2nd ed. Chicago: Univ. of Chicago Press. 272 pp.

SEQUEIRA, R. 1976. Monsoonal deposition of sea salt and air pollutants over Bombay, *Tellus* 28 (3): 275–281.

SHACKLETTE, H. T., J. C. HAMILTON, J. G. BOERNGEN, and J. M. BOWLES. 1971. *Elemental composition of surficial materials in the conterminous United States.* U.S.G.S. Prof. Paper no. 574–D. 71 pp.

SIMPSON, H. J. 1977. Man and the global nitrogen cycle. In *Global Chemical Cycles and Their Alterations by Man*, ed. W. Stumm, pp. 253–274. Berlin: Dahlem Konferenzen.

SMIC (STUDY OF MAN'S IMPACT ON CLIMATE, STOCKHOLM). 1970. *Inadvertent climate modification.* Cambridge, Mass.: MIT Press. 308 pp.

SÖDERLUND, R. 1977. NO$_x$ pollutants and ammonia emissions—a mass balance for the atmosphere over NW Europe, *Ambio* 6 (2): 118—122.

SÖDERLUND, R., and B. H. SVENSSON. 1976. The global nitrogen cycle. In *Nitrogen, phosphorus and sulphur—Global cycles*, ed. B. H. Svensson, and R. Söderlund, pp. 23–73 SCOPE Report no, 7. Stockholm: Ecol. Bull. 22.

STALLARD, R. F. 1980. Major element geochemistry of the Amazon river system. Ph.D. dissertation, MIT/Woods Hole Oceanographic Inst., WHO I-80-29. 366 pp.

STALLARD, R. F., and J. M. EDMOND. 1981. Chemistry of the Amazon, precipitation chemistry and the marine contribution to the dissolved load at the time of peak discharge, *J. Geophys. Res.* 86 (C10): 9844–9858.

STUMM, W., and J. J. MORGAN. 1981. *Aquatic chemistry,* 2nd ed. New York: Wiley Interscience. 780 pp.

SUGAWARA, K. 1967. Migration of elements through phases of the hydrosphere and atmosphere. In *Chemistry of the Earth's Crust*, vol. 2, ed. A. P. Vinogradov, pp. 501–510. Israel Program for Scientific Translation Ltd., Jerusalem. Reprinted in *Geochemistry of Water*, ed. Y. Kitano, pp. 227–237. New York: Halsted Press, 1975.

TANAKA, S., M. DARZI, and J. W. WINCHESTER. 1980. Sulfur and associated elements and acidity in continental and marine rain from north Florida, *J. Geophys. Res.* 85 (C8): 4519–4526.

THORNTON, J. D., and S. J. EISENREICH. 1982. Impact of land use on the acid and trace element composition of precipitation, *Atmos. Environ.* 16: 1945–1955.

TJEPKEMA, J. D., R. J. CARTICA, and H. F. HEMOND. 1981. Atmospheric concentration of ammonia in Massachusetts and deposition on vegetation, *Nature* 294:445–446.

TUREKIAN, K. K. 1972. *Chemistry of the earth.* New York: Holt, Rinehart & Winston. 131 pp.

U.S. EPA (United States Environmental Protection Agency). 1982. National air pollution emissions estimates, 1940–1980. Research Triangle Park, N.C.: *EPA*-450/4-82-001.

VARHELYI, G., and G. GRAVENHORST. 1981. Production rate of airborne sea salt sulfur deduced from chemical analysis of marine aerosols and precipitation. Paper presented at IAMAP-ROAC Symposium, Hamburg, Germany, Aug. 24–27, 1981.

VISSER, S. 1961. Chemical composition of rainwater in Kampala, Uganda, and its relation to meteorological and topographical conditions. *J. Geophys. Res.* 66: 3759–3766.

WENT, F. W. 1960. Organic matter in the atmosphere, *Proc. Natl. Acad. of Sciences* 46: 212–221.

WHITE, E. J., and F. TURNER. 1971. An assessment of the relative importance of several chemical sources to the waters of a small upland catchment, *J. Applied Ecology* 8: 743–749.

WHITEHEAD, H. C., and J. H. FETH. 1964. Chemical composition of rain, dry fallout and bulk precipitation, Menlo Park, Calif., 1957–1959. *J. Geophys. Res.* 69: 3319–3333.

WINNER, W. E., C. L. SMITH, G. W. KOCH, H. A. MOONEY, J. D. BEWLEY, and H. R. KROUSE. 1981. Rates of emission of H_2S from plants and patterns of stable sulfur isotope fractionation, *Nature* 289: 672–673.

WRIGHT, R. F., and E. T. GJESSING. 1976. Changes in the chemical composition of lakes, *Ambio* 5 (5–6): 219–223.

ZEHNDER, H. J. B., and S. H. ZINDER. 1980. The sulfur cycle. In *The handbook of environmental chemistry*, vol. 1A: *The natural environment and the biogeochemical cycles*, ed. O. Hutzinger, pp. 105–145. New York: Springer-Verlag.

ZOBRIST, J., and W. STUMM. 1980. Chemical dynamics of the Rhine catchment area in Switzerland: Extrapolation to the "pristine" Rhine river input into the ocean. In *River Inputs to Ocean Systems*, ed. J–M. Martin, J. D. Burton, and D. Eisma, pp. 52–63. Rome: SCOR/UNEP/UNESCO Review and Workshop, FAO.

ZVEREV, V. P., and V. Z. RUBEIKIN. 1973. The role of atmospheric precipitation in circulation of chemical elements between atmosphere, hydrosphere, and lithosphere, *Hydrogeochemistry* 1: 613–620.

4

Soil Water and Groundwater:

Weathering

INTRODUCTION AND HYDROLOGIC TERMINOLOGY

Water falling to the surface of the continents as rain, upon striking the surface, undergoes major modification of both its mode of transport and its chemical composition. The water may infiltrate the soil or it may immediately run off the surface. It may be first intercepted by vegetation and then "drip" to the ground. The water may be lost back to the atmosphere by evaporation from the ground or from trees. Finally, it may pass deep underground only to emerge at a much later date in a river or lake. In all cases it comes into contact with substances that react with it and, as a result, alter its composition.

An idea of the paths that water may take once it strikes the ground, is shown diagrammatically in Figure 4.1. Water that has been intercepted by vegetation and then drips off it is termed *throughfall*. (That running down plant stems or tree trunks is called *stemflow*). Water infiltrating the soil is called *soil water*, and that passing directly into the nearest stream is referred to as *surface runoff*. Once in the soil, the water either passes downward or is taken up by plant and tree roots. In the latter case, the water is transported up through the tree and eventually evaporated from leaf surfaces. In this way the water is returned to the atmosphere and the overall process is known as *transpiration*. Water trickling downward through the soil eventually encounters a level in the soil or underlying bedrock where all pore space is filled with water. At this point the water becomes *groundwater*, the rock or soil is said to be *saturated* with water, and the level where this occurs is known as the *water table* (Figure 4.1). (Above the water table, pore space is filled by a mixture of air and water and this region is referred to as the *unsaturated zone* or *zone of aeration*.)

Groundwater flows underground until the water table intersects the land surface and the flowing water becomes surface water in the form of springs, rivers, swamps, and lakes. The continual contribution of groundwater to rivers, important between rainstorms, is known as *base flow*. Groundwater continues to flow due to an hydrostatic head built up by the *recharge* of new rainwater at the source. Because of day-to-day, seasonal, and longer-term climatic changes, the rate of rainfall input and, consequently, the position of the water table, can fluctuate but the degree of fluctuation and its timing can be considerably damped and delayed depending upon the capacity of the subsurface rocks to store groundwater. (For further details on groundwater hydrology consult, e.g., Domenico 1972 or Freeze and Cherry 1979).

Water coming into contact with rocks (and derived soils) reacts with primary minerals contained in them. The minerals dissolve to varying extents and some of the dissolved constituents react with one another to form new, or secondary, minerals. Dissolution is brought about mainly by acids provided by plant activity and bacterial metabolism (and, in areas of pollution, by acid rain), and the overall process is called *chemical weathering*. Besides biological factors, chemical weathering is also aided by physical processes that act to break up rocks and expose additional mineral surface area to weathering solutions. This is known as *physical weathering* and the dominant process is the fracturing of rocks by expansion accompanying the freezing of water in cracks. Thus, physical weathering is most important at higher latitudes and elevations. Together chemical, biological, and physical weathering result in the breakdown of rock and the formation of soil.

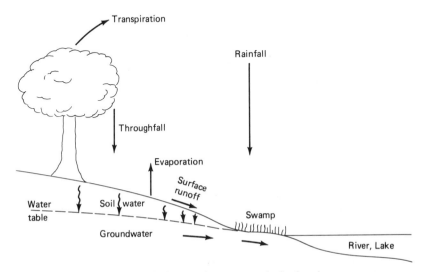

Figure 4.1 Pathways of water near the land surface.

Besides soil formation, chemical weathering also results in a radical change in the composition of soil water and groundwater. These changes reflect both the composition of the primary minerals and the degree of biological activity acting to bring about mineral dissolution. Although limited dissolution can occur by reactions of primary minerals with pure rainwater, it is safe to say that most weathering and consequent change in water composition is brought about, directly or indirectly, by biological activity, and it is this intimate interplay between rocks, water, and life that we shall emphasize in this chapter. In fact, we shall begin our coverage of weathering with a discussion of biological cycling or, in simple terms, a "tree's-eye view of weathering."

BIOGEOCHEMICAL CYCLING IN FORESTS

In forested areas, natural water chemistry is influenced by the uptake, storage, and release of nutrients by the vegetation. (*Nutrients* are elements needed by living things.) A forest can be considered a reservoir of nutrient elements between their input and their release in streamwater runoff. Thus, biogeochemical cycles exist that illustrate interaction of the hydrologic (water) cycle with plant nutrient cycles. The degree of interaction varies, and, in some cases, the biological cycle may be in large part independent of the weathering process.

The major nutrients needed by plants are shown in Table 4.1 in the order of concentrations required for growth. Life cannot exist without water and carbon dioxide, which are the principal sources of carbon, hydrogen, and oxygen. In addition, essential life components, such as proteins, nucleic acids, and ATP, also require nitrogen, phosphorus, and sulfur. Together these elements combine to produce an overall average composition for land plants (on a mole basis) of C_{1200} H_{1900} O_{900} N_{35} P_2 S_1 (based on data of Table 4.1). Other major elements include magnesium and potassium, which are essential components in the chlorophyll which is used by plants for photosynthesis. Besides major elements, trace nutrients (Table 4.1) are also required but are taken up in much lower concentrations.

Nutrient elements are taken up by plants from four *immediate* sources (Likens et al. 1977). They are: the atmosphere, dead organic matter, minerals, and soil solutions. Atmospheric nutrients include gases, which can be added directly to the plants, and dust, which is carried by the atmosphere and trapped on foliage. Dead organic matter in the soil contains nutrients that are made available to plants upon decomposition. (In tropical rain forests, which are commonly developed on heavily leached soils, dead organic matter constitutes the principal source of nutrients.) Minerals, both primary and secondary, are the principal source of many nutrients but require solubilization, via weathering, before the nutrients can become available. Finally, soil solutions carry nutrients, derived from minerals, rainwater, and organic matter, directly to the plant roots. (An intermediate state of nutrient availability also exists as adsorbed ions and molecules on the surfaces of organic matter and secondary minerals.)

TABLE 4.1 Elements Essential for Nutrition of Plants

Element	Adequate Concentration (% dry wt. of Tissue)
Carbon	45
Oxygen	45
Hydrogen	6
Nitrogen	1.5
Potassium	1.0
Calcium	0.5
Phosphorus	0.2
Magnesium	0.2
Sulfur	0.1
Chlorine	0.01
Iron	0.01
Manganese	0.005
Zinc	0.002
Boron	0.002
Copper	0.0006
Molybdenum	0.00001

Source: Zinke, P.J. 1977. "Man's activities and their effect upon the limiting nutrients for primary productivity in marine and terrestrial ecosystems," In *Global Chemical Cycles and Their Alterations by Man*, ed. W. Stumm, p. 92. Berlin: Dahlem Konferenzen.

Nutrients can also be provided by the trees themselves. Within the overall biogeochemical cycle of a forested watershed, there is an internal cycle that is at least partly independent of weathering (Likens et al. 1977; Graustein 1981). An element is taken up by a tree in solution through its roots and carried upward through the tree. A certain amount (which varies from element to element) is stored in living and dead vegetation (*biomass storage*) and the rest is released to be recycled as (1) *litterfall*, dead leaves lost to the forest floor (important for Ca and Mg); (2) *throughfall*, the washoff and leaching of an element from the leaves by precipitation (important for K); and (3) *root exudates*, element loss in solution from living and dead roots (important for Na). By these processes the litterfall, throughfall, and exudates all add nutrients to the soil that are readily taken up again by the tree through its roots.

A generalized biogeochemical cycle of an element in a forested area, including the internal cycle just discussed, is shown in Figure 4.2. The trees take up nutrients from the various sources discussed above. The primary *external* inputs are atmospheric gases, rainfall, trapped aerosol dust, and underlying rocks. Atmospheric gases and rain provide some of the sulfur, nitrogen, and phosphorus, whereas aerosols and rocks provide most of the potassium, magnesium, calcium, sodium, and silicon. Losses from the forest take place via groundwater flow and streams. If the forest is at steady state, that is, if it isn't growing and storing nutrients, inputs

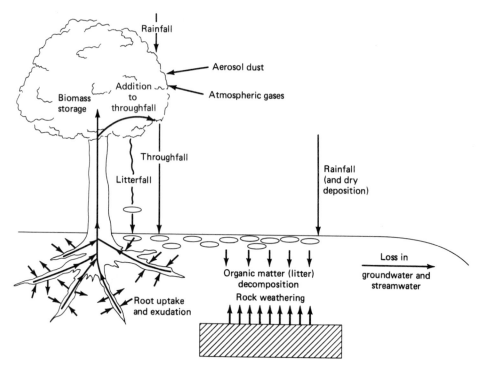

Figure 4.2   Biogeochemical cycling in forests.

equal outputs and the change in water composition, in passing from rainwater to streamwater, is due solely to the release of elements by rock weathering (and aerosol dust dissolution). For large forested regions, the assumption of steady state is probably valid but only in the absence of human deforestation (which has become a major global problem). For single, small watersheds, year-to-year variations in biological activity can have a major effect on runoff composition, and corrections for biomass storage (or release) should be made before any attempt is undertaken to estimate the degree of rock weathering from the composition of streamwater or groundwater. Unfortunately, although this correction has been emphasized by Likens et al. (1977) and Graustein (1981), it is often not done.

The relative importance of the various inputs and outputs, and resulting storage or release of elements varies both with the area considered and with the element. For example, Likens et al. (1977), in a forested New Hampshire area (Hubbard Brook Experimental Watershed) found that 80%–90% of the streamwater output of calcium, magnesium, potassium, and sodium came from rock weathering whereas nitrogen and sulfur were derived almost entirely from rain and atmospheric gases with little or none from weathering. (In this area, aerosol dust input was believed to be small because of the relatively humid climate.) Likens et al. also showed that the amount of total weathering, calculated after correction for biomass storage, was considerably greater than that indicated simply by concentrations in runoff minus those in rainfall.

Graustein (1981), working in a mountainous area in New Mexico, found a much greater input from aerosols. Spruce and fir trees trapped considerable quantities of windblown dust, which was washed off by precipitation and became part of the throughfall (in addition to elements recycled by the tree and released from the foliage). Based partly on Sr isotope studies, he concluded that tree-trapped dust accounted for about two-thirds of the calcium added to throughfall as opposed to one-third from biologic recycling. (The relative importance of dust versus botanical recycling was greater for sodium and less for Mg and K.) The amount of calcium, sodium, magnesium and potassium added by tree-trapped dust was about two to three times the amount originally present in precipitation. However, the type of tree is important; the trapping of dust by nearby aspen trees, by contrast, was found to be negligible.

In Graustein's (1981) study the flux of spruce-fir-trapped dust dissolved in throughfall was a significant portion of the flux of calcium and potassium in runoff but a minor part of sodium and magnesium losses. Thus, calculations of rates of chemical weathering for Ca^{++} and K^+ based on streamwater output minus precipitation input of these elements without considering tree-trapped dust input would have resulted in too large an estimate for the amount of bedrock weathering.

Graustein (1981) also points out the importance of biological internal cycling on soil water composition. Most of the dissolved K^+ in soil waters of his watersheds was derived from throughfall additions as a result of leaf exudation (not dust). This increased K^+ should affect the rate and type of weathering of potassium minerals as well as estimates of the importance of weathering on streamwater runoff composition.

If one measures biological activity as the ratio of an element stored annually in living and dead vegetation to the annual loss from the area in streamflow, the elements in order of decreasing biological activity are

$$P > N > K > Ca > S > Mg > Na$$

The most biologically affected element is thus phosphorus, with nitrogen coming second. K and Ca are considerably affected by vegetation while Na is little affected. Thus, in New Hampshire (Likens et al. 1977) the annual increase in biomass storage for potassium was much greater than streamwater losses while that for calcium was somewhat less than streamwater runoff. In addition, the flux of potassium cycled internally by the vegetation was around 25 times the flux lost in streamflow whereas that for Ca and Mg was about 3–6 times the loss via runoff.

In temperate climates with deciduous forests, the biological control over elements such as potassium and nitrogen is strongly seasonal. During the forest growing season in summer, potassium input in precipitation is greater than potassium loss in streamflow; apparently the vegetation takes up potassium from both the soil and from rainwater. Only during the dormant season, when leaves are on the ground, is the streamflow output greater than the precipitation input. This is in contrast to Ca, Mg, Na, and Al, which show greater output than input regardless of the season.

Phosphorus input in precipitation in many forests is greater than output in streamflow because tremendous amounts of phosphate are stored in vegetation. Only in agricultural areas, where large amounts of waste and fertilizer phosphate enter the runoff, does stream runoff exceed precipitation input. Weathering is not as important a source for phosphate as it is for Ca, Mg, K, and Na.

Besides nutrient cycling, trees exert an indirect control on water composition. This is brought about by transpiration. During transpiration pure water is lost via evaporation to the atmosphere from leaf surfaces. As a result the remaining salts become concentrated in soil water. Likens et al. (1977) have shown that at the New Hampshire location, where transpiration (plus evaporation) accounts for about 40% of the yearly water loss, the concentration of dissolved substances in runoff may be increased by as much as 60% over that which would have prevailed in the absence of this process.

SOIL WATER AND MICROORGANISMS: ACID PRODUCTION

The chemical composition of soil water is affected by inputs from throughfall and rainfall, additions from rock weathering, and inputs and removals due to biological activity. In addition to the nutrient cycling discussed in the previous section, there are important biologically induced changes brought about by soil microorganisms, most notably the production of soil acids. These acids constitute the principal agents of rock weathering (Carroll 1970; Alexander 1961). Acids provide hydrogen ions, which replace cations on mineral surfaces, thus bringing about disintegration of the minerals. Also, some organic acids react with specific elements contained in minerals to form *chelates* or soluble, multiply bonded, metal-organic complexes. Acid production is accomplished by a variety of organisms including bacteria, fungi, actinomycetes, and algae. (For a detailed discussion of soil microbiology consult, e.g., Alexander 1961.)

The principal acids produced by soil microorganisms are carbonic and sulfuric acids and a whole host of various organic acids. Carbonic acid (H_2CO_3) is produced by the oxidation of organic matter by microbes to CO_2. In other words, representing organic matter as CH_2O:

$$CH_2O + O_2 \rightarrow CO_2 + H_2O$$

This CO_2 then combines with water to form carbonic acid, which further partly dissociates (it is a weak acid) to H^+ and HCO_3^-:

$$CO_2 + H_2O \rightarrow H_2CO_3$$

$$H_2CO_3 \rightarrow H^+ + HCO_3^-$$

Sulfuric acid (H_2SO_4) is produced by the bacterially catalyzed oxidation of sulfide minerals and, in soils developed on rocks with high sulfide mineral contents, high concentrations of this strong acid and, consequently, low pH values can result. This is especially true where mining has brought about enhanced oxidation of sulfides. (Further details on sulfuric acid weathering are provided later in this chapter.)

Organic acids are formed by the partial breakdown of organic matter. Chief among these is a variety of high molecular weight, poorly defined compounds collectively referred to as humic and fulvic acids (see, e.g., Schnitzer and Khan 1972). These common acids, which impart characteristic brown and yellow colors to soil solutions, attack minerals and chelate and solubilize several metallic elements (e.g., Fe, Al). Precipitation of the acids brings about the formation of humus. In addition, certain organisms secrete specific, well-defined acids. For example, fungi living in the litter layer that accumulates on the soil surface exude oxalic acid ($H_2C_2O_4$), which reacts with iron and aluminum minerals to form dissolved iron and aluminum oxalate chelates (Graustein 1981). Iron and aluminum, which are otherwise highly insoluble, are carried downward in the soil by oxalate until the oxalate itself is broken down to CO_2 and HCO_3^- by additional microorganisms. As a result, the iron and aluminum are precipitated in lower portions of the soil. The net effect of the oxalate is to move the iron and aluminum downward. This movement, which is also accomplished by humic, fulvic, and other soil acids, is one of the main processes that bring about the differentiation of soils into specific horizons (see section on soil formation).

As a result of the attack on minerals by soil acids, the input from throughfall, the removal by plant roots, and vertical transport by organic chelates, concentrations of principal elements dissolved in soil water can vary from place to place and with depth in the soil. Some idea of depth variation, taken from the work of Graustein (1981) is shown in Table 4.2. Note that K^+ added at the top by throughfall is greatly depleted with depth due to root uptake. Continual increase of sodium with depth is due to rock weathering. High dissolved Al concentration in the shallow soil water is due to organic chelation and decrease in Al with depth is due to precipitation upon downward transport and chelate decomposition. Although data of this sort are sparse, these profiles are probably typical of many forest soils.

CHEMICAL WEATHERING

Minerals Involved in Weathering

The chemical weathering of rocks and minerals has been alluded to throughout the previous discussion without detailing exactly what rocks and minerals are involved. As a guide we present in Tables 4.3 and 4.4 a list of the most common minerals involved in weathering. Primary minerals (Table 4.3) are those undergoing destruction by weathering, while secondary minerals (Table 4.4) are those formed

TABLE 4.2   Concentrations of Na+, K+, Ca+, and Al in Soil Waters of a Watershed Forested with Aspen Trees, Sangre de Cristo Mountains, New Mexico, and in Rainfall and Throughfall in the Same Area

Water	Depth (cm)	Concentration ($\mu g/\ell$)			
		Na+	K+	Ca++	Al
Rainfall	—	67	120	360	5
Throughfall	—	85	2800	780	10
Soil water	30	710	2200	4400	350
Soil water	100	1600	430	1300	30
Soil water	150	3100	350	1050	38
Soil water	200	3800	510	2450	11

Source: From W.C. Graustein. "The Effect of Forest Vegetation on Solute Aquisition and Chemical Weathering: A Study of the Tesuque Watersheds near Santa Fe, New Mexico," pp. 44, 118, 293. 1981. Unpublished Ph.D. dissertation, Yale University, New Haven, Conn.

by weathering.   (Technically speaking, all minerals can undergo destruction by weathering but the secondary minerals are the most resistant.)   In interpreting Table 4.3 those with no geological background may find it helpful to visualize rocks simply as aggregates of minerals.   There are three basic types of rock: igneous, sedimentary, and metamorphic.   Igneous rocks are formed by crystallization from a melt at high temperature and include the common rock types, granite (Na-plagioclase feldspar, K-feldspar, quartz, biotite) and basalt (Ca-plagioclase feldspar, pyroxenes, olivine).   Sedimentary rocks are laid in water at the earth's surface and include eroded debris from preexisting rocks (e.g., quartz and feldspar), as in sandstones; fine-grained secondary minerals formed by weathering (e.g., iron oxides and clays), as in shales; the skeletal remains of organisms (mainly $CaCO_3$), as in limestones; and precipitates from seawater (gypsum and halite), as in evaporites.   Metamorphic rocks form by the recrystallization and alteration of sedimentary and igneous rocks at elevated temperatures and pressures (but without melting) and contain many different minerals, including amphiboles, muscovite, biotite, quartz and feldspar.

Also included in Table 4.3 are the principal weathering reactions that each primary mineral undergoes.   Weathering reactions are classified here according to the nature of the attacking substance and whether the primary mineral simply dissolves or whether a portion of it reprecipitates to form a secondary mineral or minerals.   Simple dissolution is referred to as *congruent dissolution*, and dissolution with reprecipitation of some of the components of the mineral is called *incongruent dissolution*.   Attacking substances are separated into soil acids, dissolved oxygen, and water itself.   Dissolved oxygen only attacks those minerals that contain reduced forms of elements, principally iron and sulfur, and that undergo oxidation to form new minerals.

Although most minerals are attacked mainly by soil acids, a few very soluble ones simply dissolve in water. This is shown in Table 4.3. In addition, these soluble minerals may also reprecipitate under arid conditions. This is why gypsum, for example, appears in Tables 4.3 and 4.4 both as a primary and a secondary mineral.

Minerals can be listed in order of their degree of resistance to weathering. In other words, if two minerals are present in the same soil and attacked by the same acids for the same length of time, one will be destroyed faster than the other. On the basis of observations of partly weathered rocks and soils and of responses to different climatic conditions, a table has been prepared (Table 4.5) of minerals listed in order of increasing resistance to weathering. This table is similar to those

TABLE 4.3 Common Primary Minerals that Undergo Weathering

Mineral	Generalized Composition	Rock Type(s)	Main Weathering Reaction
Olivine	$(Mg,Fe)_2SiO_4$	Igneous	Oxid. of Fe Cong. diss. by acids
Pyroxenes	$Ca(Mg,Fe)Si_2O_6$ or $(Mg,Fe)SiO_3$	Igneous	Oxid. of Fe Cong. diss. by acids
Amphiboles	$Ca_2(Mg,Fe)_5Si_8O_{22}(OH)_2$ (also some Na and Al)	Igneous Metamorphic	Oxid. of Fe Cong. diss. by acids
Plagioclase feldspar	Solid solution between $NaAlSi_3O_8$ (albite) and $CaAl_2Si_2O_8$ (anorthite)	Igneous Metamorphic	Incong. diss. by acids
K-feldspar	$KAlSi_3O_8$	Igneous Metamorphic Sedimentary	Incong. diss. by acids
Biotite	$K(Mg,Fe)_3(AlSi_3O_{10})(OH)_2$	Metamorphic Igneous	Incong. diss. by acids Oxid. of Fe
Muscovite	$KAl_2(AlSi_3O_{10})(OH)_2$	Metamorphic	Incong. diss. by acids
Volcanic glass (not a mineral)	Ca,Mg,Na,K,Al,Fe-silicate	Igneous	Incong. diss. by acids and H_2O
Quartz	SiO_2	Igneous Metamorphic Sedimentary	Resistant to diss.
Calcite	$CaCO_3$	Sedimentary	Cong. diss. by acids
Dolomite	$CaMg(CO_3)_2$	Sedimentary	Cong. diss. by acids
Pyrite	FeS_2	Sedimentary	Oxid. of Fe and S
Gypsum	$CaSO_4 \cdot 2H_2O$	Sedimentary	Cong. diss. by H_2O
Anhydrite	$CaSO_4$	Sedimentary	Cong. diss. by H_2O
Halite	$NaCl$	Sedimentary	Cong. diss. by H_2O

Note: cong. = congruent; incong. = incongruent; diss. = dissolution; oxid. = oxidation.

TABLE 4.4   Common Secondary Minerals Formed by Weathering in Soils

Mineral	Composition
Hematite	Fe_2O_3
Goethite	$HFeO_2$
Gibbsite	$Al(OH)_3$
Kaolinite	$Al_2Si_2O_5(OH)_4$
Smectite	$(1/2 \ Ca, Na) \ Al_3MgSi_8O_{20}(OH)_4 \cdot nH_2O$ (average composition)
Vermiculite	Basically biotite or muscovite composition with K^+ replaced by hydrated cations
Calcite	$CaCO_3$
Opaline silica (not a mineral)	$SiO_2 \cdot nH_2O$
Gypsum	$CaSO_4 \cdot 2H_2O$

prepared by others (e.g., Goldich 1938; Loughnan 1969; Carroll 1970) and is based largely on this older work. Although some reordering can occur in different soils, overall the order shown is considered to be well established. Goldich (1938) noted that the order shown for igneous silicate minerals parallels their temperature of formation from molten magma. In other words, the silicate minerals that weather fastest (e.g., olivine) are those that originally formed at the highest temperatures. The reason for this correlation is not clear, but a common explanation (e.g., Goldich 1938) is that minerals formed under conditions more distantly removed from those at the earth's surface are less stable there and thus weather faster. This explanation agrees with the position of the common secondary minerals at the bottom of the list but does not account for the high weatherability of nonsilicates, such as halite, gypsum, calcite, and pyrite, which also form under earth surface conditions.

Mechanism of Silicate Dissolution

Of all mineral groups, silicates have received the most attention in weathering studies because they make up the most abundant rock types. How primary silicates dissolve during weathering, however, is not well known. A prevalent theory during the past decade has been that silicate dissolution occurs by means of the formation of a protective surface layer of altered composition on each mineral grain (e.g., Luce, Bartlett, and Parks 1972; Pačes 1973; Busenberg and Clemency 1976; Chou and Wollast 1984). This layer is assumed to be so tight that it severely inhibits the migration of dissolved species to and from the surface of the primary mineral, and in this way is protective. The layer forms from components of the underlying primary mineral and, as weathering proceeds, it increases in thickness. It was invoked originally to explain the results of laboratory dissolution experiments (simulating weathering) where rates of dissolution were seen to decrease with time, due presumably to the thickening of a protective surface layer.   However, attempts to prove the existence of a layer with any appreciable thickness, using both electron microscopic and surface chemical techniques, have proven to be negative (e.g., Berner and Holdren 1979; Berner and Schott 1982).

What really appears to happen during weathering is that soil solutions penetrate through *permeable* (nonprotective) clay layers right to the bare surfaces of the primary silicate mineral grains and there react with them.   Dissolution does not occur at all places on the surface so as to produce general rounding of the grains, as predicted by the protective surface layer theory, but instead affects only those portions of the surface where there is excess energy such as at outcrops of dislocations.   (Dislocations are rows of atoms in a crystal that are slightly out of place and therefore more energetic.)   As a result of selective etching, distinct crystallographically controlled *etch pits* form on the mineral surface and, upon growth and coalescence, form interesting features.   Some examples taken from our studies of soil feldspars and pyroxenes are shown in Figure 4.3.   These etch pits reflect the crystal structure of the underlying mineral and therefore are regular in shape and aligned in certain directions.

Dissolution of primary minerals, therefore, occurs via etch pit formation and growth.   If the pits are located primarily at outcrops of dislocations, then a fun-

TABLE 4.5   Mineral Weatherability (Decreasing from top-to-bottom)

Halite
Gypsum-anhydrite
Pyrite
Calcite
Dolomite
Volcanic glass
Olivine
Ca-plagioclase
Pyroxenes
Ca-Na plagioclase
Amphiboles
Na-plagioclase
Biotite
K-feldspar
Muscovite
Vermiculite, smectite
Quartz
Kaolinite
Gibbsite, hematite, goethite

Note: Minerals are listed in order of increasing resistance to weathering. (Exact positions for some minerals can change one or two places due to effects of grain size, climate, etc.) See also Goldich (1938), Loughnan (1969), and Carroll (1970).

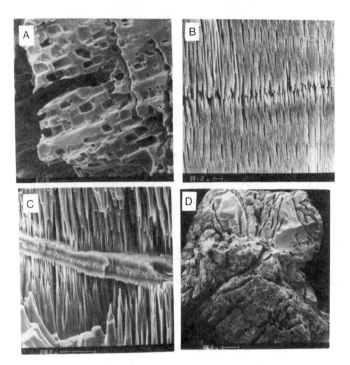

Figure 4.3 Scanning electron photomicrographs of partly weathered primary silicate mineral grains taken from soils. (A) Square-shaped ("prismatic") etch pits developed on dislocations in plagioclase feldspar (oligoclase from Piedmont, North Carolina). (x3000) (After R. A. Berner & G. R. Holdren. "Mechanism of Feldspar Weathering: Some Observational Evidence," *Geology*, 5, p. 372. © 1977 by The Geological Society of America. All rights reserved.) (B) Lens-shaped etch pits developed on dislocations in amphibole (hornblende from Ashe County, North Carolina). (x3000) (C) Clay-filled crack and surrounding "teeth" in hornblende formed by the coalescence of lens-shaped etch pits. (x1000) (D) Compound double pyroxene grain consisting of a single crystal of augite, $Ca(Mg,Fe)Si_2O_6$, in lower portion of photo and a single crystal of hypersthene, $(Mg,Fe)SiO_3$, in upper portion. Note greater degree of pitting of augite showing that it weathers faster because of a higher dislocation density. (x50) (B, C, & D. After R. A. Berner and J. Schott. "Mechanism of Pyroxene and Amphibole Weathering II: Observations of Soil Grains," *American Journal of Science*, 282, pp. 1219, 1222, 1223. © 1982 by the American Journal of Science, reprinted by permission of the publisher.)

damental controlling factor on dissolution of a *given* mineral during weathering is the density of dislocations. (Dissolution of *different* minerals is still controlled by differences in chemical composition.) This can help to explain different rates of dissolution of different occurrences of the same mineral under similar soil conditions. For example, the work of Berner and Holdren (1979) has shown that adularia, $KAlSi_3O_8$, reacts in the laboratory with hydrofluoric acid (as a simulator of soil acids) much more slowly than does microcline, also $KAlSi_3O_8$. The major difference between the two minerals is the presence of numerous dislocations in microcline and very few in adularia. In addition, electron microscope studies (Berner and Schott 1982) show that coexisting hypersthene and augite (two pyroxene minerals in the same soil) weather at different rates because of differences in dislocation density. This is illustrated in Figure 4.3.

Silicate Weathering Reactions: Secondary Mineral Formation

As noted in Table 4.3 some weathering reactions involve simple congruent dissolution by water or acids. In the case of silicate minerals, congruent dissolution is rare and confined only to relatively iron-free olivine, amphiboles, and pyroxenes. In this case we have the following reactions, assuming that attack is by carbonic acid:

$$Mg_2SiO_4 + 4H_2CO_3 \rightarrow 2Mg^{++} + 4HCO_3^- + H_4SiO_4 \qquad (4.1)$$

$$2H_2O + CaMgSi_2O_6 + 4H_2CO_3 \rightarrow Ca^{++}$$
$$+ Mg^{++} + 4HCO_3^- + 2H_4SiO_4 \qquad (4.2)$$

(Note that dissolved silica is represented here as H_4SiO_4, which closely represents the *actual form* found in solution. It is sometimes also represented by the alternative formula $Si(OH)_4$. We shall not represent silica in solution by SiO_2, as is commonly done, because it can be confused with quartz.) Most other silicate minerals, especially those containing aluminum, dissolve incongruently with the consequent formation of iron oxides and/or clay minerals. (*Clay minerals* is a common term applied to fine-grained aluminosilicates and includes kaolinite, smectite, and vermiculite as well as other minerals, such as chlorite, not discussed here). Thus, because silicate minerals constitute the fundamental components of most major rock types, it is important to inquire in detail into how they weather and to what they weather.

The most abundant silicate mineral in the earth's crust, which also readily undergoes weathering, is plagioclase feldspar. On this basis we begin our discussion of weathering reactions using Na-plagioclase or albite. Let us suppose that albite, $NaAlSi_3O_8$, is attacked by an organic acid, here represented as oxalic acid, $H_2C_2O_4$. First of all the oxalic acid dissociates to form H^+ ions:

$$H_2C_2O_4 \rightarrow 2H^+ + C_2O_4^{--} \qquad (4.3)$$

These H^+ ions then attack albite, liberating its constituent elements to solution:

$$4H^+ + 4H_2O + NaAlSi_3O_8 \rightarrow Al^{+++} + Na^+ + 3H_4SiO_4 \qquad (4.4)$$

Since oxalate ion, $C_2O_4^{--}$ readily reacts with Al^{+++} to form a chelate (Graustein 1981) we also have

$$Al^{+++} + C_2O_4^{--} \rightarrow Al\,(C_2O_4)^+ \qquad (4.5)$$

If we combine reactions 4.3–4.5 so as to cancel H^+ ion, we obtain

$$2H_2C_2O_4 + 4H_2O + NaAlSi_3O_8 \rightarrow Al\,(C_2O_4)^+$$
$$+ Na^+ + C_2O_4^{--} + 3H_4SiO_4 \qquad (4.6)$$

This is a dissolution reaction characteristic of the uppermost, highly acid portion of most temperate soils.

Aluminum oxalate and oxalate ion in solution are not stable, however, due to bacterial decomposition (Graustein 1981). On passing downward in migrating soil water, the oxalate is oxidized by bacteria and the liberated Al^{+++}, which is unstable in solution at most pH values found in soils, precipitates to form $Al(OH)_3$ or clay minerals. Let us assume here that the common clay mineral, kaolinite, $Al_2Si_2O_5(OH)_4$, is formed. The reactions are

$$2C_2O_4^{--} + O_2 + 2H_2O \rightarrow 4HCO_3^- \qquad (4.7)$$
$$2Al\,(C_2O_4)^+ + O_2 + 2H_4SiO_4 \rightarrow Al_2Si_2O_5\,(OH)_4$$
$$+ 4CO_2 + H_2O + 2H^+ \qquad (4.8)$$

Multiplying reaction 4.6 by 2, and adding to reactions 4.7 and 4.8, we obtain the overall reaction:

$$4H_2C_2O_4 + 2O_2 + 9H_2O + 2NaAlSi_3O_8 \rightarrow Al_2Si_2O_5\,(OH)_4$$
$$+ 2Na^+ + 4HCO_3^- + 2H^+ + 4CO_2 + 4H_4SiO_4 \qquad (4.9)$$

As written, this reaction is not quite complete. Since H^+ and HCO_3^- rapidly react with one another, we must take into account:

$$H^+ + HCO_3^- \rightarrow CO_2 + H_2O \qquad (4.10)$$

Doubling reaction 4.10 and adding to 4.9, we obtain the final overall reaction for the weathering by oxalic acid of albite to kaolinite:

$$4H_2C_2O_4 + 2O_2 + 7H_2O + 2NaAlSi_3O_8 \rightarrow Al_2Si_2O_5\,(OH)_4$$
$$+ 2Na^+ + 2HCO_3^- + 4H_4SiO_4 + 6CO_2 \qquad (4.11)$$

Note that reaction 4.11, even though the attacking acid is oxalic acid, ends up with the production of only Na^+, HCO_3^-, and H_4SiO_4 in solution. (The CO_2 is readily lost as a gas from the soil.) These are the dissolved species that would be found if the water were to pass out of the soil zone to become groundwater and eventually stream and river water. There is no memory of the oxalate and it is as though the albite had actually reacted with carbonic acid:

$$2H_2CO_3 + 9H_2O + 2NaAlSi_3O_8 \rightarrow Al_2Si_2O_5(OH)_4$$

$$+ 2Na^+ + 2HCO_3^- + 4H_4SiO_4 \qquad (4.12)$$

In fact, recalling that H_2CO_3 forms by the reaction

$$H_2O + CO_2 \rightarrow H_2CO_3$$

the only difference between reactions 4.11 and 4.12 is:

$$4H_2C_2O_4 + 2O_2 \rightarrow 8CO_2 + 4H_2O \qquad (4.13)$$

which is the reaction for the oxidative decomposition of oxalic acid.

What has been said above for oxalic acid and albite is true of any organic acid and silicate (and carbonate) mineral. Thus, even though the actual acid attacking a mineral is organic, the overall reaction, as far as most groundwater composition is concerned, can be represented as though the only attacking acid were H_2CO_3. In other words, we find HCO_3^- and not $C_2O_4^{--}$ in most ground and river waters. (In some rivers draining swampy areas of heavy organic matter production, organic acids and their anions can resist bacterial oxidation and survive to become carried some distance in the rivers; see Chapter 5). The simplification provided by this reasoning, enables the prediction of the origin of ions in ground-waters without concern for the type of organic acid actually attacking the primary minerals. Thus, the assumption that silicate weathering consists solely of attack by carbonic (and sulfuric) acids (Garrels 1967) is justified even though, when looked at in detail, it is not correct. The organic acids really do much of the attacking but they disappear. From now on in our discussion of weathering we shall adopt the convention of writing reactions in terms only of H_2CO_3, but it should be remembered that this is only a convention and a shorthand way of representing a series of chemical weathering reactions that are far more complex.

In the above example, we assumed that all aluminum liberated by feldspar dissolution was precipitated to form a secondary mineral (in this case, kaolinite). This is a reasonable assumption for most soils (e.g., Loughnan 1969). Except for localized redistribution accompanying chelate transport, aluminum does not migrate any appreciable distance in solution and normally accumulates in soils as weathering proceeds. (However, in unusually acid soils containing appreciable H_2SO_4, Al can be removed in solution; see Chapter 6). In fact the change in the ratio of other elements to aluminum in soils is often used as a measure of the degree of removal by weathering (eg., Goldich 1938). Iron, because of its insolubility in the presence of dissolved O_2, also accumulates in soils, as ferric oxides. Because of their lack of removal in solution we shall continue in all *overall* weathering reactions, as is the custom, to assign Al and Fe only to secondary minerals and allow none to appear in solution.

Although all Al is reprecipitated to form a secondary mineral, it need not always be kaolinite. Two other common aluminous weathering products are gibbsite, $Al(OH)_3$, and smectite, a complex cation Al-silicate (see Table 4.4). (Vermiculite forms by the loss of K^+ from biotite and muscovite and constitutes a special case of structural inheritance which will not be discussed here.) Conditions under which each of these minerals would be expected to form can be deduced on the basis of some rather simple reasoning: Weathering of primary aluminosilicates to gibbsite, kaolinite, or smectite should occur under different conditions because of fundamental differences in the composition of the three phases. Smectite contains Al, Si, and various cations; kaolinite contains only Al and Si; and gibbsite, only Al (see Table 4.4). In addition, the ratio of Si/Al is higher in smectite than it is in kaolinite. Thus, we would expect that increase of H_4SiO_4 in solution would favor the formation of kaolinite over gibbsite and, at higher values, smectite over kaolinite. In addition, increase in cations, for example, Na^+, should favor smectite. Quantitative representation of these ideas is shown in Figure 4.4. Here a plot of log $[Na^+]/[H^+]$ versus log $[H_4SiO_4]$, where brackets represent molar concentrations in solution, shows the regions of stability for the various secondary minerals as well as that of albite in aqueous solution. If a natural water has concentrations of $[Na^+]$, $[H^+]$, and $[H_4SiO_4]$, that fall in, let us say, the field for kaolinite, we would expect to find kaolinite forming from this water, and so forth for the other minerals. Note that as $[H_4SiO_4]$ is increased and $[Na^+]$ is increased (for constant pH), kaolinite and smectite, respectively, become the favored phases. By use of this diagram the compositional evolution of a water reacting with albite can be predicted (for details consult Helgeson, Garrels, and Mackenzie 1969).

With reference to Figure 4.4, following the reasoning of Helgeson, Garrels, and Mackenzie (1969), imagine the sequence of events as albite (again used to represent a typical primary aluminosilicate) undergoes dissolution during weathering at constant pH. If we start with a very dilute solution (low Na^+; low H_4SiO_4) its composition will fall in the field of stability of gibbsite (marked x on Figure 4.4). The weathering reaction in this case is

$$7H_2O + H_2CO_3 + NaAlSi_3O_8 \rightarrow Al(OH)_3$$
$$+ Na^+ + HCO_3^- + 3H_4SiO_4 \qquad (4.14)$$

If the water does not leave the rock, concentrations of Na^+, HCO_3^-, and H_4SiO_4 will build up and the solution composition will move to the northeast on the diagram following the arrow-marked path. When the boundary between gibbsite and kaolinite is reached, gibbsite begins to convert to kaolinite by the addition of silica:

$$2H_4SiO_4 + 2Al(OH)_3 \rightarrow Al_2Si_2O_5(OH)_4 + 5H_2O \qquad (4.15)$$

During this reaction all silica released by albite dissolution is used to convert gibbsite

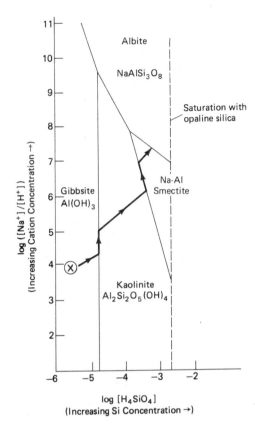

Figure 4.4 Stability fields of gibbsite, kaolinite, smectite, and albite as a function of log $[Na^+]/[H^+]$ and log $[H_4SiO_4]$ in solution. (Brackets denote concentration in moles per liter.) Smectite is represented by its pure Na-Al end-member, Na-beidellite, and analcite (and other zeolite) formation is ignored. (Albite is metastable with respect to analcite). A typical weathering reaction path taken for a closed system is shown by the heavy line with arrows. (Modified and recalculated from Bricker and Garrels 1965; and Helgeson, Garrels, and Mackenzie 1969.)

to kaolinite and this is why the solution path on the diagram turns abruptly northward. After all gibbsite is converted to kaolinite, we proceed again along a northeasterly trend until the boundary between kaolinite and smectite is reached. Again we follow the kaolinite-smectite boundary until all kaolinite is converted to smectite by the addition of silica and cations. We then proceed again to the northeast as smectite forms from albite until we reach the albite-smectite boundary. At this point, the solution is saturated with albite; it cannot continue to dissolve, and therefore weathering ceases.

The scenario described above is what would be expected if the water always stayed in contact with the albite (and there were no kinetic problems with the precipitation of secondary minerals). In other words, this is analogous to adding water and albite to a beaker and letting them react until albite solubility is reached. It is a closed system. Soils, on the other hand, are open to flow of water through them. Thus, concentrations will build up during contact of the water with albite (or other primary minerals), but further buildup will cease when the water leaves the rocks. The faster the rock is flushed with water, the shorter the time of contact with the primary minerals, and the lower the dissolved concentrations in the exiting waters. A steady state is attained between rate of addition by dissolution and rate of water flow, so that the water composition for any given soil containing albite may fall anywhere along the reaction path of Figure 4.4 depending on the relative magnitude of these two rates. Gibbsite formation should represent a high degree of flushing with removal of both cations and silica; kaolinite should represent less rapid flushing where less silica is removed; and smectite should represent rather stagnant conditions of water flow so that appreciable buildup of both silica and cations can occur. Also, for a given rate of flushing, more rapidly reacting minerals providing more silica and cations to solution should favor the formation of smectite or kaolinite over gibbsite.

These predictions are borne out when actual soils are examined. In general, gibbsite forms only in areas where there is very rapid flushing due to a combination of high rainfall and good drainage due to high relief. An example is the island of Jamaica where valuable deposits of bauxite (an ore of aluminum consisting largely of gibbsitelike minerals) are formed as a result of intense weathering accompanying high rainfall in a mountainous terrain. Less strong flushing, but still enough to remove all cations, is found in most tropical and subtropical soils and here, as predicted, the characteristic secondary mineral is kaolinite.

Smectite is the characteristic mineral of soils of semiarid regions and where volcanic glass is the primary material undergoing weathering. In semiarid regions, rainfall is low and water adheres to soil grains for long periods before being replaced by new water. Volcanic glass is the most reactive silicate known and is weathered very rapidly. Again, both observations are in agreement with our predictions.

A nice demonstration of the effects of flushing by water on the weathering of a single rock type is shown by the studies of Sherman (1952) and Mohr and van Baren (1954). Sherman found that clay mineralogy of the soils developed on the basalts of the island of Hawaii are correlated very well with mean annual rainfall. (Rainfall varies considerably on Hawaii because of the effects of rain shadowing by mountains.) This is shown in Figure 4.5. Mohr and van Baren found that for the same rock type and same rainfall, soils on islands of Indonesia showed different clay minerals depending on the degree of drainage. Upland soils, where drainage

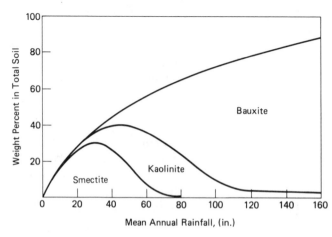

Figure 4.5 Weathering products of basalt in Hawaii. Note the excellent correlation of clay mineral type with rainfall in agreement with predictions based on the degree of flushing of the soil with water. (After G. D. Sherman. "The Genesis and Morphology of the Alumina-rich Laterite Clays," In *Problems in Clay and Laterite Genesis*, p. 159. © 1952 by the Amer. Inst. Min. Metal. Eng., reprinted by permission of the publisher.)

was good, consisted of kaolinite, whereas those in poorly drained or swampy depressions consisted of smectite. This is again what would be expected from our predictions.

Silicate Weathering: Soil Formation

Soil can be defined as "a complex system of air, water, decomposing organic matter, living plants and animals, in addition to the residues of rock weathering, organized into definite structural patterns as dictated by the environmental conditions" (Loughnan 1969: 115). The dominant control, on a worldwide basis, of soil formation is climate. In fact most classifications of soils (zonal soils) are based primarily on climatic differences. A common classification is that shown in Figure 4.6. (For a more detailed discussion of weathering as it relates to soil formation, consult Loughnan 1969; Keller 1957; Carroll 1970; Paton 1978; and Bohn, McNeal, and O'Connor 1979.) In addition to the role of rainfall as it affects flushing of the rock (as discussed above), climate exerts an important influence on weathering via its control on vegetation and the organic content of soils. High rainfall, which enhances plant growth, and low temperature, which retards bacterial destruction, both favor the accumulation of organic matter in soils. Organic matter is important, as mentioned earlier, in that it is the source of carbonic and organic acids and chelating compounds that react with silicate minerals. It is because of the accumulation of organic matter that the most acid soils in Figure 4.6 are those formed under cool, humid conditions, in other words, the podzols. (Tundra soils are even richer in organic matter, but, due to lack of water circulation caused by

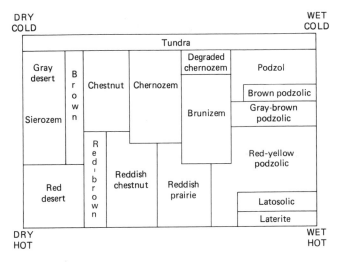

Figure 4.6 A climatic classification of major soil groups. (After C. E. Millar, L. M. Turk, and H. D. Foth, *Fundamentals of Soil Science*, p. 137. Copyright © 1958. John Wiley & Sons, Inc. Reprinted by permission of John Wiley & Sons, Inc.)

permanently frozen ground at depth, little weathering takes place in them.) By contrast, less organic matter is formed in the warm climate latosols and laterites, due to almost complete destruction of organic matter by microorganisms, and in desert and semiarid soils due to a lack of appreciable plant growth.

Since our concern is mainly with water chemistry, we needn't go into great detail about various soil types and classifications. However, a few additional comments concerning some of the soils of Figure 4.6 are in order. The characteristic soil of humid temperate regions is the podzol. *Podzols* are characterized by a strong vertical zonation, an example of which is shown in Figure 4.7. The top of the soil (O and A_1 horizons) consists of leaf litter, which becomes decayed to structureless humus upon burial. The downward flow of water, which becomes highly acidic upon passing through the organic layer, results in intense leaching and removal of the immediately underlying material, resulting in a residue consisting mainly of resistant quartz (A_2 horizon). Cations, silica, organic matter, and even Al and Fe are removed. The Al and Fe, however, as well as some silica and dissolved organic matter (humic and fulvic acids), do not leave the rock and are precipitated at depth in the so-called B horizon. Underlying the B horizon is the C horizon (Figure 4.7) or zone where the rock is only slightly weathered and little affected by biological activity.

In other soil types one also encounters vertical zonation but often different from that exhibited by podzols. For example, in chernozems (the characteristic soil of subhumid temperate grasslands), as well as in soils of more arid climates, one finds the precipitation of $CaCO_3$ at depth within the B horizon due to the dominance, during dry periods, of evaporation over downward percolation and

Horizon Designation	Soil Character
O	Leaf litter (partly decomposed)
A_1	Dark brown: humic organic matter with some minerals
A_2	Light gray; residual quartz; intense leaching, including Fe, Al
B	Dark brown humic layer underlain by red-brown to yellow-brown accumulation of clays and iron oxides; blocky structure
C	Slightly altered parent rock

Figure 4.7 Simplified vertical profile of a typical podzolic soil of a humid temperate climate.

leaching. In tropical and subtropical soils, there is little organic matter but still there is intense leaching of cations and silica, due to heavy rainfall, resulting in the buildup of residual iron and aluminum within the B horizon. In desert soils or in soils that have been subjected to weathering for only short periods (immature soils), vertical zonation is often poorly developed due to a relative lack of weathering. Examples of immature soils are those developed on young glacial deposits or on relatively recent volcanic flows.

Carbonate Weathering

Compared to silicates, carbonate minerals weather far more simply. Dissolution is normally congruent and the overall reaction, accomplished by carbonic acid (H_2CO_3) attack (or organic acid attack, which can be represented as H_2CO_3; see above) is for calcite:

$$H_2CO_3 + CaCO_3 \rightarrow Ca^{++} + 2HCO_3^- \qquad (4.16)$$

and for dolomite:

$$2H_2CO_3 + CaMg\,(CO_3)_2 \rightarrow Ca^{++} + Mg^{++} + 4HCO_3^- \qquad (4.17)$$

Carbonate minerals are not as abundant as silicate minerals but they exert a dominant influence on groundwater and river water composition. Most of the dissolved Ca^{++} and HCO_3^- in river water, on a worldwide basis, arises from carbonate dissolution via reactions 4.16 and 4.17.

Calcium carbonate dissolution, under certain circumstances, can be followed by reprecipitation. In soil water and groundwater where carbonic (and organic) acid concentrations are high, the resulting concentrations of Ca^{++} and HCO_3^- can build up to high values. If the waters then encounter conditions where degassing of dissolved CO_2 occurs, the waters may become supersaturated with respect to $CaCO_3$. Carbonic acid constantly maintains equilibrium with dissolved CO_2:

$$H_2CO_3 \rightleftarrows H_2O + CO_2 \qquad (4.18)$$

and if CO_2 is lost from solution, reaction 4.18 shifts to the right, using up H_2CO_3, and consequently, reaction 4.16 shifts to the left to replace H_2CO_3. If Ca^{++} and HCO_3^- are present in high enough concentrations, $CaCO_3$ precipitation may then take place.

An outstanding example of the reprecipitation of $CaCO_3$ occurs in limestone caves. Limestone is a rock consisting largely of calcite. It undergoes congruent dissolution during weathering along joints, cracks, and other avenues of water flow by carbonic acid derived from organic matter decomposition in the overlying soils. The Ca^{++} and HCO_3^- are normally removed from the rock and, as a result, enlargement of cracks and ultimately the formation of caves takes place. If a cave is sufficiently connected, via cracks, to the atmosphere, a low, atmospheric value of CO_2 gas is maintained in the cave air. In this case, water emerging in the cave from above and containing high levels of H_2CO_3, CO_2, and HCO_3^-, can lose CO_2 to the cave atmosphere. As a result, supersaturation with $CaCO_3$ is attained and calcite is precipitated to form stalactites, stalagmites, and other cave deposits.

Cave deposits also form in dolostones (dolomite rocks) but in this case dolomite is not reprecipitated. There are severe problems in precipitating dolomite so that, in solutions supersaturated with respect to dolomite and calcite, calcite is invariably the precipitate. Thus, in dolostones, we find caves that contain deposits of calcite, not dolomite (see, e.g., Holland et al. 1964).

Calcite can also precipitate in soils. (The Ca^{++} and HCO_3^- may come from the weathering of silicates rather than carbonates.) This comes about by a similar process of degassing but the precipitation occurs much closer to the ground surface. In arid climates, downward percolating soil waters undergo degassing and loss of CO_2 at a depth of only a few tens of centimeters in the soil. (The CO_2 is picked up from organic decay at even shallower depths.) As a result, calcite precipitation takes place and the resulting deposit (Loughnan 1969) is referred to as *caliche*.

Sulfide Weathering

Sulfide minerals occur in minor quantities in many different rock types and locally in major quantities in ore deposits. The most abundant mineral, by far, is pyrite (FeS_2), which is found mainly in organic-rich, fine-grained sedimentary rocks known as black shales, and in coal. Upon exposure to dissolved oxygen during weathering, pyrite (and other sulfides) are chemically unstable and rapidly undergo oxidative decomposition. This decomposition is important in that it results in the production of sulfuric acid, which can be used, in turn, to bring about further weathering of silicate and carbonate minerals. Sulfuric acid forms as follows:

$$4FeS_2 + 15O_2 + 8H_2O \rightarrow 2Fe_2O_3 + 8H_2SO_4 \tag{4.19}$$

$$H_2SO_4 \rightarrow 2H^+ + SO_4^{--} \tag{4.20}$$

(Various other reactions can be written involving Fe^{++} or Fe^{+++} in solution, etc., but in all cases there is a distinct lowering of pH due to the formation of H_2SO_4.) Almost always the oxidation is catalyzed by bacteria (Stumm and Morgan 1981).

The acidity of water draining rocks where sulfide oxidation is taking place depends on the content of sulfides and the presence of other minerals, especially carbonates, which can readily neutralize acid. During coal mining, large expanses of pyrite-bearing coal are suddenly exposed to oxygenated water, and because associated rocks generally contain little carbonate, the waters draining coal-mining localities are very acid. Values of pH of less than 3, resulting from bacterial catalysis of pyrite oxidation, are common for such mine drainage waters (Stumm and Morgan 1981; Kleinmann and Crerar 1979).

By contrast, calcareous rocks that also contain pyrite do not weather to produce highly acid water. Instead the sulfuric acid is neutralized by $CaCO_3$.

Reactions are

$$4FeS_2 + 15O_2 + 8H_2O \rightarrow 2Fe_2O_3 + 16H^+ + 8SO_4^{--} \qquad (4.21)$$

$$16H^+ + 16CaCO_3 \rightarrow 16Ca^{++} + 16HCO_3^- \qquad (4.22)$$

which, added together, give

$$4FeS_2 + 15O_2 + 8H_2O + 16CaCO_3 \rightarrow 2Fe_2O_3$$
$$+ 16Ca^{++} + 8SO_4^{--} + 16HCO_3^- \qquad (4.23)$$

In this way Ca^{++}-SO_4^{--}-HCO_3^- groundwaters and streamwaters can arise. In arid and semiarid regions where evaporative concentration of soil waters is common, the Ca^{++} and SO_4^{--} concentrations often reach the point where the precipitation of gypsum, $CaSO_4 \cdot 2H_2O$ takes place. The common occurrence of gypsum crystals in weathered outcrops of calcareous black shales in the western interior of the United States is an example of this process.

Sulfuric acid is also partially neutralized by silicate minerals. It is this neutralization, in fact, that constitute one of the more important weathering reactions for silicates. For example, for albite

$$H_2SO_4 + 9H_2O + 2NaAlSi_3O_8 \rightarrow Al_2Si_2O_5(OH)_4$$
$$+ 2Na^+ + SO_4^{--} + 4H_4SiO_4 \qquad (4.24)$$

Sometimes, if there is sufficient pyrite (or other sulfides), acidity can be so high that common secondary minerals, such as kaolinite and iron oxides, become unstable and dissolve:

$$6H^+ + Al_2Si_2O_5(OH)_4 \rightarrow 2Al^{+++} + 2H_4SiO_4 + H_2O \qquad (4.25)$$

$$6H^+ + Fe_2O_3 \rightarrow 2Fe^{+++} + 3H_2O \qquad (4.26)$$

In this case, minerals such as alunite, $KAl_3(SO_4)_2(OH)_6$, or jarosite, $KFe_3(SO_4)_2(OH)_6$, may form. Finding these minerals in soils is a good indication that the soils are highly acidic (van Breemen 1976).

GROUNDWATERS AND WEATHERING

Upon percolating downward below the water table, soil waters become groundwaters. Once the zone dominated by biological activity is left, weathering and rock dissolution are much slower but by no means negligible. *Given sufficient time,* silicate rock decomposition at depth can be extensive, giving rise to the formation of thick accumulations of clays, iron oxides, etc. known as *saprolites.*

An example of extensive saprolitization is provided by the deep weathering profiles found in the southeastern United States. Such saprolitization takes place by means of the attack on primary minerals of carbonic acid and by water itself, and little *direct* biological activity is involved (Carroll 1970). (However, the carbonic acid is derived indirectly from microbiological organic matter decomposition in the overlying soil.) This deep weathering takes place and affects water composition both above and below the water table (Cleaves 1974; Velbel 1984). Below the water table the groundwater is out of contact with the atmosphere and, as a result of continuing reactions, may become anoxic (O_2-free). In this case, dissolved Fe^{++} and Mn^{++} appear in solution because of a lack of dissolved O_2, which would otherwise remove them by oxidation plus precipitation. Such groundwater, when pumped to the surface, undergoes rapid O_2 uptake from the atmosphere and consequent ferric hydroxide and manganese hydroxide precipitation. This explains the "rusty colored" water often encountered when using well water (see, e.g., Crerar, Knox, and Means 1979).

The results of chemical reaction taking place at depth, such as Fe^{++} liberation, as well as the many weathering reactions occurring in overlying soils, are manifested together in the composition of groundwater. In many circumstances, these reactions can be deduced from the chemical composition of the groundwater, and because of the abundance of chemical data (see, e.g., White, Hem, and Waring 1963), various schemes can be constructed for deducing the origin of groundwater composition. This will be pursued here (see also Drever 1982). Throughout the discussion it should always be kept in mind, however, that we are assuming that differences in water composition between groundwater and rainwater are due only to rock weathering. This may not be true if there are large effects on groundwater composition of soil-water alterations due to biomass increases or decreases or the dissolution of windblown dust as discussed at the beginning of this chapter. Fortunately, for several elements such effects can often be ignored as shown by good agreement between predictions based on groundwater composition and actual observations of soils.

Studies of groundwater and springwater draining silicate rocks have been used to construct a mobility series for elements reminiscent of the mineral weatherability series shown in Table 4.5. One can define element mobility as the weight fraction of total dissolved matter that is made up by a given element divided by the weight fraction of the same element in the primary rock undergoing weathering. Examination of much groundwater data by Feth, Roberson, and Polzer (1964) has led to the following order of mobility:

$$Ca \geq Na \geq Mg > Si \geq K >> Al \approx Fe$$

In other words, Ca, Na, and Mg are the most mobile and most easily liberated by weathering; Si and K are intermediate; and Al and Fe are essentially immobile and remain in the soil. This order is in agreement with results of mineralogical studies in that the most rapidly weathered silicate minerals are Na-Ca silicates (plagioclase feldspars) and Mg-containing silicates (e.g., pyroxenes, amphiboles), whereas Al, Fe, and Si (the latter to a lesser degree) form secondary minerals and remain in the soils, and K is contained in less rapidly weathered minerals, specifically biotite, muscovite, and K-feldspar.

Garrels' Model for the Composition of Groundwaters from Igneous Rocks

Since plagioclase is the most abundant mineral in the earth's crust, and is weathered rapidly, it is reasonable to assume that much of the composition of natural groundwaters can be explained in terms of plagioclase weathering. This has been done by Garrels (1967) and his model for the origin of groundwater composition will be discussed next.

In examining the composition of groundwaters draining through igneous rocks, Garrels concluded that the principal dissolved species deserving explanation are Ca^{++}, Na^+, HCO_3^-, and H_4SiO_4. (Lesser concentrations of K^+ and Mg^{++} were explained in terms of biotite and amphibole weathering and will not be discussed here.) To explain the relative concentrations of these species he made the following assumptions:

1. Water compositions (after correcting for rainfall input) result from the attack by carbonic acid on primary silicate minerals.
2. Plagioclase feldspar is the sole mineral source of Na^+ and Ca^{++} because it is abundant and readily weathered. (Also, there is no $CaCO_3$ present.)
3. Dissolved silica is derived almost entirely from the weathering of plagioclase (a small amount comes from Fe-Mg minerals).
4. Rainfall correction for Na^+ is made by subtracting an amount equivalent to the measured Cl^- concentration from the total measured Na^+ concentration.

First, Garrels constructed a series of weathering reactions of plagioclase reacting to form gibbsite, kaolinite, or smectite. For example, for weathering to kaolinite of a plagioclase with 50% $NaAlSi_3O_8$ and 50% $CaAl_2Si_2O_8$:

$$4Na_{0.5}Ca_{0.5}Al_{1.5}Si_{2.5}O_8 + 6H_2CO_3 + 11H_2O \rightarrow 3Al_2Si_2O_5(OH)_4$$
$$+ 2Na^+ + 2Ca^{++} + 6HCO_3^- + 4H_4SiO_4 \qquad (4.27)$$

As a result of this reaction one would expect to find a molar ratio in solution of $Na:Ca:HCO_3^-:H_4SiO_4$ of 1:1:3:2. For reactions of plagioclase, of different Na-Ca compositions, to kaolinite, different ratios result, and likewise for the weathering of the same or different compositions of plagioclase to gibbsite, or to smectite. As a result of writing many reactions of the type shown above, Garrels was able to construct a diagram, reproduced here in Figure 4.8, which shows solution compositions, in terms of $[HCO_3^-]/[H_4SiO_4]$ versus $[Na^+]/[Ca^{++}]$, predicted for the weathering of plagioclase separately to gibbsite, kaolinite, and smectite.

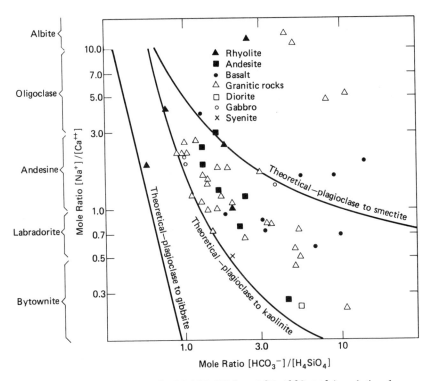

Figure 4.8 Mole ratios of $[HCO_3^-]/[H_4SiO_4]$ and $[Na^+]/[Ca^{++}]$ in solution for groundwaters from various types of igneous rocks. (Brackets denote concentration in moles per liter.) Rock types are shown by different symbols. Nomenclature for plagioclase feldspar, composition from which the solution Na^+/Ca^{++} ratio is derived, is shown on the left. Note that most of the water compositions plot between the theoretical curves for the weathering of plagioclase to kaolinite and to smectite. (After R. M. Garrels. "Genesis of Some Ground Waters From Igneous Rocks," In *Researches in Geochemistry*, ed. P. H. Abelson, p. 412. Copyright © 1967. John Wiley & Sons, Inc. Reprinted by permission of John Wiley & Sons, Inc.)

By plotting natural groundwater compositions on the plagioclase weathering diagram, Garrels was able to deduce the major plagioclase weathering reactions taking place in nature. The data (Figure 4.8) indicate that the dominant reactions are those involving the formation of either kaolinite, smectite, or a mixture of both. This prediction is in accord with mineralogical observations of most soils and indicates the basic validity of the Garrels model.

Garrels also constructed plots of various parameters versus HCO_3^- concentration. Bicarbonate concentration is a measure of the degree of ion buildup in solution and, thus, the contact time of water with plagioclase. Garrels found that at low HCO_3^- concentrations, water appeared to have HCO_3^-/H_4SiO_4 ratios expected for the weathering of plagioclase (and other minerals) to kaolinite whereas, at high concentrations of HCO_3^-, the HCO_3^-/H_4SiO_4 ratios suggested weathering to smectite. In addition, at low HCO_3^- concentrations solutions behaved as though they were undersaturated with respect to smectite, and at high concentrations they appeared to attain a constant solubility product expected if smectite were present. These observations led Garrels to conclude that at low contact times of water with plagioclase ("initial weathering") the primary product is kaolinite and at high contact times both kaolinite and smectite form. This is in full agreement with what has been said earlier in this chapter. Moderately rapid flushing prevents cation buildup and consequent smectite formation, which explains its relative absence in well-drained soils in regions of moderate to heavy rainfall.

Garrels also found that there was an interesting correlation of the concentration product $[Ca^{++}][CO_3^{--}]$ versus HCO_3^- concentration. This is shown in Figure 4.9. At high HCO_3^- levels, the product approaches that (10^{-8}) expected for calcite saturation, in other words, the solubility product for $CaCO_3$. This means that high water-plagioclase contact time, as reflected by high HCO_3^- concentrations, should lead to calcite precipitation. Again, this prediction is borne out. In semiarid regions where water contacts soil grains for long periods before being replaced by new rainwater, concentrations in solution build up, due to prolonged weathering and evaporation, and often $CaCO_3$ precipitation in the soil results, sometimes in the form of caliche (see section on carbonate weathering).

These calculations show that chemical modelling of groundwater composition can prove to be a useful tool for studying chemical weathering. However, it must

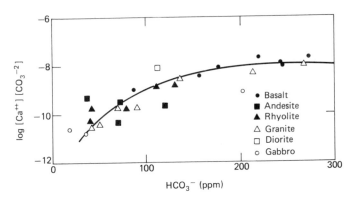

Figure 4.9 Plot of the concentration product of dissolved Ca^{++} and CO_3^- versus HCO_3^- concentration for groundwaters from different igneous rocks. (Brackets around Ca^{++} and CO_3^- denote concentration in moles per liter.) Note that above about 200 ppm HCO_3^- the product approaches a constant value, which is that expected for saturation with calcite. (After R. M. Garrels. "Genesis of Some Ground Waters From Igneous Rocks," In *Researches in Geochemistry*, ed. P. H. Abelson, p. 415. Copyright © 1967. John Wiley & Sons, Inc. Reprinted by permission of John Wiley & Sons, Inc.)

TABLE 4.6   Origin of Major Components of
Groundwater (Major Processes Only)

Component	Origin
Na^+	NaCl dissolution (some pollutive)[a] Plagioclase weathering Rainwater addition
K^+	Biotite weathering K-feldspar weathering Biomass decreases Dissolution of trapped aerosols
Mg^{++}	Amphibole and pyroxene weathering Biotite (and chlorite) weathering Dolomite weathering Olivine weathering Rainwater addition
Ca^{++}	Calcite weathering Plagioclase weathering Dolomite weathering Dissolution of trapped aerosols Biomass decreases
HCO_3^-	Calcite and dolomite weathering Silicate weathering
SO_4^{--}	Pyrite weathering (some pollutive)[a] $CaSO_4$ dissolution Rainwater addition
Cl^-	NaCl dissolution (some pollutive)[a] Rainwater addition
H_4SiO_4	Silicate weathering

Note: Order presented is approximate order of decreasing
importance. For further information consult Chapter 5.

[a] For a discussion of pollutive contributions see Chapter 5.

be remembered that in the Garrels model it is assumed that all Ca^{++} arises from
the weathering of plagioclase.   This is reasonable only if the groundwater associated
with a specific igneous rock type has contacted only the major minerals of the rock
and has not encountered any carbonate minerals.   Even if trace amounts of $CaCO_3$
were present, say as fracture fillings in a basalt, the origin of the Ca^{++} in an
igneous groundwater would be in doubt.   Calcium carbonate dissolves much more
rapidly than silicate minerals, and small amounts in a rock can dominate the composition of associated ground water.   Thus, caution must be employed in selecting
groundwaters for use in chemical modelling of silicate weathering reactions to be
sure that problems associated with carbonate dissolution can be excluded.

Origin of Major Constituents of Groundwater: A Summary

By way of summary, the minerals and types of reactions involved in the production of the various common components of groundwater are shown in Table 4.6.   Included also are inputs due to rainfall and biological effects where pertinent.   Quantitative estimation of the relative importance of the various sources for each element is done in Chapter 5 when discussing riverwater and will not be repeated here. Table 4.6 is offered merely as a summary of what has been said in the present chapter.   As can be seen, there are a variety of sources and each element exhibits its own idiosyncrasy.

REFERENCES

ALEXANDER, M.   1961.   *Introduction to soil microbiology.*   New York: John Wiley.   472 pp.

BERNER, R. A., and G. R. HOLDREN.   1977.   Mechanism of feldspar weathering: Some observational evidence, *Geology* 5: 369–372.

————.   1979.   Mechanism of feldspar weathering II: Observations of feldspars from soils, *Geochim. Cosmochim. Acta* 43: 1173–1186.

BERNER, R. A., and J. SCHOTT.   1982.   Mechanism of pyroxene and amphibole weathering II: Observations of soil grains, *Amer. J. Sci.* 282: 1214–1231.

BOHN, H. L., B. L. MCNEAL, and G. A. O'CONNER.   1979.   *Soil chemistry.*   New York: Wiley-Interscience.   329 pp.

BRICKER, O. P., and R. M. GARRELS.   1965.   Mineralogic factors in natural water equilibria.   In *Principles and applications of water chemistry,* ed. S. Faust and J. V. Hunter, pp. 449–469.   New York: John Wiley.

BUSENBERG, E., and C. V. CLEMENCY.   1976.   The dissolution kinetics of feldspars at 25° C and 1 atm. CO_2 partial pressure, *Geochim. Cosmochim. Acta* 40: 41–50.

CARROLL, D.   1970.   *Rock weathering.*   New York: Plenum Press.   203 pp.

CHOU, L., and R. WOLLAST.   1984.   Study of the weathering of albite at room temperature and pressure with a fluidized bed reactor, *Geochim. Cosmochim. Acta* 48: 2205–2218.

CLEAVES, E. T.   1974.   *Petrologic and chemical investigations of chemical weathering in mafic rocks, eastern piedmont of Maryland.*   Maryland Geol. Survey, Report of Investigations no. 25.   28 pp.

CRERAR, D. A., G. W. KNOX, and J. L. MEANS.   1979.   Biogeochemistry of bog iron in the New Jersey Pine Barrens, *Chem. Geol.* 24: 111–135.

DOMENICO, P. A.   1972.   *Concepts and models in groundwater hydrology.*   New York: McGraw-Hill.   405 pp.

DREVER, J. I.   1982.   *The chemistry of natural waters.*   Englewood Cliffs, N.J.: Prentice-Hall.   388 pp.

FETH, J. H., C. E. ROBERSON, and W. L. POLZER. 1964. *Sources of mineral constituents in water from granitic rock, Sierra Nevada, California and Nevada.* U.S.G.S. Water Supply Paper no. 1535-I.

FREEZE, R. A., and J. A. CHERRY. 1979. *Groundwater.* Englewood Cliffs, N.J.: Prentice-Hall. 604 pp.

GARRELS, R. M. 1967. Genesis of some ground waters from igneous rocks. In *Researches in geochemistry*, ed. P. H. Abelson, pp. 405–420. New York: John Wiley.

GOLDICH, S. S. 1938. A study on rock weathering, *J. Geology*, 46: 17–58.

GRAUSTEIN, W. C. 1981. The effect of forest vegetation on solute aquisition and chemical weathering: A study of the Tesuque watersheds near Santa Fe, New Mexico. Ph.D. dissertation, Yale University, New Haven, Conn. 645 pp.

HELGESON, H. C., R. M. GARRELS, and F. T. MACKENZIE. 1969. Evaluation of irreversible reactions in geochemical processes involving minerals and aqueous solutions II: Applications, *Geochim. Cosmochim. Acta* 33:455–482.

HOLLAND, H. D., T. V. KIRSIPU, J. S. HUEBNER, and U. M. OXBURGH. 1964. On some aspects of the chemical evolution of cave waters, *J. Geology* 72:36–67.

KELLER, W. D. 1957. *The principles of chemical weathering.* Columbia, Mo.: Lucas Bros. 111 pp.

KLEINMANN, R. L. P., and D. A. CRERAR. 1979. *Thiobacillus ferrooxidans* and the formation of acidity in simulated coal mine environments, *Geomicrobiol. J.* 1:373–388.

LIKENS, G. E., F. H. BORMANN, R. S. PIERCE, J. S. EATON, and N. M. JOHNSON. 1977. *Biogeochemistry of a forested eco-system.* New York: Springer-Verlag. 146 pp.

LOUGHNAN, F. C. 1969. *Chemical weathering of the silicate minerals.* New York: American Elsevier. 154 pp.

LUCE, R. W., R. W. BARTLETT, and G. A. PARKS. 1972. Dissolution kinetics of magnesium silicates, *Geochim. Cosmochim. Acta* 36:35–50.

MILLAR, C. E., L. M. TURK, and H. D. FOTH. 1958. *Fundamentals of soil science.* New York: John Wiley. 526 pp.

MOHR, E. J. C., and F. A. VAN BAREN. 1954. *Tropical soils,* New York: Interscience. 498 pp.

PAČES, T. 1973. Steady-state kinetics and equilibrium between ground water and granitic rock, *Geochim. Cosmochim. Acta* 37:2641–2663.

PATON, T. R. 1978. *The formation of soil material.* London: Allen & Unwin. 143 pp.

SCHNITZER, M., and S. U. KHAN. 1972. *Humic substances in the environment.* New York: Marcel Dekker. 327 pp.

SHERMAN, G. D. 1952. The genesis and morphology of the alumina-rich laterite clays. In *Problems in clay and laterite genesis*, Amer. Inst. Min. Metal. Eng., pp. 154–161.

STUMM, W., and J. J. MORGAN. 1981. *Aquatic chemistry.* New York: John Wiley. 780 pp.

VAN BREEMEN, N. 1976. *Genesis and solution chemistry of acid sulfate soils in Thailand.* Centre for Agricultural Publishing and Documentation, Wageningen, the Netherlands. Agricultural Research Report no. 848. 263 pp.

VELBEL, M. A. 1984. Mineral transformations during rock weathering and geochemical mass-balances in forested watersheds of the southern Appalachians. Ph.D. dissertation, Yale University, New Haven, Conn. 175 pp.

WHITE, D. E., J. D. HEM, and G. A. WARING. 1963. *Chemical composition of subsurface waters: Data of geochemistry*, 6th ed. U.S.G.S. Prof. Paper no. 440-F.

ZINKE, P. J. 1977. Man's activities and their effect upon the limiting nutrients of primary productivity in marine and terrestrial ecosystems. In *Global cycles and their alterations by man*. ed. W. Stumm, pp. 89–98. Berlin: Dahlem Konferenzen.

5

Rivers

INTRODUCTION

Having considered water vapor in the atmosphere, rain, and the reactions of rain-water with rocks and soil (weathering), we come next in the water cycle to rivers, which serve as the major routes by which continental rain and the products of continental weathering reach the oceans. Rivers do not contain (on the scale of the water cycle) a very large percentage of the total water on the earth (see Chapter 2). Their importance derives from the major role they play in the transport of water as well as dissolved and suspended matter. In this they dwarf the only other transport agent from the continents to the oceans, the atmosphere.

Rivers have played an important role in human development. Historically, settlement along rivers has occurred because the rivers provided both water supply and transportation. As a result of human proximity, rivers have been considerably affected by activities ranging from agriculture and flood control to the input of human and industrial wastes (Lerman 1980). These effects, which are not only of recent origin, have a considerable impact on the transport of water and dissolved and suspended matter.

River-Water Components

River transport to the oceans can be thought of as being made up of a number of components:

1. Water.
2. Suspended inorganic matter. Major elements include Al, Fe, Si, Ca, K, Mg, Na, and P.
3. Dissolved major species: HCO_3^-, Ca^{++}, SO_4^{--}, H_4SiO_4, Cl^-, Na^+, Mg^{++}, and K^+. These can be further divided into:
   a. Elements with no gaseous phases in the atmosphere (for which an input and output balance within the water cycle can be more easily done). These include Ca^{++}, Cl^-, H_4SiO_4, Na^+, Mg^{++}, and K^+.
   b. Elements with gasous phases, SO_4^{--} and HCO_3^-, which are derived from atmospheric gases (e.g., SO_2 and CO_2, respectively) as well as from rocks.
4. Dissolved nutrient elements, N and P (and to a certain extent Si), which are used biologically and whose concentrations vary due to this.
5. Suspended and dissolved organic matter.
6. Trace metals, both dissolved and suspended, which will not be discussed in any detail here (however, see Martin and Whitfield 1981, and Garrels, Mackenzie and Hunt 1975).

River Runoff

Rivers result from the *runoff* of water from the continents. River water itself comes ultimately from precipitation, some of which is evaporated from the land, some of which passes through the ground to the river at shallow depths (surface flow) and some of which remains in the ground much longer and reaches greater depths before entering the river (groundwater). On a global scale, continental runoff, which includes dominantly river runoff and a small amount of direct groundwater discharge to the oceans (Meybeck 1984), can be thought of as being equal to the excess of oceanic evaporation over precipitation. This excess ocean evaporation results in water vapor transport to the continents where water vapor is converted to rain and the nonevaporated portion is ultimately lost as river runoff, thereby balancing the water cycle (see Chapter 2). Thus, the initial amount of rainfall on land is determined by the evaporation rate over the oceans (where 85% of total evaporation occurs), and by the global circulation of water vapor, which is driven by meridional differences in solar heating (and temperature).

On a continental scale, in order to have river runoff, there must be *net* precipitation on land (i.e., precipitation must exceed continental evaporation). Average continental evaporation rates are temperature dependent and decrease rapidly with increasing latitude (see Chapter 2). Rainfall is not uniformly distributed because it is dependent upon the atmospheric circulation of water vapor. The result is that there are two major belts where precipitation exceeds evaporation (i.e., where there is net positive precipitation), and these are the areas where most of the large rivers occur. One belt is near the equator (10°N–10°S) where there is both high rainfall (due to high concentrations of water vapor) and high evaporation (due to high temperatures) but with rainfall exceeding evaporation. This results in large rivers such as the Amazon and Zaire. The second belt is in the temperate zone ($30°$–$60°$N and S). Here there is generally adequate rainfall (and a good supply of water vapor) as well as lower evaporation rates (due to lower average surface temperature). Two major rivers originating in this area are the Mississippi and the Yangtze. In between these two belts, there is a region of low river runoff in the subtropics ($15°$–$30°$N and S) where evaporation exceeds precipitation and deserts form. This area has fairly high surface temperatures and high evaporation.

Superimposed on the continental-scale factors influencing the amount of river runoff are more local effects resulting from the distribution of rainfall in both space and time. Geographic differences in rainfall distribution are due primarily to relief, with the windward sides of mountains receiving large amounts of rain and the leeward sides very little. Geographic heterogeneity can increase the runoff by as much as 20% (Holland 1978). For two continents having the same *average* annual rainfall, the one with greater geographically induced variation in the rainfall rate will have less loss of rainfall due to evaporation and thus more runoff. Seasonality of rainfall also increases runoff because, again, there is less annual loss by evaporation. An extreme example of this is the monsoonal climate of southern Asia where an "average" monthly rainfall of 17 cm varies seasonally from about 1 cm to 69 cm per month (Miller et al. 1983). This gives rise to such rivers as the Ganges and Brahmaputra, which undergo large seasonal fluctuations in flow rates and flood often.

When we discussed atmospheric CO_2 in Chapter 2, we noted that an increase in the average temperature of the earth as a whole would greatly affect runoff. Overall, worldwide temperature increases should speed up the hydrologic cycle and produce more runoff. The oceans would be warmer and there would be more evaporation and more water vapor to produce more rain and runoff. However, the geographic distribution of belts of excess precipitation would be different, with the result that certain areas, such as the interior of the United States, might have less runoff.

The *runoff ratio* is the ratio of average river runoff (per unit area) to average rainfall (per unit area). The world average value is about 0.46, suggesting that about 50% of rainwater that reaches the ground is returned directly to the atmosphere by evaporation and never reaches rivers. However, there is considerable variation in the runoff ratio by continent due to the factors discussed above, from a high of 0.54 in Asia (with its areas of high mountains and monsoon climate) to a low of 0.28 in Africa (which has large areas of desert and where most of the rainfall occurs in low-lying areas). South America (0.41), Europe (0.42), and North America (0.38) have intermediate runoff ratios (See Table 5.6).

Major World Rivers

Table 5.1 lists some major world rivers that empty directly into the sea in approximate order of water discharge (with water data from Milliman and Meade. 1983).

TABLE 5.1 Major Rivers That Flow to the Sea Listed in Order of Discharge

River	Location	Annual Discharge				
		Water (km³/yr)	Dissolved Solids (Tg/yr)	Suspended Solids (Tg/yr)	Dissolved/ Suspended Ratio	Drainage Area (10⁶km²)
1. Amazon	S. America	6300	223	900	0.25	6.15
2. Zaire (Congo)	Africa	1250	36	43	0.84	3.82
3. Orinoco	S. America	1100	39	210	0.19	0.99
4. Yangtze (Chiang)	Asia (China)	900	226	478	0.47	1.94
5. Brahmaputra	Asia	603	61	(see Ganges & Brahmaputra)		0.58
Ganges & Brahmaputra	Asia	971	136	1670	0.08	1.48
6. Mississippi	N. America	580	125	210	0.60	3.27
7. Yenisei	Asia (USSR)	560	65	13	5.0	2.58
8. Lena	Asia (USSR)	514	70	12	5.8	2.50
9. Mekong	Asia (Vietnam)	470	70	160	0.44	0.79
10. La Plata (Parana is a tributary)	S. America	470	16	92	0.17	2.83
11. Ganges	Asia	450	75	(see Ganges & Brahmaputra)		0.975
12. Irrawaddy	Asia (Burma)	428	92	265	0.35	0.43
13. St. Lawrence	N. America	447	59	4	14.8	1.03
15. Mackenzie	N. America	306	64	100	0.64	1.81
17. Columbia	N. America	251	35	8	4.4	0.67
20. Indus	Asia (India)	238	41	100	0.41	0.97
Huangho (Yellow)	Asia (China)	49	22	1080	0.02	0.77
Red (Hungho)	Asia (Vietnam)	123	?	160	?	0.12

Sources: Water and suspended solids from Milliman and Meade 1983. Dissolved solids calculated from Meybeck 1979; Martin and Whitfield 1981; and Ming-hui, Stallard, and Edmond 1982.

Note: Tributaries are excluded. Tg = 10⁶ tons = 10¹² g.

The first 13 rivers (with a total discharge of 14,000 km³ water per year) account for about 38% of the total water discharge to the oceans (37,400 km³/yr). Most remarkably, one river, the Amazon, accounts for 17% of the total water discharge to the oceans, and has over ten times the discharge of the Mississippi River. Also, most of the world's largest rivers are in underdeveloped countries and as a result have not been very well studied.

SUSPENDED MATTER IN RIVERS

Total Amount of Suspended Matter

The total suspended load carried by rivers to the oceans is about 13.5×10^9 metric tons/yr with an additional $1-2 \times 10^9$ tons from bedload and floods (Milliman and Meade 1983) or about 152 tons/km²/yr for the drained land area (88.6×10^6 km²). Other estimates of suspended load are somewhat larger: 15.5×10^9 tons/yr (Martin and Meybeck 1979) to 18.4×10^9 tons/yr (Holeman 1968). The total suspended sediment load can be used to calculate the average mechanical denudation rate for the continents, which is about 5.6 cm of elevation reduction in 1000 years (assuming an average rock density of 2.7 and ignoring isostatic uplift, which occurs over longer geologic time periods). However, as pointed out by Milliman and Meade (1983), the sediment transport rate from the continents to the oceans by rivers is *not the same* as the total rate of soil erosion. This is because much sediment is eroded from upland areas and deposited in lowland areas without reaching the sea. The sediment discharge rate from the conterminous United States to the oceans is only about 8% of the total erosion rate over the same area. (This number is based on an erosion rate of 5.3×10^9 tons/yr from Holeman 1980, and a sediment discharge rate of 0.445×10^9 tons/yr from Curtis, Culbertson, and Chase 1973.)

Seventy percent of the total world suspended load is from southern Asia and the large Pacific and Indian Ocean islands (particularly Taiwan, New Guinea, and New Zealand) (Milliman and Meade 1983). This is despite the fact that no river from the large Pacific islands ranks among the top 25 rivers in water discharge. For example, approximately the same amount of suspended sediment is removed from Taiwan as from a major drainage area, 575 times as large, on the African continent. Figure 5.1 shows the total suspended sediment load and suspended sediment yield (yield = load/drainage basin area) from the various world river basins and Table 5.2 gives the sediment yield and drainage area for the various continents.

The individual rivers carrying the largest sediment load shown in Table 5.1 are, in order, the Ganges and Brahmaputra rivers, the Huangho (Yellow) River, the Amazon River, and the Yangtze River. The Huangho River, whose sediment load is 12% of the world total, carries only 0.1% of the total river water discharge. The reason for this lies in the nature of the sediment source; for the Huangho River it is the easily eroded yellow loess of north central China (Holeman 1968).

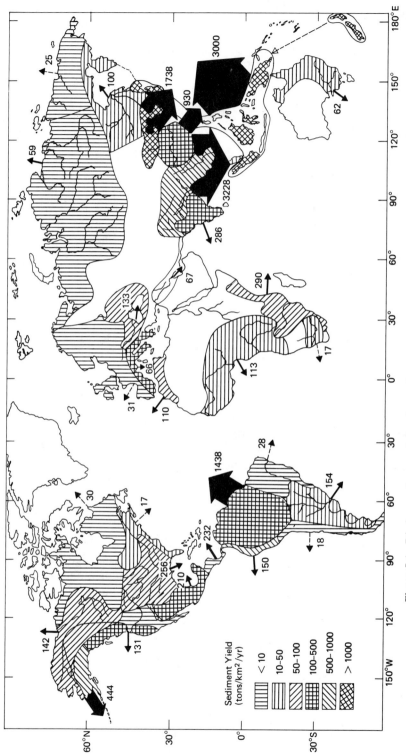

Figure 5.1 Discharge of suspended sediment from world drainage basins (in 10⁶ tons/yr) as indicated by arrows. Sediment yield (tons/km²/yr) for various drainage basins is also shown by appropriate pattern (see legend). Open pattern indicates essentially no sediment discharges to the oceans. (After J. D. Milliman and R. H. Meade. "World-wide Delivery of River Sediment to the Oceans," *Journal of Geology*, 91(1), p. 16. Copyright © 1983 by The University of Chicago Press, reprinted by permission of the publisher.)

TABLE 5.2   Suspended Sediment (in metric tons) Carried by Rivers to the Ocean

Continent	Drainage Area Contributing Sediment to Ocean (10^6 km²)	Sediment Discharge (10^6 tons/yr)	Sediment Yield (tons/km²/yr)	Mean Continental Elevation (km)
North America	15.4	1020	66	0.72
Central America[a]	2.1	442	210	—
South America	17.9	1788	97	0.59
Europe	4.61	230	50	0.34
Eurasian Arctic	11.17	84	8	~0.2
Asia	16.88	6349	380	0.96
Africa	15.34	530	35	0.75
Australia	2.2	62	28	0.34
Pacific & Indian Ocean islands[b]	3.0	~3000	~1000	~1.0
World total	88.6	13,505	152	

Sources: After J. D. Milliman and R. H. Meade. "World-wide Delivery of River Sediment to the Oceans," *Journal of Geology*, 91(1), p. 14. Copyright © 1983 by The University of Chicago Press, reprinted by permission of the publisher. Elevations from Fairbridge 1968.

[a] Includes Mexico.

[b] Japan, Indonesia, Taiwan, Philippines, New Guinea, and New Zealand.

The Ganges and Brahmaputra rivers carry large sediment loads because they drain the readily erodible foothills of the Himalaya Mountains, the highest in the world, in an area of heavy seasonal (monsoon) rainfall and large runoff. The Amazon River gets most of its sediment load only from the Andes Mountains (Gibbs 1967), not from the Brazilian lowlands; as a result, for such a large river, the Amazon does not have a particularly high sediment yield per unit area.

There are a number of natural factors that control the suspended sediment load of rivers (Milliman 1980): (1) relief of the drainage basin, (2) drainage basin area, (3) amount of water discharge, (4) geology of the river basin, (5) climate, and (6) presence of lakes along the river length. The suspended load of most rivers is influenced by more than one of these factors in combination. In addition, human activities such as deforestation, agriculture, and the building of dams, have had a very large effect.

Relief (local difference in elevation) exerts a major influence. The greatest sediment yield comes from mountainous areas (high elevation) located a short distance from the ocean, which results in steep slopes and little area for sediment storage in flood plains. The large islands of the Pacific and Indian oceans are the most conspicuous example of this effect, particularly Taiwan, which has the largest suspended sediment yield in the world (10,000 tons/km²/yr) (Milliman and Meade 1983) and an average elevation of around 1000 m. The rivers draining the Himalaya Mountains, such as the Ganges-Brahmaputra in India and the Irawaddy in Burma, have a very high sediment yield. In South America, the rivers draining the steep Andes Mountain slopes into the Pacific also have a high sediment yield (Figure 5.1). Likewise, erosion on the Andean foothills is believed to be the dominant control on both the amount and composition of suspended material in the Amazon River (Gibbs 1967). The U.S. river with the largest known sediment yield (1750 tons/km^2/yr) is the Eel River of northern California, which drains the Coastal ranges (reaching elevations of more than 2000 m) for a short distance into the Pacific. Conversely, rivers in areas of low relief have a very low sediment yield. The Eurasian Arctic rivers (Yenisei and Lena, the seventh and eighth largest rivers) in a low-lying area have a very low sediment yield (5 tons/km^2/yr) and the Congo River (the second largest river), also in an area of low relief, carries only 10 tons/km^2/yr.

Garrels and Mackenzie (1971), Hay and Southam (1977), and Holland (1978) point out that the suspended sediment yield seems to increase exponentially with mean continental elevation. This is shown in Figure 5.2 in which Milliman and Meade's (1983) data for suspended load per unit area of external drainage for the continents are plotted against mean continental elevation (Fairbridge 1968). (Hay and Southam [1977] point out that *external* drainage area, i.e., area draining to the oceans, should be more relevant than total drainage area, since suspended load is based on rivers that enter the ocean.) Also included in Figure 5.2 are the Pacific islands and the Eurasian Arctic (with a rough estimate of their elevation), which tend to confirm the correlation. Africa has a somewhat lower sediment yield than might be expected based on relief. Hay and Southam (1977) suggest that this is due to the fact that the continental slope of Africa starts 400 m higher than that of most continents; in other words, Africa is a high plateau. If the mean elevation of Africa is measured as being that above 400 m, then the sediment yield plots

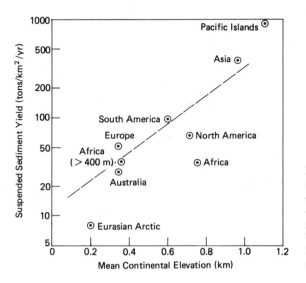

Figure 5.2 Suspended sediment yield (tons/km^2/yr) versus mean continental elevation (km). (Suspended sediment data from Milliman and Meade 1983; mean continental elevation from Fairbridge 1968, except for the Pacific Islands and Eurasian Arctic which are our rough estimates.)

near Europe and Australia in Figure 5.2 and the empirical relation provides a better fit to the data.

Milliman and Meade (1983) point out that there is an *inverse* correlation between drainage area and sediment yield for all major sediment-discharging rivers (carrying more than 10^7 tons of suspended sediment per year). However, much of this correlation can be attributed to the fact that the small drainage basins with large sediment yield considered by Milliman and Meade were ones with higher relief. The wide range of values for the larger basins can also be traced to differences in relief. The greater sediment yield in small basins is due to their inability to store sediment because of steep slopes and the absence of flood plains. The decrease in sediment yield with increasing drainage area of streams is attributed by Trimble (1977) at least partly to human activities. The inability of large-drainage-area streams to carry artificially increased solid erosion loads due to human activities may result in the deposition of the excess sediment in their channels and flood plains.

The amount of water discharged by a river (either total annual discharge or runoff per unit area) seems to have less effect on the suspended load, either on a local basis (Milliman and Meade 1983; Holland 1978) or on a continental basis (Hay and Southam 1977), than other factors. Although the Amazon (the largest river) does carry a large sediment load, its sediment yield per unit area is only about average for world rivers (150 tons/km^2/yr). Certainly some of the largest rivers (Congo, Yenisei, and Lena) have abnormally low sediment yields (less than 10 tons/km^2/yr) and others (Ganges-Brahmaputra) have abnormally high yields. Most major rivers have an average suspended sediment *concentration* between 100 mg/ℓ and 1000 mg/ℓ (Milliman 1980). Dividing the total suspended load (13.5 $\times$ 10^9 tons/yr) by the total water discharge (37,400 km^3/yr) gives an average concentration for all rivers of 360 mg/ℓ.

The geology of the river basin can have a very important effect on the sediment load. The most obvious case is the Huangho (Yellow) River, with one of the largest of all sediment loads, which flows through vast areas of *loess* (windblown glacial dust), a material that is extremely easily eroded. The presence of active glaciers also increases the sediment load of rivers (Milliman and Meade 1983) because of the production of easily eroded rock debris (moraine, drift, etc.) from glacial grinding. Examples of this are Alaskan rivers draining from glaciers, which average suspended sediment yields of 1000 tons/km^2/yr. (The glacial Copper River, for example, carries a greater sediment load than the Yukon River, which has four times its drainage area.) Similarly, the southern European rivers draining Alpine glacial areas (Rhone and Po rivers) carry a much larger sediment load than most European rivers. On the other hand, areas from which the rock debris has been removed by glaciers, northeastern North America and the Eurasian Arctic, tend to have low sediment yields. A lot of tectonism also tends to increase sediment load because of increased relief and the production of easily erodible volcanic material, such as ash.

Climatic effects on the suspended load are due to a combination of mean annual temperature, total rainfall, and rainfall seasonality. Heavy rainfall is obviously important in producing runoff but, in particular, heavy *seasonal* rainfall such as the monsoon climate in southern Asia contributes to a large suspended sediment load. Heavy rainfall alone, such as in the Congo Basin, will not necessarily give large sediment loads. Conversely, rivers in desert areas have high suspended sediment *concentrations* but (because of a lack of water) low total sediment yields. Thus the Australian and African continents with their large percentage of deserts have fairly low sediment yields.

When lakes are present along a river they trap suspended sediment before it reaches the ocean, greatly reducing the river sediment load (Milliman and Meade 1983). Examples are the effects of the Great Lakes on the St. Lawrence River and lakes along the Zaire River. On a large scale, the Black Sea traps more than half of the sediment draining off nonarctic Europe, and the Mediterranean Sea traps much of the rest, preventing it from reaching the Atlantic Ocean.

To summarize, the greatest suspended sediment yields are from mountainous tropical islands, areas with active glaciers, mountainous areas near coasts, and areas draining loess (Figure 5.1).

Human Influence

Human activities may increase or decrease the suspended sediment load. Increases are due to (1) deforestation and cultivation of the land, and (2) overgrazing. Decreases in the suspended load result from (1) building of dams and reservoirs, which trap sediment, (2) bank stabilization of rivers, and (3) soil conservation practices.

The initial effect of humans was to increase the suspended sediment load through deforestation and conversion of the land into cropland. These effects can be seen as far back as Roman times when human activity resulted in large increases in sediment deposition in an Italian lake (Judson 1968). Similar effects have been noted in the eastern United States (Trimble 1975) corresponding to the arrival of European settlers. However, there is a difference between the increased *erosion* rate (rate at which soil is removed from the upland land surfaces) due to human activities and the stream *transport* rate of suspended sediment to the oceans. In the southeastern United States, apparently streams were unable to handle the increased erosional loads due to European settlement, and only 5% of the material eroded from upland slopes has been exported to the oceans as suspended sediment. Most of the eroded sediment has been deposited instead as colluvium and alluvium in valleys and valley streams (Trimble 1975). Many major rivers that transport large amounts of sediment, such as those in India and China have been affected by agriculture for centuries (Milliman and Meade 1983). Overgrazing by animals can also cause unusually large suspended sediment yields. One such case noted by Milliman and Meade (1983) is the Tana River in Kenya.

More recently, humans have been building huge dams that trap large amounts of river suspended sediment. Milliman and Meade (1983) point out that the suspended sediment loads normally transported to the oceans by the Nile and

TABLE 5.3 U.S. (excluding Alaska) Soil Erosion Rates by Land Type

Land Type	U.S. Area (10^6 km^2)[a]	Percent of Total U.S. External Drainage Area[b]	Percent of Total Erosion Load[c]	Erosion Load (10^9 tons/yr)[c]	Calculated Erosion Rate (tons/km^2/yr)	Relative Erosion Rate[d]
Arable (cropland and pasture)	1.90	28				
Cropland	1.43	21	37	1.96	1371	$8n$
Pasture	0.47[e]	7	7	0.37	787	$4n$
Forest	2.38	34	8	0.42	176	$\equiv n$
Rangeland	2.61[f]	38	26	1.38	529	$3n$

[a] From *World Almanac* 1982.

[b] Total U.S. external drainage area is 6.89×10^6 km^2 (Curtis, Culbertson, and Chase 1973).

[c] Total soil erosion load (5.3×10^9 tons/yr) and percentages of this load from Holeman 1980.

[d] See text for calculation.

[e] Calculated as difference between arable and cropland.

[f] Equal to $6.89 - (1.90 + 2.38)$.

Colorado rivers have been reduced to nearly nothing by dams. Dams and soil conversion have reduced the load on the Mississippi River by one-third (Milliman and Meade 1983), and the load of the Zambezi River has been similarly reduced. The loss in suspended sediment transported to the ocean due to dams on large rivers, according to Milliman and Meade (1983), is about 0.5×10^9 tons of sediment a year or around 4% of the total river suspended load. Trimble (1977) notes that in the southeastern United States the erosion rate is being reduced due to declining agriculture and better soil conservation; however, as a result, excess sediment stored in streams from past human-induced erosion may now be transported. Nevertheless, new dams will trap 80%–90% of this sediment before it reaches the ocean. Soil conservation plus bank stabilization on the Orange River in South Africa has also greatly reduced the suspended sediment load there.

What is the net effect of the various human influences on the suspended load in both U.S. and world rivers? In the United States, deforestation and the conversion of land to cropland and pasture since the arrival of European settlers has greatly increased the soil erosion rate (Judson 1968). From a knowledge of the present total U.S. erosion rate, and the percentage of it contributed by various land types (forests, rangeland, cropland, pasture) combined with the areal extent of these land types, we have estimated the pre-European suspended sediment load. First, we note from Table 5.3 that the present erosion rate from cropland (1371 tons/km^2/yr) is about eight times the rate from forested land, that from pasture about four times, and that from rangeland about three times the forest rate. The United States was originally one-half forest and one-third grassland (*Encyclopedia Americana* 1959, vol. 27: 311) with the latter, by analogy with rangeland, assumed to have a natural erosion rate three times that of forests. (The remaining one-sixth of the United States is and was desert, which presumably has few rivers and contributes little sediment to the sea.) From these assumptions, we calculate that the pre-European total erosion rate for the United States was roughly equal to 1.5 times the present forest erosion rate $[1/2n + 1/3(3n) = 1.5n$; see Table 5.3].

The rates given in Table 5.3 are rates for *erosion* of soil from the land surface, but what we want is the *transport* (yield) of sediment from drainage basins to the oceans as river suspended sediment load, which, as pointed out above, can be a considerably lower number. If we assume that annual river suspended sediment yield for each land type and for the total continent are in the same proportion as is found for relative erosion rates (given above and in Table 5.3), then we can calculate the present sediment yield rate for forested land only. Let n equal this rate; from Table 5.3 the yield rate, on the above assumption, then is about $8n$ for cropland, $4n$ for pasture, and $3n$ for rangeland. Multiplying these rates by their respective areas, we obtain the following equation by summing all four types and equating the result to the total U.S. suspended load (Curtis, Culbertson, and Chase 1973):

$$1.43 \times 10^6 \text{ km}^2 (8n) + 0.47 \times 10^6 \text{ km}^2 (4n) + 2.61 \times 10^6 \text{ km}^2 (3n)$$

$$+ 2.38 \times 10^6 \text{ km}^2 (n) = 445 \times 10^6 \text{ metric tons/yr}$$

TABLE 5.4 World Areas of Different Land
Types, and Estimated Sediment Yield

Land Type	Area[a] $(10^6 \ km^2)$	Relative Suspended Sediment Yield[b] $(tons/km^2/yr)$
Cropland	14	$10n$
Grassland	24	$3n$
Forests	51	n

[a] After Zehnder and Zinder (1980) and Logan (1983).

[b] Total world exterior drainage area $= 89 \times 10^6 \ km^2$; Total suspended sediment load $= 14,000 \times 10^6$ tons/ yr (Milliman and Meade 1983).

Solution of this equation results in $n = 18.9$ tons/km^2/yr, the value for the present forested land sediment yield rate. Multiplying this by 1.5, we obtain the value 28 tons/km^2/yr for the pre-European sediment yield rate for all types. By comparison the present value is 65 tons/km^2/yr. Thus, the effect of European settlement of the United States has been to increase the sediment loss to the oceans by a factor of about 2.3.

By a similar procedure, one can calculate a prehuman sediment yield for the world. We start with the present-day sources of sediment. We assume that the sediment yield rates for different types of land for the world have approximately the same relative proportions as those for the United States. In other words, if the forest sediment yield rate is n, then the rate for grassland is $3n$ and that for cropland is $10n$. (The latter value is raised from $8n$ to account for the poorer worldwide soil conservation on agricultural land.) The worldwide areas of various land types are given in Table 5.4 along with the total world exterior drainage area (area over which rivers flow to the ocean rather than dry up by evaporation), and the world suspended sediment load. (The suspended sediment load has been increased from 13.5×10^9 tons/yr to 14.0×10^9 tons/yr to account for sediment removal by humans in reservoirs on big rivers.) The equation for the world analogous to that for the United States is

$$51 \times 10^6 \ km^2 \ (n) + 14 \times 10^6 \ km^2 \ (10n) + 24 \times 10^6 \ km^2 \ (3n)$$

$$= 14.0 \times 10^9 \ tons/yr$$

Solving for n, we get a forested land suspended sediment yield rate of 53 tons/km^2/ yr. Assuming that the prehuman land distribution for the world was the same as for the pre-European United States, we multiply, as above, this value by 1.5 to obtain a prehuman yield rate of 80 tons/km^2/yr. Over a total world exterior drainage area of $89 \times 10^6 \ km^2$, this equals 7.1×10^9 tons/yr. The present world suspended sediment load is 14.0×10^9 tons/yr, so that the effect of agriculture worldwide has caused an increase in sediment yield by a factor of about 2. Thus, our results for both the United States and the world indicate that human activity has had a profound influence on the suspended sediment yield. This is in overall qualitative agreement with the results of Judson (1968), who used a somewhat different method of calculation and obtained an even greater value for human influence.

Chemical Composition of Suspended Matter

Table 5.5 shows the average concentrations of the major elements in river suspended (particulate) material and dissolved material compared with the concentration in average surficial rocks and soils. Because of chemical weathering and the dissolution of soluble elements combined with reprecipitation of insoluble elements in secondary weathering products, the river suspended matter is enriched in the less soluble elements (such as Al and Fe) relative to the parent rock and strongly depleted in the most soluble elements (Na and Ca). The ratio of concentrations of various elements in river suspended matter to average surficial rock is also shown in Table 5.5. A ratio greater than 1.0 indicates enrichment in river suspended matter as is the case for Al and Fe. Although silicon has about the same concentration in river suspended matter as in surficial rocks, it is somewhat enriched and left behind in soils; this is partly due to the resistance of quartz (SiO_2) to weathering. Na and Ca in river suspended material are strongly depleted to about half their concentrations in surficial rocks and Mg and K are also fairly strongly depleted. Phosphorus, which is a nutrient, is enriched greatly over surficial rocks but this is probably a biological and possibly a pollutional effect.

The relative size of the particulate load to the total load (dissolved and suspended) is also shown in Table 5.5. As expected from their enrichment in the suspended load, Al and Fe are carried almost entirely in the suspended load, and Si is largely carried there. At the other extreme, Na and Ca, which are the most depleted in the river particulate load, have only 40% of their total load in the particulate form. Potassium and magnesium are interesting in that they are depleted similarly in the river particulate load over rock concentrations (enrichment = 0.7) but a larger part of the Mg (40%) appears in the dissolved load than does K (14%). This may have to do with the fact that K is retained in clay minerals in soils while Mg, although it is found in clay minerals, also occurs in soluble carbonate rocks. In addition, more K is stored in the biosphere.

Martin and Meybeck (1979) believe that the geographic variations in major concentrations in river particulate matter can be explained by differences in *climate* and resultant weathering regimes between river basins. They distinguish between intensely weathered tropical river basins, and temperate and arctic river basins which have less intense weathering. The tropical rivers have high Al and Fe concentrations because their particulates originate from soil material enriched in insoluble elements left behind when soluble elements were leached away. Temperate and arctic rivers on the other hand have lower concentrations of Al and Fe in the suspended matter relative to soluble elements because smaller amounts of

TABLE 5.5 Concentrations of Major Elements in Continental Rocks and Soils and in River Dissolved and Particulate Matter

| | Continents | | Rivers | | | | Element Weight Ratio | |
Element	Surficial Rock Concentration (mg/g)	Soil Concentration (mg/g)	Particulate Concentration (mg/g)	Dissolved Concentration (mg/ℓ)	Particulate Load (10^6 tons/yr)	Dissolved Load (10^6 tons/yr)	River Particulate/ Rock	Particulate/ (Partic. + Dissolved)
Al	69.3	71.0	94.0	0.05	1457	2	1.35	.999
Ca	45.0	15.0	21.5	13.40	333	501	0.48	.40
Fe	35.9	40.0	48.0	0.04	744	1.5	1.33	.998
K	24.4	14.0	20.0	1.30	310	49	0.82	.86
Mg	16.4	5.0	11.8	3.35	183	125	0.72	.59
Na	14.2	5.0	7.1	5.15	110	193	0.50	.36
Si	275.0	330.0	285.0	4.85	4418	181	1.04	.96
P	0.61	0.8	1.15	0.025	18	1.0	1.89	.82

Sources: After Martin and Meybeck 1979; Martin and Whitfield 1981; and Meybeck 1979, 1982.

Note: Elements with no gaseous phase only. Particulate and dissolved loads based, respectively, on the total loads, 15.5 × 10⁹ tons solids/yr and 37,400 km³ water/yr.

soluble elements have been removed.   Their suspended load comes from rock debris and poorly weathered material, especially in mountainous areas, and the composition of the suspended matter is closer to the average composition of surficial rocks.   Martin and Meybeck also state that Ca and Na concentrations are lower in tropical river particulate matter than in temperature and arctic river particulate matter.   Based on their data (for 15 rivers), particulate Al does seem generally to be higher and Ca lower in tropical rivers, but there are a number of exceptions and the effect of variations in the original lithology of the rocks and relief of the river basin certainly must be important.   In fact, those rivers that have higher suspended matter Ca concentrations also have high dissolved Ca concentrations, suggesting that the rocks in their river basins contain more Ca than other basins, probably in the form of limestone.   Also, in general, sedimentary rocks are more abundant in temperate areas than in tropical areas (Meybeck 1986).   Tropical rivers with high particulate Al concentrations also tend to have high dissolved SiO_2 concentrations, which is suggestive of both a siliceous and aluminous lithology in the source regions and higher weathering rates associated with higher temperatures (Meybeck 1986).

For most of the major elements listed in Table 5.5, nearly 90% of the total transport is in the particulate load.   However, HCO_3^-, which is the dominant dissolved ion, is not included in the elements listed.   In addition, as far as the availability of major elements for chemical reactions in the oceans is concerned, obviously the dissolved elements will be more important and the suspended load much less so.   A large part of the suspended load is merely dumped and buried upon reaching the oceans.   There are, nevertheless, some changes involving the suspended load, such as ion exchange, when it goes from river water to the ocean and this can affect ocean water chemistry.   (Discussion of this topic is deferred to Chapters 7 and 8 on estuaries and the oceans.)

CHEMICAL COMPOSITION OF RIVERS

World Average River Water

The chemical composition of world average river water according to Meybeck (1979) is given in Table 5.6.   This represents river transport *to the oceans* of dissolved components and does not include transport to internal basins.   "Natural" world river water is corrected for pollution and "actual" river water includes pollution.   Meybeck's estimate for natural river water is similar to earlier estimates (Livingstone 1963).

In estimating natural (prehuman) world river-water composition, Meybeck attempted to avoid conspicuously polluted river data by using early (pre-1900) data for such rivers as the Mississippi, St. Lawrence, and Rhine.   In addition, he made a further correction for pollution by estimating anthropogenic inputs from the changes with time in the composition of five large river basins in industrial areas

TABLE 5.6 Chemical Composition of Average River Water

By Continent	River Water Concentration[a] (mg/ℓ)									Water Discharge (10³ km³/yr)	Runoff Ratio[b]
	Ca^{++}	Mg^{++}	Na^+	K^+	Cl^-	SO_4^{--}	HCO_3^-	SiO_2	TDS		
Africa:											
Actual	5.7	2.2	4.4	1.4	4.1	4.2	26.9	12.0	60.5	3.41	0.28
Natural	5.3	2.2	3.8	1.4	3.4	3.2	26.7	12.0	57.8		
Asia:											
Actual	17.8	4.6	8.7	1.7	10.0	13.3	67.1	11.0	134.6	12.47	0.54
Natural	16.6	4.3	6.6	1.6	7.6	9.7	66.2	11.0	123.5		
S. America:											
Actual	6.3	1.4	3.3	1.0	4.1	3.8	24.4	10.3	54.6	11.04	0.41
Natural	6.3	1.4	3.3	1.0	4.1	3.5	24.4	10.3	54.3		
N. America:											
Actual	21.2	4.9	8.4	1.5	9.2	18.0	72.3	7.2	142.6	5.53	0.38
Natural	20.1	4.9	6.5	1.5	7.0	14.9	71.4	7.2	133.5		
Europe:											
Actual	31.7	6.7	16.5	1.8	20.0	35.5	86.0	6.8	212.8	2.56	0.42
Natural	24.2	5.2	3.2	1.1	4.7	15.1	80.1	6.8	140.3		
Oceania:											
Actual	15.2	3.8	7.6	1.1	6.8	7.7	65.6	16.3	125.3	2.40	—
Natural	15.0	3.8	7.0	1.1	5.9	6.5	65.1	16.3	120.6		

TABLE 5.6 (continued)

	Ca++	Mg++	Na+	K+	Cl-	SO4--	HCO3-	SiO2	TDS	Water Discharge (10³ km³/yr)	Runoff Ratio[b]
						River Water Concentration[a] (mg/ℓ)					
World Average											
World actual	14.7	3.7	7.2	1.4	8.3	11.5	53.0	10.4	110.1	37.4	0.46
World natural (unpolluted)	13.4	3.4	5.2	1.3	5.8	8.3 (6.6)[c]	52.0	10.4	99.6	37.4	0.46
Pollution	1.3	0.3	2.0	0.1	2.5	3.2 (4.9)[c]	1.0	0	10.5		
World % pollutive	9%	8%	28%	7%	30%	28% (43%)[c]	2%	0%	—	—	

Source: All river water concentrations and discharge values from Meybeck 1979 except "actual" concentrations by continent which were calculated from Meybeck's data. (M. Meybeck. "Concentrations des eaux fluviales en éléments majeurs et apports en solution aux océans," *Rev. Géol. Dyn. Geogr. Phys.*, 21 (3), pp. 220, 227. Copyright © 1979. Reprinted by permission of the publisher.)

[a] *Actual* concentrations include pollution. *Natural* concentrations are corrected for pollution.

[b] Runoff ratio = average runoff per unit area/average rainfall (calculated from Meybeck).

[c] We have raised the pollutive contribution; see Table 5.11. (Our values are in parentheses.)

and from direct measurements of river pollution. He corrected the various continents differently for pollution based on their population and stage of industrial development. The main constituents affected are Cl^-, SO_4^{--}, and Na^+ and, to a lesser extent, Ca^{++} and Mg^{++}. As can be seen in Table 5.6, Meybeck estimated that about 30% of the Na^+, Cl^-, and SO_4^{--} concentrations in actual river water can be considered as arising from pollution. (However, we estimate that sulfate pollution is somewhat higher; see Table 5.11 and our detailed discussion of river pollution later in this chapter.) Meybeck gives only the natural river-water values for each continent. We here calculate the actual (polluted) values by continent based on his correction scheme.

Looking at world average river water, we can see first that the *total concentration* of dissolved major ions (TDS = Total Dissolved Solids) in river water is about 100 mg/ℓ or 20 times greater than the concentration in rain. Additional ions are added to rainwater on the continents before it becomes river water. However, river water would be more concentrated than rain even if no additional ions were added, due to the loss of water, through evaporation, after it reaches the ground. Using the runoff ratio of 0.46 given in Table 5.6, we calculate that the concentration of TDS in river water due solely to evaporation should be 2.2 times greater than the concentration in rainwater. This is a considerably lower concentration factor than the value of 20 actually found and the difference is due mainly to contributions from rocks during weathering. There is also considerable anthropogenic input, particularly for ions like Na^+, Cl^-, and SO_4^{--}.

Some examples are instructive. Figure 5.3 compares the North American average natural river water concentration of various ions from Table 5.6 to 2.6 times the average U.S. rainwater concentration from Chapter 3. (The value 2.6 is the inverse of the North American runoff ratio of 0.38.) The estimated pollutive input to North American rivers is also shown. On the order of 10%–15% of the Ca, Na, and Cl in U.S. river water, one quarter of the K, and nearly half of the sulfate (most of which is pollutive) comes from rain. By comparison, SiO_2 and HCO_3^- are essentially entirely from rock weathering (0% from rain).

Considering the effect of the runoff ratio on river concentration by continent, one might expect on the basis of the runoff ratio that African surface water (runoff ratio = 0.28) would be more concentrated and Asian surface water (runoff ratio = 0.54) more dilute than other continents, but this is not the case. African and South American river waters (which have a greater influence of crystalline rocks) are both more dilute (TDS = 61 and 55, respectively) than Asian, North American and European river waters, which are more influenced by sedimentary rocks. Thus, for the continents as a whole, the runoff ratio does not seem to be the dominant influence on river concentrations.

The above discussion emphasizes the importance not only of total concentration of dissolved ions but also of chemical composition. World average river-water composition is dominated by Ca^{++} and HCO_3^-, both of which are derived predominantly from limestone weathering. Meybeck (1979) found that 98% of all river water was of the calcium carbonate type (i.e., had Ca^{++} and HCO_3^- as

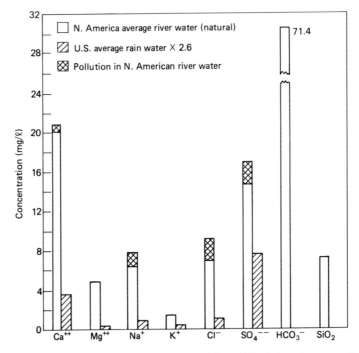

Figure 5.3 Comparison of dissolved composition of North American natural and polluted river water (data from Meybeck 1979) with U.S. rainwater (concentrations in mg/ℓ). Rainwater concentrations are multiplied by 2.6 to correct for evaporation from the continents (see text).

the principal ions). Less than 2% of surface waters had Na^+ (linked with Cl^-, SO_4^{--}, or HCO_3^-) as the principal ion. In the next section, we shall discuss the natural chemical composition of river water in an effort to show the origin of various major dissolved ions and how they are affected by (1) the amount and nature of rainfall and evaporation, (2) geology of the river basin and weathering history, (3) average temperature, (4) relief, and (5) vegetation and biologic uptake. The chemical composition of some major world rivers is given in Table 5.7.

Chemical Classification of Rivers

Several attempts have been made to classify rivers on the basis of their chemistry. The reason for classifying rivers is to determine which of a number of *natural* environmental factors that affect river-water chemistry (i.e., those listed above) are most important. By studying well-known rivers, it is possible to extrapolate results to those whose environments are less well known. As we have seen in the previous section, there are also many human influences that tend to increase the total dissolved solids (TDS) in many rivers and particularly increase Cl^-, SO_4^{--}, and certain cations relative to HCO_3^- and SiO_2.

TABLE 5.7 Major Ion (Dissolved Only) Chemical Composition of Principal World Rivers

River	Concentration (mg/ℓ)									Discharge (km³/yr)	Drainage Area (10⁶ km²)	Reference
	Ca^{++}	Mg^{++}	Na^+	K^+	Cl^-	SO_4^{--}	HCO_3^-	SiO_2	TDS			
North America												
Colorado (mean 1960s)	83	24	95	5.0	82	270	135	9.3	703	20	0.64	Meybeck 1979
Columbia	19	5.1	6.2	1.6	3.5	17.1	76	10.5	139	250	0.67	Meybeck 1979
Mackenzie	33	10.4	7.0	1.1	8.9	36.1	111	3.0	211	304	1.8	Meybeck 1979
St. Lawrence (1870)	25	3.5	5.3	1.0	6.6	14.2	75	2.4	133	337	1.02	Meybeck 1979
Yukon	31	5.5	2.7	1.4	0.7	22	104	6.4	174	195	0.77	Meybeck 1979
Mississippi (1905)	34	8.9	11.0	2.8	10.3	25.5	116	7.6	216	580	3.27	Meybeck 1979
Mississippi (1965–1967)	39	10.7	17	2.8	19.3	50.3	118	7.6	265	580	3.27	Meybeck 1979
Frazer	16	2.2	1.6	0.8	0.1	8.0	60	4.9	93	100	0.38	Meybeck 1979
Nelson	33	13.6	24	2.4	30.2	31.4	144	2.6	281	110	1.15	Meybeck 1979
Rio Grande (Laredo, Tex.)	109	24	117	6.7	171	238	183	30	881	2.4	—	Livingstone 1963
Ohio	33	7.7	15	3.6	19	69	63	7.9	221	—	—	Livingstone 1963
Europe												
Danube	49	9	(9)	(1)	19.5	24	190	5	307	203	0.8	Meybeck 1979
U. Rhine (unpolluted)	41	7.2	1.4	1.2	1.1	36	114	3.7	307	—		Zobrist and Stumm 1980
L. Rhine (polluted)	84	10.8	99	7.4	178	78	153	5.5	256	68.9	0.145	Zobrist and Stumm 1980
Norwegian rivers	3.6	0.9	2.8	0.7	4.2	3.6	12	(3.0)	31	383	0.34	Meybeck 1979
Black Sea— Azov rivers	43	8.6	17.1	1.3	16.5	42	136	—	265	158	1.32	Meybeck 1979
South America												
U. Amazon (Peru)	19	2.3	6.4	1.1	6.5	7.0	68	11.1	122	1512		Stallard 1980
L. Amazon (Brazil)	5.2	1.0	1.5	0.8	1.1	1.7	20	7.2	38	7245	6.3	Stallard 1980

River												Reference
L. Negro (Amazon trib.)	0.2	0.1	0.4	0.3	0.3	0.2	0.7	4.1	6	1383		Stallard 1980
Madeira (Amazon trib.)	5.6	0.2	2.6	1.6	0.8	5.6	28	9.4	53	1550		Stallard 1980
Parana	5.4	2.4	5.5	1.8	5.9	3.2	31	14.3	69	567	2.8	Meybeck 1979
Magdalena	15.0	3.3	8.3	1.9	(13.4)	14.4	49	12.6	118	235	0.24	Meybeck 1979
Guyana rivers	2.6	1.1	2.6	0.8	3.9	2.0	12	10.9	36	240	0.24	Meybeck 1980
Orinoco	3.3	1.0	(1.5)	(0.65)	2.9	3.4	11	11.5	34	946		Meybeck 1979
Africa												
Zambeze	9.7	2.2	4.0	1.2	1	3	25	12	58	224	1.34	Meybeck 1979
Congo (Zaire)	2.4	1.3	1.7	1.1	2.9	3	11	9.8	33	1230	4.0	Meybeck 1979
Niger	4.1	2.6	3.5	2.4	1.3	(1)	36	15	66	190	1.12	Meybeck 1979
Nile	25	7.0	17	4.0	7.7	9	134	21	225	83	3.0	Meybeck 1979
Orange	18	7.8	13.4	2.3	10.6	7.2	107	16.3	183	10	0.8	Meybeck 1979
Asia												
Brahmaputra	17	2.9	3.7	3.1	3.8	9.6	54	7.3	101	[603]	0.58	Ming-hui, Stallard, and Edmond 1982; [Milliman and Meade 1983]
Ganges	24.5	5.0	4.9	3.1	3.4	8.5	105	12.8	167	450	0.975	Meybeck 1979
Mekong	14.2	3.2	3.6	2.0	(5.3)	3.8	58	8.9	99	577	0.795	Meybeck 1979
Japanese rivers	8.8	1.9	6.7	2.2	5.8	10.6	31	19	86	550	0.37	Meybeck 1979
Indonesian rivers	5.2	2.5	3.8	1.0	3.9	5.8	26	10.6	58		1.23	Meybeck 1979
New Zealand rivers	8.2	4.6	5.6	0.7	5.8	6.2	50	7	88	400	0.27	Meybeck 1979
Yangtze (Chiang)	45	6.4	4.1	1.2	4.1	17.9	148	5.8	232	1063	1.95	Ming-hui, Stallard, and Edmond 1982
Yellow (Hwang Ho)	50	22	49	2.4	40	73	226	5.8	467	[49]	0.745	Ming-hui, Stallard, and Edmond 1982; [Milliman and Meade 1983]
Philippines	31	6.6	10.4	1.7	3.9	13.6	131	30.4	228	332	0.3	Meybeck 1979

According to the classification of Gibbs (1970), the major natural mechanisms controlling world surface-water chemistry are (1) atmospheric precipitation, both composition and amount; (2) rock weathering; and (3) evaporation and fractional crystallization. A boomerang-shaped diagram resulted when he plotted the ratio of two major cations in world surface waters, Ca^{++} and Na^+, plotted as $Na/(Na + Ca)$, versus total dissolved salts (TDS) (see Figure 5.4). Rivers plot either in areas dominated by each of the three mechanisms, in other words in the three corners of the "boomerang," or in intermediate areas where more than one mechanism controls their chemistry. The TDS axis is roughly inversely proportional to rainfall or runoff. Atmospheric precipitation–controlled rivers are in areas of high rainfall, and evaporation-crystallization-controlled rivers are in arid areas, while rock-dominated rivers are in areas of intermediate rainfall. Thus, this classification is based to a considerable extent upon the amount of rainfall and, thus, runoff. Gibbs also found that a practically identical location for almost all rivers was found if a similar diagram was plotted which used $Cl/(Cl + HCO_3)$ in place of $Na/(Na + Ca)$.

According to Gibbs, rivers whose composition is controlled mainly by atmospheric precipitation are those whose composition resembles that of rainfall; that is, they have low total dissolved solids and high Na relative to Ca (or high Cl relative to HCO_3). In Figure 5.4 they plot in the lower right corner of the boomerang. These are generally tropical rivers of South America and Africa and Atlantic Coastal Plain rivers of the United States in areas of high rainfall, low relief, and heavy weathering or draining sandy rocks with a resulting low supply of dissolved salts. Gibbs states that the ratios of the major dissolved ions in atmospheric precipitation–controlled Amazon Basin rivers are similar to sea salt except for H_4SiO_4 and K, which are derived from rock weathering.

The middle portion of the boomerang where values of TDS are intermediate and values of $Na/(Na + Ca)$ are low, is the location of rivers whose composition is controlled, according to Gibbs, by rock weathering. This includes most of the major world rivers (Mississippi, Ganges, Nile, etc.). For these rivers, rock weathering supplies most of the dissolved salts. Since sedimentary rocks occupy about 75% of the earth's surface and their weathering is dominated by the dissolution of $CaCO_3$ (see Chapter 4), one would expect that rivers controlled by rock weathering would consist mainly of $Ca^{++} + HCO_3^-$ resulting from carbonate dissolution. This is why such rivers plot at low values of $Na/(Na + Ca)$ and low values of $Cl/(Cl + HCO_3)$.

The upper right corner of the boomerang in Figure 5.4 is the domain of rivers whose composition is controlled by evaporation and fractional crystallization. These rather atypical rivers have a high concentration of total dissolved salts and high Na (or Cl^-) relative to Ca^{++} (or HCO_3^-). Two examples given by Gibbs are the Rio Grande and Pecos. The composition evolves in a sequence of river samples taken in a downstream direction from rock-weathering type waters found in headwaters to more concentrated and sodic waters as a result of evaporation and concentration of salts accompanied by the crystallization of $CaCO_3$ (see arrows in Figure 5.4). As evaporation continues, the concentrations of Na and Cl increase

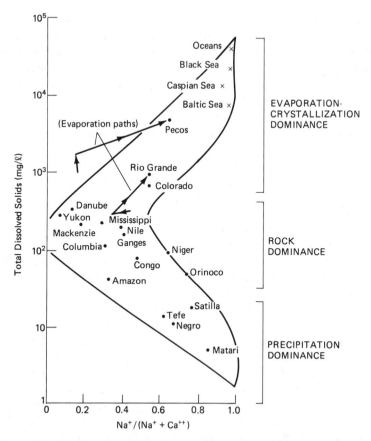

Figure 5.4 Variation of the weight ratio, $Na^+/(Na^+ + Ca^{++})$, as a function of total dissolved solids, of world rivers (·) and other water bodies (×). The "evaporation paths" in a downstream sequence of samples from the Pecos and Rio Grande rivers are discussed in the text. (Redrawn after R. J. Gibbs. "Mechanisms Controlling World Water Chemistry," *Science*, 170, p. 1088. Copyright © December 4, 1970 by The Amer. Assoc. for the Advancement of Science, reprinted by permission of the publisher.)

but those of Ca and HCO_3 do not because the latter two are constantly removed by precipitation of $CaCO_3$. (For a further discussion of evaporation and fractional crystallization, see the section on saline lakes in Chapter 6.)

Gibbs' classification has proved to be controversial. The main points of contention have to do with the two high Na/(Na + Ca) corners of the boomerang. Stallard and Edmond (1983) point out that several of the Amazon tributaries, used by Gibbs as prototypes of rainfall-controlled rivers, upon resampling and reanalysis of both rivers and nearby rainwater, prove to have compositions rather different from that of sea salt. In fact, Stallard and Edmond disagree with Gibbs' assumption that rivers in the "precipitation dominance" field, with low concentration of dissolved salts (TDS) and high Na^+ relative to Ca^{++}, are dominated by atmospheric precipitation.   When Stallard and Edmond made a correction for cyclic salts in their data for coastal and lowland Amazon rivers they found that it made only a minor change in the position of these rivers on Gibbs' diagram (except for one river, the Matari).   They reasoned that if these rivers were dominated by atmospheric precipitation of cyclic salts (i.e., marine sea salt), then correcting for cyclic salt should reduce their TDS and Na^+ concentrations and they should be shifted toward the less sodic part of the diagram.   However, this was not the situation. Only in the coastal Matari River was the cyclic contribution to TDS large (42%). The Negro River, which is about 1500 km inland and which falls in Gibbs' atmospheric precipitation–dominated field, contains only on the order of 10% cyclic salt.   Gibbs (1970) states that 81% of the Na^+, K^+, Ca^{++}, and Mg^{++} in the coastal Rio Tefe (which also falls in his precipitation-dominated field) is from precipitation, but *using Stallard's (1980) data* we calculate that only 10% of the Na^+, K^+, Ca^{++}, and Mg^{++} and 3%–5% of the TDS consists of cyclic salt.

Stallard and Edmond (1983) believe that the chemical composition of these dilute, high Na/Ca rivers results from their geology and erosional regime rather than from atmospheric precipitation.   Their high Na/Ca ratio arises because the rocks are mainly siliceous (as compared to calcareous), and the low TDS arises because the sediments and soils of the drainage basins have already undergone intense weathering and are, thus, relatively unreactive.   Thus, they would consider these rivers to be rock dominated (in Gibbs' terminology).

A river that might better fit Gibbs' "atmospheric precipitation–dominated" category is the Satilla River in Georgia (Beck, Reuter, and Perdue 1974).   The Satilla drains a swampy area of the U.S. coastal plain, is very dilute, has a low pH, and its organic constituents are greater (30 ppm C) than its inorganic constituents (15 ppm).   Sodium and chloride dominate the inorganic ions in the Satilla River and the river ratios of Na^+/Cl^- and SO_4^{--}/Cl^- are very similar to sea salt. Beck, Reuter, and Perdue concluded that rainfall input is the main source of the major element ions in the river except for minor additions of K and Ca from the soil.   If all the river Cl^- is assumed to be from sea salt, then we calculate that about 50% of the TDS in the Satilla River and most of the Na^+ and SO_4^{--} (see Table 5.14) are also derived from sea salt.   Pine Barrens rivers in the New Jersey coastal plain are also atmospheric-controlled, that is, they receive a large part of their composition from precipitation (Yuretich et al. 1981).

The other major controversial points of Gibbs' classification concerns the saline rivers.   Feth (1971) states that the evaporation-crystallization process is not the controlling mechanism for the composition and concentration for two of Gibbs' major examples, the Pecos and Rio Grande rivers.   Feth agrees that the river water becomes progressively more concentrated downstream due to evapotranspiration in this arid area and by the addition of irrigation return water.   However, he feels that the *main* increase in dissolved salts downstream in the Pecos River is from groundwater flow into the river of NaCl brines from near-surface halite deposits (see also Gibbs 1971 and Feth 1981).

Stallard (1980), based on Amazon Basin data, also feels that high Na/Ca and high TDS rivers are mainly the result of the weathering of evaporites (which have a high Na/Ca ratio) rather than the concentration and evolution of typical "rock weathering" river water.   Stallard found that three-fourths of the chloride in the main Amazon River comes from evaporites (the rest is cyclic salt) with 90% of this evaporite-derived chloride coming from the weathering of halite-bearing salt diapirs in the Peruvian Andes.   Buried evaporites there result in the formation of salt springs and saline rivers.   The salt domes do not occur in an arid area (precipitation of 150 cm/yr); thus, highly saline rivers can result from evaporite weathering in the *absence* of aridity.

Stallard and Edmond (1983) have created their own classification of Amazon Basin rivers emphasizing the role of geology and erosional regime as the major control on river-water composition.   They group rivers in terms of total cation charge, which for the sake of simplification we have roughly converted to mg/ℓ total dissolved solids (TDS).   Stallard and Edmond's categories and the Amazon Basin geology and erosional regime to which they correspond, are shown in Table 5.8.   Although the classification is based on study of rivers of the Amazon Basin, where there is rapid tropical weathering and a lack of evaporative concentration and little pollution, the categories can be expanded to cover other areas.   Thus, we have also included here other world rivers that fit these categories.   The first category ($<$ 20 mg/ℓ TDS) comprises rivers draining intensely weathered material—cation-poor siliceous rocks and deeply weathered soils and saprolites—and includes Gibbs' "atmospheric precipitation control" rivers.   The second category (20–40 mg/ℓ TDS) represents rivers draining (cationic) siliceous terrains (igneous and metamorphic rocks and shales lacking major marine deposits) and it falls between Gibbs' "atmospheric precipitation control" and "rock (weathering) dominance."   The third category (40–250 mg/ℓ TDS) includes rivers draining marine sedimentary rocks.   Such rivers have high Ca^{++} and HCO_3^- concentrations as a result of the weathering of carbonate minerals and high sulfate from the weathering of gypsum and pyrite in shale.   This corresponds to Gibbs' rock dominance field and, if extrapolated to world rivers, includes most major world rivers.   The fourth category (TDS $>$ 250 mg/ℓ) represents rivers draining evaporites.   As pointed out above, this corresponds to Gibbs' "evaporation-crystallization control" rivers.

As further evidence of the geologic or rock-weathering control of Amazon rivers, Stallard (1980) and Stallard and Edmond (1983) show that rivers in the first two categories (i.e., those draining cationic siliceous rocks and heavily weathered areas) generally have a molar ratio of 2:1 between Si and (Na + K) after (minor) cyclic salt corrections are made (see Table 5.8).   This is the ratio predicted for water draining areas dominated by the weathering of primary silicate minerals (such as Na-plagioclase and K-feldspar) to kaolinite:

$$2NaAlSi_3O_8 + 2H^+ + 9H_2O \rightarrow Al_2Si_2O_5(OH)_4 + 2Na^+ + 4H_4SiO_4$$
plagioclase   kaolinite

TABLE 5.8 Stallard and Edmond's River Classification

Total Cationic Charge (μeq/ℓ)	Approximate TDS (mg/ℓ)	Predominant Source-Rock Type	Characteristic Water Chemistry[a]	Examples	Gibbs Category
<200	<20	Intensely weathered (cation-poor) siliceous rocks and soils (thick regolith)	Si-enriched; low pH; Si/(Na + K) = 2; high Na/(Na + Ca)	Amazon tributaries (Matari, Tefe, Negro)	Atmosphere-precipitation controlled
200–450	20–40	Siliceous (cation-rich); igneous rocks and shales (sedimentary silicates)	Si-enriched; (higher Si from igneous and metamorphic rocks); Si/(Na + K) = 2; intermediate Na/(Na + Ca)	L. Amazon, Orinoco, Zaire	Between atmosphere-precipitation controlled and rock-dominated
450–3000	40–250	Marine sediments; carbonates, pyrite; minor evaporites	Na/Cl = 1; (Ca + Mg)/($\frac{1}{2}$HCO$_3$ + SO$_4$) = 1; low Na/(Na + Ca)	Most major rivers	Rock-weathering dominated
>3000	>250	Evaporites; CaSO$_4$ and NaCl	Na/Cl = 1; (Ca + Mg)/($\frac{1}{2}$HCO$_3$ + SO$_4$) = 1; high Na/(Na + Ca)	Rio Grande, Colorado	Evaporation-crystallization

Source: Data from Stallard 1980; and Stallard and Edmond 1983.

[a] Element ratios given refer to mole ratios.

or

$$2KAlSi_3O_8 + 2H^+ + 9H_2O \rightarrow Al_2Si_2O_5(OH)_4 + 2K^+ + 4H_4SiO_4$$

K-feldspar   kaolinite

The ion ratio of the weathering products of these reactions is 2Na:4Si and 2K:4Si, or in other words, 1(Na + K):2Si. (For further discussion see Chapter 4.) Similarly, rivers in Stallard's third and fourth groups (draining carbonate and evaporite terrains) have molar ratios of 1:1 for Na:Cl and 1:1 for (Ca + Mg):($\frac{1}{2}$ $HCO_3^- + SO_4^{--}$). These are the ratios expected for waters draining areas dominated by carbonate and evaporite weathering (Table 5.9); (see also Chapter 4).

Drever (1982) and Garrels and Mackenzie (1971) also stress the importance of rock type in determining river composition. Drever's description of the river water produced by different rock types is similar to Stallard's. Drever characterizes water from shales as having variable TDS, and a lower ratio of Si to total cations than igneous rocks because illite and quartz (the predominant mineral sources of Si in shales) do not weather readily. Sulfate from pyrite weathering and Cl^- from trapped NaCl (probably originally seawater) are the primary shale-derived anions. From these considerations, waters from shales should fall in Stallard's categories 2 and 3. (Because black shales contain pyrite, which readily weathers to H_2SO_4, they tend to produce waters that extend over more than one category.)

Meybeck (1984) has studied French rivers draining single rock types and from these rivers has derived representative river compositions for each rock type, again emphasizing the importance of rock type on river composition. Reeder, Hitchon, and Levinson (1972) in a study of the Mackenzie River drainage in Canada, have concluded that salinity is largely controlled by lithology with higher salinities resulting from carbonates and evaporites. In general, the total dissolved solids (TDS) in rivers draining sedimentary rocks (including carbonates) tend to be at least twice the TDS in rivers draining only crystalline (igneous and metamorphic) rocks (Holland 1978).

Stallard plotted the composition of Amazon Basin rivers (corrected for cyclic salt) on a ternary diagram with the three vertices represented by siliceous rock weathering (Si), carbonate weathering (HCO_3^-), and evaporite (plus pyrite) weathering ($Cl^- + SO_4^{--}$). He found that the total dissolved solids (represented by total cation charge) increases from the Si vertex to the HCO_3^- vertex and then from the HCO_3^- vertex to the $Cl^- + SO_4^{--}$ vertex.

As an extension of Stallard's work (and also that of Ming-hui, Stallard, and Edmond 1982), we have plotted world rivers on a similar ternary diagram (Figure 5.5). Here the rivers are not corrected for cyclic salt, so those few rivers with considerable cyclic salt are shifted toward the right as compared to Stallard's diagram. The data for Figure 5.5 are from Table 5.7 given in the previous section. The diagram also includes world average river water and certain other rivers for comparison. "Natural world average river water" (Meybeck 1979), which has been corrected for pollution (denoted as World Average—N), plots near a number of major world rivers (Orange, Columbia, Brahmaputra, and Upper Amazon).

TABLE 5.9 Carbonate and Evaporite Weathering Reactions

Reaction	Molar Ratio of Products in Solution
1. $NaCl \rightarrow Na^+ + Cl^-$ halite	$Na^+:Cl^- = 1:1$
2. $CaSO_4 \rightarrow Ca^{++} + SO_4^{--}$ anhydrite	$(Ca^{++} + Mg^{++}):(\frac{1}{2}HCO_3^- + SO_4^{--}) = 1:1$
3. $CaCO_3 + H_2CO_3 \rightarrow Ca^{++} + 2HCO_3^-$ calcite carbonic acid	$(Ca^{++} + Mg^{++}):(\frac{1}{2}HCO_3^- + SO_4^{--}) = 1:1$
4. $2CaCO_3 + H_2SO_4 \rightarrow 2Ca^{++} + SO_4^{--} + 2HCO_3^-$ calcite sulfuric acid	$(Ca^{++} + Mg^{++}):(\frac{1}{2}HCO_3^- + SO_4^{--}) = 1:1$
5. $CaMg(CO_3)_2 + 2H_2CO_3 \rightarrow Ca^{++} + Mg^{++} + 4HCO_3^-$ dolomite carbonic acid	$(Ca^{++} + Mg^{++}):(\frac{1}{2}HCO_3^- + SO_4^{--}) = 1:1$
6. $2CaMg(CO_3)_2 + 2H_2SO_4 \rightarrow 2Ca^{++} + 2Mg^{++} + 2SO_4^{--} + 4HCO_3^-$ dolomite sulfuric acid	$(Ca^{++} + Mg^{++}):(\frac{1}{2}HCO_3^- + SO_4^{--}) = 1:1$

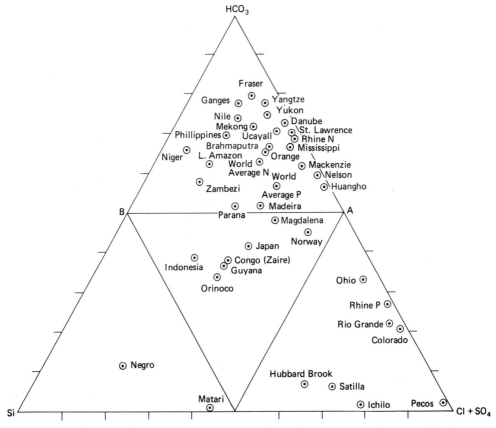

Figure 5.5 Major rivers: Percentage of Si (μ moles/ℓ); HCO_3^- (μ Eq/ℓ); and sum of Cl^- + SO_4^- (μ Eq/ℓ). For example, 100% HCO_3^- plots at the HCO_3^- vertex; 50% HCO_3^- and 50% Si plots at point B; and 50% HCO_3^- and 50% (Cl^- + SO_4^-) plots at point A. River TDS increases from Si vertex to HCO_3^- vertex to (Cl^- + SO_4^-) vertex. (Data from Tables 5.7 and 5.14.)

(The inclusion of the Orange River in this group is surprising because it is from an arid area.) Uncorrected or polluted world average river water (denoted as World Average—P) after Meybeck (1979) is also plotted and shows a shift toward the (Cl^- + SO_4^{--}) vertex since these two ions are the major pollutants. Similarly when the unpolluted Rhine River (Rhine—N) is compared to the polluted Rhine River (Rhine—P), there is a tremendous increase in Cl^- and SO_4^{--} (data from Zobrist and Stumm 1980).

From Figure 5.5 one can see also that the vast majority of major world rivers have more than 50% HCO_3^- and from 10%–30% (Cl^- + SO_4^{--}). Thus, most large rivers are dominated by sedimentary rock weathering and are mainly Ca^{++} and HCO_3^- waters derived from carbonate minerals. (To a lesser extent rivers draining areas of incomplete silicate weathering where cation-rich soils are forming will also plot in this area [Stallard and Edmond 1983].)   Arid area rivers (Rio Grande and Colorado) plot along the axis between(Cl^- and SO_4^{--}) and HCO_3^- toward the ($Cl^- + SO_4^{--}$) vertex, and the Pecos plots virtually at the vertex.   The Huangho River is interesting in that its composition is dominantly due to the weathering of carbonates but is shifted toward the Cl-SO_4 vertex.   There is extensive evaporation in the arid part of its course and there is also probably considerable input from soil salts and even carbonate precipitation (as in Gibbs' evaporation-crystallization) in this area (Ming-hui, Stallard, and Edmond, 1982).

Some other rivers shown in Figure 5.5 represent special situations.   The sulfuric acid weathering trend of pyritiferous black shales described by Stallard for the Madeira River (and in particular one tributary, the Ichilo) also includes Japanese rivers (which have thermal springs and volcanic input).   Hubbard Brook, which is a very small river receiving sulfuric acid rain on a siliceous terrain, plots as an extension of the same general trend (see Figure 5.5).   The Congo, Guayana, and Orinoco rivers, which are all tropical rivers flowing over a weathered siliceous terrain, plot together.   We have also included, on Figure 5.5, the Satilla and Matari rivers, which we mentioned earlier as having considerable atmospheric precipitation influences.   The main reason they are included in Figure 5.5 is that both are quite acidic and this results in a very low proportion of HCO_3^-.   (In the case of the Satilla, this is due to organic acidity; see later discussion.)

Other factors in determining river composition not specifically mentioned in the classifications discussed above, are relief, average temperature, and vegetation.   Relief is an important factor in Stallard and Edmond's (1983) "erosional regime" which is used to produce their classification of Amazon Basin rivers using total cation concentration or TDS.   Greater relief means greater physical erosion and faster exposure of fresh rock for chemical weathering.   Because transport away of weathering material is rapid with high relief, chemical weathering in the soil is incomplete and the amount depends upon how easily the rocks undergo decomposition.   Thus, both lithology (carbonates vs. silicates) and structure are important.   For instance, massive silicates (shield rocks) weather more slowly than more porous silicates such as sediments and volcaniclastics.   With low relief, chemical weathering is much more complete and thick soils and saprolites develop over bedrock so that variations in geology become less important.   For example, where evaporites are exposed along the lower Amazon Basin their contribution to rivers is fairly minor—more in proportion to their relative abundance (Stallard 1985), as compared to their conspicuous contributions to Andean rivers.   However, even in the Amazon Basin lowland areas geology is important; that is, igneous and metamorphic silicates weather more slowly than sedimentary silicates, and rivers draining carbonates have a greater TDS concentration than those draining silicates.

Relief in a particular area often correlates with rock type, temperature, and vegetation, and these factors are hard to separate (Drever 1982).   For example, Gibbs (1967) points to relief as the dominant factor in controlling the chemical composition of the Amazon River system.   In the Amazon River about 70% of the dissolved load is derived from the Andean highlands, which are underlain by easily weathered carbonates and evaporites (Stallard 1980). Here, high relief and easily weathered rock types tend to correlate. By contrast in North America and other areas, the headwaters of rivers are often in crystalline rock, which is resistant to chemical weathering, and the headwater streams are very dilute. These rivers become more concentrated downstream as more easily weathered rocks are encountered (Livingstone 1963). The Mississippi River is an example of this effect.

Vegetation has two opposing effects on river composition (Drever 1982). It increases chemical weathering by supplying CO_2 and organic acids. However, vegetation also prevents physical weathering by stabilizing soil and thus decreasing exposure of bedrock to chemical weathering. This can be seen in recently deforested areas where the TDS of rivers increases. (For further discussion of the role of vegetation in weathering, consult Chapter 4.)

MAJOR DISSOLVED COMPONENTS OF RIVER WATER

Chloride and Cyclic Salt

Chloride in natural waters is extremely mobile and very soluble. Once in solution, Cl^- has a very uneventful geochemistry. It reacts very little with other ions, does not form complexes, is not greatly adsorbed on mineral surfaces, and is not active in biogeochemical cycling (Feth 1981). Because it is so chemically unreactive, Cl^- is used as a tracer of natural water masses, particularly in the oceans, but also in ground and surface water.

The main sources of chloride in river water are (according to Feth 1981): (1) sea salt, that is, cyclic salt from rain and dry fallout; (2) dissolution during weathering of halite (NaCl) in bedded evaporites or dispersed in shales; (3) thermal and mineral springs in volcanic areas; (4) redissolution of saline crusts in desert basins (not a primary source); and (5) pollution such as domestic and industrial sewage, oil well brines, mining, and road salt.

"Cyclic salt," or sea salt carried inland in the atmosphere and deposited in rain or snow, has been assumed by some to be the main source of Cl in river water. However, as pointed out by Feth (1981) this may be true only for some very dilute rivers, such as those draining nonsedimentary rocks, particularly those along coasts. In rivers with higher Cl concentrations, Feth (1981) states that Cl from rain probably makes up less than 20% of the total Cl concentration. Similarly, Zverev and Rubeikin (1973) estimate the average contribution of Cl of atmospheric origin in all USSR rivers to be 21%. In the section on chemical classification of rivers, we mentioned several exceptional coastal rivers whose Cl content is presumably dominated by cyclic salt, for example, the Matari River in the Amazon Basin and the Satilla River in Georgia. Some other notable, but small, river basins whose Cl^- input is mainly from sea salt include Hubbard Brook, New Hampshire (Likens et al. 1977); Pond Branch, Maryland, (Cleaves, Godfrey, and Bricker 1970); the Pine Barrens rivers, New Jersey (Yuretich et al. 1981); and the Mattole River basin in coastal northern California (Kennedy and Malcolm 1977).

Table 5.10 gives the percentage of Cl^- (and other ions) in world average river water that comes from cyclic salt as estimated by us and by other studies. Our values are based on Stallard and Edmond's (1981) estimate of cyclic salt in the relatively unpolluted Amazon River basins. They determined the relationship between Cl^- concentrations in Amazon River basins draining terrains with no chloride-bearing rocks and the distance of the rivers from the Atlantic Ocean. The Cl^- concentration dropped rapidly from the coast, and levelled off at 2000 km inland; this inland Cl^- value was then used to estimate cyclic Cl^- in inland rivers. Overall, they found that 18% of the Cl^- in the Amazon River was cyclic and, as they suggest, this seems to be applicable to world average river water. However, this value is for unpolluted rivers and since about 30% of riverine Cl worldwide is due to pollution (see Table 5.11 and Meybeck 1979), we calculate that only 70% of Stallard and Edmond's value of 18%, or 13%, represents the fraction of *total* Cl^- carried by rivers that is due to cyclic salt. The cyclic Cl^- estimates of Holland (1978) and particularly Garrels and Mackenzie (1971) and Meybeck (1983, 1984) are much larger than our estimate of 13% (see Table 5.10). The primary reason for this is the underestimation by these authors of the large contribution of Cl^- from sedimentary halite (see below). Meybeck (1983) has an unusually large sea-salt contribution because of his emphasis on the importance of *coastal* rivers (within 100 km of the coast) returning cyclic Na^+ and Cl^- back to the ocean. However, analyses of many major rivers are taken further upstream than this so that it confuses the matter to emphasize coastal effects.

The remaining Cl^- in river water (57% of total Cl; see Table 5.11) comes dominantly from the weathering of sodium chloride. (Contributions from saline groundwater and hydrothermal springs are included here since the Cl^- comes ultimately from NaCl dissolution.) NaCl occurs in bedded evaporite rocks as the mineral halite, which is normally accompanied by $CaSO_4$ minerals, and as inclusions in sandstone and shale formed by the trapping of interstitial seawater. Since NaCl is both widespread and very soluble, it provides an abundant source of Cl^- in both ground and surface waters. For example, according to Mattox (1968), 25% of continental areas are underlain by evaporites and 15% of these areas have salts containing Cl. (In the Northern Hemisphere, more than half of the land areas are underlain by evaporites.) Evaporite NaCl occurs most commonly in the sub-surface because surface outcrops are usually dissolved away. Likewise, shales and sandstones generally contain more Cl^- in the subsurface because of removal or preferential dissolution near the earth's surface (see Feth 1981).

The halite sources of Cl in a river basin are most often quite localized. In the Amazon Basin, for example, Stallard (1980) found that 90% of the Cl (after correction for cyclic salt) comes from the Peruvian Andes where Cl is added from salt domes, salt springs from buried evaporites, and saline water migrating up fault planes from depth. The Pecos River in Texas receives tremendous quantities of

TABLE 5.10 Atmospheric Cyclic Salt (Sea Salt) as a Percentage of World Average River Water

Element	Present Study[a]	Holland 1978[b]	Garrels and Mackenzie 1971[b]	Meybeck 1983[c]
Cl^-	13 (18)	27	55	72
Na^+	8 (11)	19	35	53
SO_4^-	2 (2)	39[d]	6	19
Mg^{++}	2 (2)	≤3	7	15
K^+	1 (1)	≤14	15	14
Ca^{++}	0.1 (0.1)	1.3	0.7	2.5

[a] Values are given as a percentage of "actual" (including pollution) world average river water (Meybeck 1979) and, in parenthesis, of "natural" world river water (corrected for pollution). (Cyclic Cl is set equal to 18% which is the Amazon value given by Stallard and Edmond 1981. Sea salt contributions to other ions are set in proportion to this based on the composition of average seawater.)

[b] Based on world average river water from Livingstone 1963, which is not corrected for pollution.

[c] Based on "natural" world average river water.

[d] Based on total atmospherically derived sulfur (natural + pollution).

Cl^- from saline springs, which are leaching subsurface halite. Feth (1981) notes that two-thirds of the continental United States is underlain by groundwater with high dissolved solid concentrations (> 1000 mg/ℓ), and that many of these contain considerable NaCl. This saline groundwater is likely derived from NaCl dissolution and may constitute a source of Cl^- to rivers. Chloride is also added to surface water by redissolution of saline evaporative crusts in desert basins, but this is not a primary source—merely a secondary, concentrated one.

Hydrothermal alteration of volcanic rocks (as in New Zealand) constitutes an additional source of Cl^-. In Japan, around 5% of the Cl in river water is estimated as being from thermal and mineral springs (Sugawara 1967). Meybeck (1984) estimates a hydrothermal Cl input to world average river water equal to 8%.

Pollution can be an important Cl^- source for many rivers. Worldwide, Meybeck estimates that about 30% of Cl^- in river water arises from pollution (see Table 5.11). Domestic sewage contains considerable Cl^-. According to Feth (1981), chlorination of public water supplies to purify them adds 0.5–2.0 mg/ℓ Cl^- to the Cl^- concentration of water, and domestic uses add another 20–50 mg/ℓ to sewage. In addition to direct sewage discharge there is additional seepage from septic tanks. Other sources of Cl^- pollution include road salt, fertilizer, and industrial Cl^--containing brines. In some areas mining of salt can locally add Cl^- to river water, and in the western United States oil field brines are also a local source. Two notable examples of rivers bearing a large component of chloride pollution are the St. Lawrence and the lower Rhine.

TABLE 5.11 Sources of Major Elements in World River Water (in percent of actual concentrations)

Element	Atmospheric Cyclic Salt	Weathering			Pollution[b]
		Carbonates	Silicates	Evaporites[a]	
Ca^{++}	0.1	65	18	8	9
HCO_3^-	<<1	61[c]	37[c]	0	2
Na^+	8	0	22	42	28
Cl^-	13	0	0	57	30
SO_4^{--}	2[d]	0	0	22[d]	43
Mg^{++}	2	36	54	<<1	8
K^+	1	0	87	5	7
H_4SiO_4	<<1	0	99+	0	0

[a] Also includes NaCl from shales and thermal springs.

[b] Values taken from Meybeck 1979 except sulfate, which is based on calculations given in the text.

[c] For carbonates, 34% from calcite and dolomite and 27% from soil CO_2; for silicates, all 37% from soil CO_2; thus, total HCO_3^- from soil (atmospheric) CO_2 = 64% (see also Table 5.13).

[d] Other sources of river SO_4^{--}: natural biogenic emissions to atmosphere delivered to land in rain, 17%; volcanism, 5%; pyrite weathering, 11%.

When plotted against runoff, the concentration of Cl^- in river water from most rivers behaves in the manner expected for simple dilution of a limited number of highly concentrated sources (Holland 1978). In other words, there is an inverse proportionality between chloride concentration and runoff. This provides further evidence for the importance of localized deposits of halite and point-source pollution (i.e., cities) as major sources of this ion.

Sodium

Because sodium is a major component of seawater, it is a principal contributor to atmospheric cyclic salt, and, therefore, should be prominent in river water. However, cyclic salt is not the major source of Na in river water. Using the sea-salt value of Na/Cl, total concentrations of Na and Cl in world average river water (Table 5.6), and the proportion of Cl^- in average river water that is contributed by cyclic sea salt (13%; see Table 5.10), we can calculate that only about 8% of river-water sodium owes its origin to cyclic salt. This is shown in Table 5.10. A much more important source of Na is halite in sedimentary rocks. Since Na^+ is the major ion accompanying Cl^- in halite, everything said in the previous section concerning the rock sources of Cl^- also refers to Na^+. In fact, by knowing the relative importance of each source of Cl^-, we can calculate from the stoichiometry of NaCl the relative importance of the same source for Na. The result (Table

5.11) is that about 42% of Na comes from the weathering of halite in evaporite beds and from NaCl occluded in shales and sandstones.

The thing about Na^+ that sets it apart from Cl^- is that it is a major component of silicate rocks. As discussed previously in the chapter on soil water and groundwater, sodium in silicate rocks is present mainly as the albite component of plagioclase, with the formula $NaAlSi_3O_8$. Plagioclase is a major source of Na for groundwater and thus it also is a major source for river water. If we sum up all the sources of Na from pollution (28%), sea salt (8%), and halite (42%), the remaining (22%) must come from the weathering of plagioclase. This is all summarized in Table 5.11. The ways and means by which Na is added to soil water and groundwater, and eventually to river waters by silicate weathering, is summarized in Chapter 4, and will not be repeated here. Our point here is merely to show that an appreciable fraction (22%) of Na in river water arises from silicate weathering.

Pollution is an important source of sodium in river water (28% of the total; see Table 5.11). In fact, along with Cl^- and SO_4^{--}, Na is one of the ions most affected by pollution. Most of pollutive sodium is associated with Cl^- in NaCl, and thus, a number of the sources of Cl pollution mentioned previously are also Na sources. These include domestic sewage, mining of halite, industrial brines, and road salt. Other Na salts, such as Na_2CO_3, Na_2SO_4 and Na-borate, are mined and used industrially for paper, soaps and detergents, and other products (Skinner 1969) and sodium is also used in water softeners as sodium zeolite and Na_2CO_3, replacing Ca^{++} and Mg^{++} in industrial and domestic water by Na^+. These provide an additional source of Na in sewage.

Potassium

Potassium in river water comes predominantly (nearly 90%; see Table 5.11) from the weathering of silicate minerals, particularly potassium feldspar, as orthoclase and microcline, and mica, as biotite. These silicate minerals are found in both sedimentary, and metamorphic and igneous rocks. Meybeck (1984) estimates that about three-fourths of silicate weathering K comes from sedimentary silicates and one-fourth from igneous and metamorphic silicates. Potassium is not dissolved and released as quickly during weathering as most other cations because the potassium-containing primary minerals weather more slowly than those containing Na, Ca, and Mg (see Chapter 4), and therefore, considerable amounts of potassium remain in the soil. Holland (1978) estimates that on the average only about 50% of rock potassium is released to solution during silicate weathering. This would agree with the fact that the soil concentration of potassium is about half that of average surficial rocks (Table 5.5). From Table 5.5, we can also see that only 15% of the river transport of K is in the dissolved load while the rest is particulate. Thus, in behavior, K is intermediate between the rapidly weathering Ca, Na, and Mg and the resistant Si and Al. (We have also assumed that 5% of weathering

K comes from evaporates as KCl, after Meybeck 1984.   This is minor compared to silicate weathering.)

Nonweathering sources of potassium include pollution (7%) and minor contributions from cyclic salt (1%).   Rare evaporite deposits of KCl and similar salts are mined for K fertilizer and river pollution of potassium results from such mining as well as from the use of the K fertilizers themselves.   An example is the mining of KCl in Alsace (France) which is a contributor to K pollution in the Rhine River (as well as to NaCl pollution; Meybeck 1979).   Since the world's consumption of K salts for fertilizers is increasing at the rate of 10% a year (Skinner 1969), potassium pollution of rivers should be expected to increase with time.

Although K in river water is ultimately dominantly from silicate weathering, it is a very biogenic element due to its utilization by growing vegetation.   For example, Dion (1983) found that 60% of K^+ from groundwater in the Connecticut River Basin was taken up by biological activity.   In temperate drainage basins, there is a seasonal variation in stream K concentrations, with K being lower during summer, when plants are growing and taking up K, than during periods of plant dormancy (winter).   In Hubbard Brook, for example, Likens et al. (1977) found K concentrations to vary by about ±33% from this effect.   Potassium is concentrated in leaves of trees so when trees lose their leaves in the fall, there tends to be a rise of K concentration in streamwater due to the leaching of K from the leaf litter (Cleaves, Godfrey, and Bricker 1970; Likens et al. 1977).   Also, unlike most other elements, potassium tends to increase in concentration during increases in discharge (e.g., from flood flow, spring runoff, or heavy rains) due to the dissolution of soluble salts from trees, leaf litter, and the top of the soil (Cleaves, Godfrey, and Bricker 1970; Miller and Drever 1977).   Because there is a net accumulation of potassium in vegetation (the net yearly biomass accumulation in the Hubbard Brook area is three times the stream output [Likens et al. 1977; see also Chapter 4]), forest cutting results in a large increase in stream K concentrations.

Overall, worldwide, there is little variation in potassium concentrations between major rivers and it is always the least abundant of the four major cations (Meybeck 1984).   (The average potassium concentration is 1.3 mg/ℓ and the range 0.5–4.0 mg/ℓ; Meybeck 1980).   Highest concentrations are found in high TDS rivers from arid areas (Nile, Colorado, and Rio Grande), but, in general, potassium is the least variable of the major dissolved ions in river water (Meybeck 1980).   There are a number of factors that probably contribute to the low variability of potassium concentrations in major world rivers.   First, there is not much difference between the average concentration of K in sedimentary rocks (2.0% K_2O; Holland 1978) and in crystalline igneous rocks (3.2% K_2O; Holland 1978), so the lithology of various drainage basins should be less important.   Secondly, K is released much more slowly and less completely during weathering than many other major dissolved ions.   Thirdly, because K, released by silicate weathering, is taken up to a considerable degree by the biomass, its release to streamwater is partly controlled by organic decay, and over a year, stream release should be related to the balance between biologic uptake and decay.   In established vegetation, which presumably is adjusted for K uptake and release, this should help make K release more uniform on an annual basis.

Calcium and Magnesium

Calcium and magnesium are contributed to river water almost entirely from rock weathering. Only 9% of Ca and 8% of Mg arise from pollution (Table 5.11) and <1% of Ca and 2% of Mg^{++} from cyclic sea salt (Table 5.10). The sources of Ca consist mainly of carbonate rocks containing calcite, $CaCO_3$, and dolomite, $CaMg(CO_3)_2$, with a lesser proportion derived from Ca-silicate minerals, chiefly calcian plagioclase, and a minor amount from $CaSO_4$ minerals. Mg-silicate minerals, chiefly amphiboles, pyroxenes, olivine, and biotite (for compositions see Chapter 4), as well as dolomite, constitute the main sources of Mg.

The relative proportions of different minerals contributing Ca and Mg to river water have been calculated by Holland (1978) and more recently by Berner, Lasaga, and Garrels (1983) and Meybeck (1984). The results of the Berner, Lasaga, and Garrels study are included in Table 5.12. As pointed out by Holland, these types of calculations rest upon a variety of assumptions having to do with such things as the area of the continents underlain by sedimentary (as opposed to igneous and metamorphic) rocks, the contribution of Ca in river water from Ca in sedimentary rocks, the rate of weathering of carbonates relative to silicates, the average Mg/Ca ratio for both carbonates and silicate rocks, and the weathering flux of sulfate from $CaSO_4$ minerals. Thus, the calculation is open to a variety of potential errors; nonetheless, we feel that the values given in Table 5.12 are the best that can be obtained from presently available data. At least there is essential agreement between the three independent studies.

The most important finding from the data of Table 5.12 is the predominance

TABLE 5.12 Sources of Ca and Mg in
World Average River Water

Source	Percent of Total Ca	Percent of Total Mg
Weathering of calcite, $CaCO_3$	52	—
Weathering of dolomite, $CaMg(CO_3)_2$	13	36
Weathering of $CaSO_4$ minerals	8	—
Weathering of Ca-silicates	18	—
Weathering of Mg-silicates	—	54
Cyclic sea salt	<<1	2
Pollution	9	8
Total	100	100

Source: Data for rock sources from Berner, Lasaga, and Garrels 1983.
Cyclic sea salt from Table 5.10 and pollution from Meybeck (1979).

of carbonate minerals (calcite and dolomite), which occur almost entirely in sedimentary rocks, as the main source (65%) of Ca^{++} in river water. Since the most abundant cation in world average river water is Ca^{++}, this result further emphasizes the importance of the weathering of sedimentary carbonate rocks to the composition of natural waters. As a crude, first-order approximation, average river water can be characterized as a $Ca(HCO_3)_2$ solution derived from the dissolution of limestone.

Bicarbonate (HCO₃⁻)

Like Ca and Mg, almost all bicarbonate in average river water is derived from rock weathering. Pollution contributes only 2% (Meybeck 1979) and cyclic sea salt far less than 1%. (Where acid rain is important, pollution can even be considered as destroying HCO_3^-—see below under acid rivers.) The weathering-derived bicarbonate, as discussed in Chapter 4, comes from two sources. One source is the carbon in carbonate minerals, such as calcite and dolomite. The other arises as a result of the reaction of carbon dioxide dissolved in soil water and groundwater with carbonate and silicate minerals. The carbon dioxide is derived almost entirely from the bacterial decomposition of soil organic matter. Two representative weathering reactions (see Chapter 4) are

$$CO_2 + H_2O + CaCO_3 \rightarrow Ca^{++} + 2HCO_3^-$$

$$2CO_2 + 11\ H_2O + 2NaAlSi_3O_8 \rightarrow 2Na^+$$
$$+ 2HCO_3^- + Al_2Si_2O_5(OH)_4 + 4H_4SiO_4$$

As these reactions demonstrate, half of the HCO_3^- resulting from carbonate weathering and all of the HCO_3^- from silicate weathering is derived from soil CO_2. It should be noted that soil CO_2 from the decomposition of organic matter was originally atmospheric CO_2 since organisms photosynthetically fix CO_2. Thus, the *ultimate* source of much of the HCO_3^- in river water is the atmosphere.

If all weathering were accomplished only by dissolved CO_2 (which reacts by way of the intermediate formation of carbonic acid), then the amount of HCO_3^- added to river water from each weathering source could be obtained by calculating that accompanying the release each of Na^+, K^+, Ca^{++}, and Mg^{++} according to the stoichiometry of reactions like the two given above. However, some silicate and carbonate weathering is also brought about by sulfuric acid formed in soils by the oxidation of pyrite (and other sulfides). Typical reactions are

$$H_2SO_4 + 2CaCO_3 \rightarrow 2Ca^{++} + 2HCO_3^- + SO_4^{--}$$

$$H_2SO_4 + 9\ H_2O + 2\ NaAlSi_3O_8 \rightarrow 2Na^+$$
$$+ SO_4^{--} + Al_2Si_2O_5(OH)_4 + 4\ H_4SiO_4$$

Note that, in this case, HCO_3^- arises only from the carbon in carbonate minerals and none comes from soil CO_2.

To take into account H_2SO_4 weathering we calculate the percentage contribution of HCO_3^- from various weathering sources as follows: from the total measured concentration of HCO_3^- (Table 5.6) is subtracted the HCO_3^-, added by weathering, which was originally contained within carbonate minerals. The latter is equivalent to the amount of Ca and Mg added by carbonate weathering (Table 5.12). The remaining HCO_3^- must be derived from soil CO_2. This HCO_3^- is then subdivided and assigned to each of Na^+, K^+, Ca^{++}, and Mg^{++} in the same proportions as these cations are given off to river water from silicate plus carbonate weathering (see Tables 5.11 and 5.12). However, the assignment is not simply to balance the charge of each cation, according to CO_2-type weathering reactions, but rather the proportioning is done so that the total HCO_3^- added from all four cations equals the remaining CO_2-derived HCO_3^- calculated above. In this way correction is automatically made for H_2SO_4 weathering. (The relative proportioning of H_2SO_4 weathering to each cation is assumed to be the same as that for CO_2-weathering.)

Results of our calculations of the sources of HCO_3^- in river water are shown in Table 5.13. Note that most HCO_3^- (64%) is derived from soil CO_2 with roughly equivalent proportions coming from carbonate and silicate weathering. Only about half as much (34%) comes from carbon originally contained within carbonate minerals. (These proportions are in excellent agreement with similar calculations by Holland [1978].) If it weren't for the fact that HCO_3^- in river water can exchange carbon with dissolved CO_2 obtained from the atmosphere, the proportions calculated here could be checked using the stable isotopes of carbon. The $^{13}C/^{12}C$ ratio of soil CO_2 is distinctly different from that of carbonate minerals, and knowledge of the $^{13}C/^{12}C$ ratio of river-water bicarbonate could, in principle, be used to calculate the relative importance of each carbon source. Unfortunately, because of isotopic exchange with the atmosphere, and consequent change in river-water $^{13}C/^{12}C$, this sort of calculation is much more difficult than it, at first, appears.

TABLE 5.13 Sources of Rock Weathering-Derived HCO_3^- in World Average River Water

Weathering Source	Percent of Total HCO_3^- from Soil CO_2	Percent of Total HCO_3^- from Carbonate Minerals
Calcite + dolomite	27	34
Ca-silicates	13	—
Mg-silicates	15	—
Na-silicates	6	—
K-silicates	3	—
Total	64	34

Note: For method of calculation see text. (An additional 2% of total HCO_3^- is added by pollution; see Table 5.11.)

What limited isotopic work has been done along these lines suggests that our calculated proportions of soil CO_2 and carbonate-derived HCO_3^- are essentially correct.

Since the advent of acid rain (see Chapter 3), one might expect that HCO_3^- in rivers might have decreased. However, most major rivers show little change in HCO_3^- concentration during the last century (Meybeck 1979; Zobrist and Stumm 1980). Decreased HCO_3^- concentrations in rivers could arise by titration of HCO_3^- by H^+ to carbonic acid (H_2CO_3), and because sulfuric and nitric acid weathering tends to replace carbonic acid weathering in affected land areas. Apparently these effects are still to be noted on a worldwide basis.

Silica

Dissolved silica in river water comes almost entirely from silicate weathering. Silicate minerals are very plentiful, but because carbonate and evaporite minerals weather much more readily, the ions resulting from silicate weathering are usually swamped by those from carbonate weathering. For example, in Figure 5.5, relatively few rivers have more than 20% dissolved Si relative to HCO_3^- and (Cl^- + SO_4^{--}) expressed in equivalents.

Meybeck (1980) states that the dissolved silica content of rivers is determined dominantly by the average temperature of the drainage basin and also by the geology of the river basin. The correlation between the mean temperature of the drainage basin and the dissolved silica concentration of rivers (expressed as mg SiO_2 per liter) is shown in Figure 5.6. For nonvolcanic rivers, arctic rivers (average temperature $<4°$ C) have a silica concentration of about 3 ± 2 mg/ℓ, temperate rivers (average temperature $4°$ to $19°$ C) have an average SiO_2 concentration of about 6 ± 3 mg/ℓ and tropical rivers (average temperature $>20°$ C) contain about 13 ± 4 mg/ℓ SiO_2. Thus the increase from the Arctic to the tropics is around four times for nonvolcanic rivers. Meybeck attributes the temperature effect on dissolved silica to the differences in silicate weathering products (clays) formed in different climatic zones. In the tropics, chemical weathering is more complete and silicate minerals tend to weather to kaolinite, which releases 1.5 times as much dissolved silica as weathering to smectite, which is more common in temperate areas. The formation of gibbsite, which is the result of even more intense tropical weathering, would release twice the amount of silica as weathering to smectite.

The geology of the river basin controls the differences in silica concentration within any temperature zone. The weathering of volcanic rocks (from recent volcanic activity) releases twice the amount of silica as that of crystalline plutonic and metamorphic rocks (see Figure 5.6) and nearly four times the amount released by sedimentary silicates. Stallard (1980) also found greater silica concentrations in igneous and metamorphic rivers than in sedimentary silicate rivers of the Amazon Basin. The reason that sedimentary silicates (sandstone and shale) release less silica is that they consist largely of detrital minerals that are basically resistant to weathering (quartz, micas, and clays). The large silica release from volcanic rocks

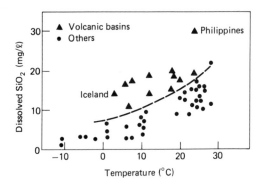

Figure 5.6 Variation in dissolved silica content of world rivers (mg/ℓ SiO₂) with average temperature of drainage basin (°C) for nonvolcanic and volcanic basins. (After Meybeck 1980.)

is due to the presence in them of easily weathered minerals, such as pyroxenes and Ca-plagioclase, and to the fact that volcanic glass, a major constituent found only in volcanics, weathers more rapidly than any silicate mineral (see Chapter 4). In addition, volcanic rocks are often porous and permeable, which speeds up weathering by enhancing the area of contact between minerals and water. Meybeck (1980) points to the Philippine rivers, which have a very high silica concentration (30 mg/ℓ), as an example of the effect of volcanic rocks on silica weathering. Because these rivers are also influenced by carbonate weathering they do not show a particularly high percentage of Si (Figure 5.6) relative to HCO_3^-, but their concentration (and dissolved load) of silica is very high due to the weathering of volcanic rocks, which is nearly as rapid as carbonate weathering. Other high-silica-content volcanic rivers include Japanese rivers (19 mg/ℓ SiO₂), New Guinea rivers (19 mg/ℓ SiO₂), and those from Iceland, which average 14 mg/ℓ SiO₂ despite their arctic location.

Silica is a biogenic element and is used by diatoms in the formation of their tests. Oceanic silica concentrations are dominated by biogenic controls, but the effect is much less strong in rivers. However, silica concentrations in rivers can be affected by the presence of diatoms in lakes along their course or in the river itself. Meybeck (1980) points out that the presence of diatoms in the Great Lakes may reduce the amount of silica that reaches the St. Lawrence River. Ming-hui, Stallard, and Edmond (1982) suggest that diatoms in reservoirs along the Huangho River in China may lower the Si concentration, and a similar diatom influence on Pine Barrens rivers of New Jesey is suggested by Yuretich et al. (1980). Dion (1983) found a strong seasonal variation in silica concentrations in the Connecticut River, which is wide with a low gradient toward its mouth. Concentrations of dissolved silica in the winter were around 6–7 mg/ℓ SiO₂ as opposed to concentrations of 4 mg/ℓ in the summer, when diatom populations were high. However, because diatom tests dissolve when the organisms die, returning dissolved silica to the river, such silica removal is not permanent. The question is how much of the silica is not returned by diatom dissolution; failure to correct for this can result in incorrect river flux data. Meybeck (1984) estimates that silica removal in lakes might amount to as much as 8% of the total river flux of silica.

Sulfate

The sources of sulfate in river water are shown in Table 5.11 (page 208). The amount of sulfate from cyclic salt (Table 5.10) is quite low (2%). The other, more significant sources, are rock weathering (33%), pollution (43%), and natural biogenically derived sulfate in rain (17%), with a minor fraction (5%) coming from volcanic activity. Rock weathering sources include the two major forms of sulfur in sedimentary rocks: sulfide sulfur in pyrite (FeS_2) and sulfate sulfur as gypsum ($CaSO_4 \cdot H_2O$) and anhydrite ($CaSO_4$).

Pyrite weathers rapidly by oxidation to sulfuric acid (H_2SO_4), which then reacts with silicate and carbonate minerals in the enclosing rock to release cations (particularlay Ca^{++}) and SO_4^{--} into river water (see section in Chapter 4 on sulfide weathering). Since pyrite occurs as a minor dispersed phase, in order to release SO_4^{--} the rock must be continually broken down. With this in mind we have calculated the percentage of pyrite-derived sulfate in river water by the following method: the ratio of pyrite-S to Ca in sedimentary rocks (times three to convert to SO_4^{--}) is multiplied by the amount of Ca in river water derived from the weathering of sedimentary rocks:

$$(SO_4^{--})_{\substack{\text{river from} \\ \text{pyrite}}} = \left[\frac{S_{\text{pyrite}}}{Ca} \right]_{\substack{\text{sedimentary} \\ \text{rocks}}} \times 3 \times (Ca^{++})_{\substack{\text{river from} \\ \text{sedimentary rocks}}}$$

In this expression we substitute S_{pyrite} = 0.3% and Ca in sedimentary rocks = 8.5% (Holland 1978). The amount of Ca^{++} in river water from the weathering of all rocks (excluding Ca from $CaSO_4$) is 13.6 mg/ℓ of which 86%, or 11.7 mg/ℓ, is derived from sedimentary rocks (Berner, Lasaga, and Garrels 1983). Then

$$(SO_4^{--})_{\text{river from pyrite}} = \left[\frac{0.3}{8.5} \right] \times 3 \times (11.7 \text{ mg}/\ell)$$

$$= 1.24 \text{ mg}/\ell$$

The amount of sulfate in river water from pyrite weathering thus amounts to 11% of the total concentration of river sulfate.

Gypsum and anhydrite are the other main rock-weathering source of sulfate in river water. They occur mainly as beds within evaporite deposits associated sometimes with halite but more often with dolomite and calcite. Gypsum should weather faster than pyrite because (1) it goes into solution faster (see Table 4.5) and (2) it occurs in discrete beds, which can be selectively removed during weathering, as compared with pyrite, which is disseminated and requires breakdown of the surrounding rock before sulfate is released. Assuming gypsum weathers twice as fast as pyrite (Berner and Raiswell 1983) and the abundance of sedimentary sulfate-sulfur (gypsum and anhydrite) is approximately equal to the abundance of sedimentary sulfide-sulfur (Holland 1978), then the percentage of river sulfate from gypsum and anhydrite weathering should be about twice that from pyrite weathering or 22%. (Note: This also gives the minor amount of Ca from $CaSO_4$ in river water used in the pyrite calculation above so that this calculation has been recycled.)

The influence of different rock weathering sources of sulfate to rivers can be seen in some examples. Since gypsum and pyrite occur in sedimentary rocks, sedimentary rock weathering tends to produce much larger sulfate concentrations in unpolluted rivers than does the weathering of igneous and metamorphic rocks. An example for Canadian rivers is a comparison of the sulfate concentration of 36 mg/ℓ in the Mackenzie River (sedimentary and igneous rocks) versus the Northwest Territory rivers (crystalline rocks) where SO_4 is 2 mg/ℓ.

Another natural source of sulfate in river water (Table 5.11) is volcanism, which yields 15 Tg SO_4/yr directly to rivers (Friend 1973) and another 6 Tg SO_4/yr to rivers via the atmosphere (Table 3.12) with the total representing 5% of total river sulfate (430 Tg SO_4/yr). The influence of volcanism on sulfate concentrations can be seen in the rivers of some volcanic islands such as New Guinea and New Zealand where sulfate is fairly high ($\sim$6 mg/ℓ SO_4) even in the absence of pollution (Meybeck 1982).

If our estimates above are correct, weathering (33%), volcanism (5%), and cyclic salt (2%) account for 40% of river sulfate. The remaining 60% or 258 Tg SO_4 must come from pollution and natural biogenic sulfur emissions to the atmosphere. In Chapter 3 we estimate that coastal and land biogenic sulfur emissions are 60 Tg SO_4/yr plus another 12 Tg SO_4 from the transport of marine biogenic sulfate. Assuming that all of this biogenic sulfate deposited on land (72 Tg SO_4) ends up in rivers, 17% of river sulfate would be from natural biogenic sulfur emissions to the atmosphere.

Pollutive sulfate in river water was estimated to amount to about 28% (or 120 Tg SO_4/yr) by Berner (1971) and Meybeck (1979). The sources of direct sulfate pollution to rivers are fertilizers, industrial and municipal wastes, and enhanced pyrite weathering from mining activities. Nriagu (1978) gives a 1975 world anthropogenic sulfate consumption of 144 Tg/yr about half from the use of sulfuric acid in the production of fertilizer and half from other industrial uses. We assume 80% of fertilizer sulfate ends up in rivers or (0.80 × 72 =) 58 Tg SO_4/yr (Möller 1984), along with most of the industrial sulfate (72 Tg SO_4/yr). Another potential river sulfate source is atmospheric sulfate production from fossil fuel burning and forest burning (216–231 Tg SO_4/yr), of which we have estimated on the order of 75% or 162–173 Tg SO_4 is deposited on the continents (see Table 3.12). Thus, the total contribution to land of anthropogenic sulfate from industry (maximum value of 72 Tg SO_4/yr) and from fossil fuel burning could amount to as much as 234–245 Tg SO_4/yr. If one assumes a large part of this ends up in rivers, along with fertilizer sulfate (58 Tg), anthropogenic river sulfate could amount to considerably more than Meybeck's (1979) estimate of 120 Tg of pollutive river sulfate. For this reason, in Table 5.11 instead of using Meybeck's estimate of river sulfate, which amounts to 28%, we have assigned the *remainder* of river sulfate, after allowing for weathering, volcanism, cyclic salt, and natural biogenic sulfate, to pollution. This results in 43% pollutive sulfate or 185 Tg SO_4. This is within the range that Husar and Husar (1985) estimate for human-derived sulfate pollution in rivers of 138–255 Tg SO_4/yr.

It may be that rock weathering makes a larger contribution to river sulfate and there is considerable uncertainty in the estimate of natural biogenic sulfur emissions (see Chapter 3). If either rock weathering or biogenic sulfur emissions are larger, then the estimate of pollutive sulfur in river water would be reduced. Since riverine sulfate pollution is so extensive and affects the pH of rivers, we shall discuss it further in detail in a separate section below.

Several authors have noted that the rain input of sulfate to the river basins that they studied seemed to be considerably larger than the river dissolved sulfate output. The explanations offered to account for the difference include transport of sulfur in river organic matter, biogenic storage on land of organic sulfur, and release back to the atmosphere of biogenic sulfur gases. In the Satilla River in Georgia, Beck, Reuter, and Perdue (1974) concluded that the rain input of sulfate is greater than the river dissolved sulfate output. Since natural organic matter contains considerable sulfur, they suspect that this river, which has high concentrations of dissolved organic matter, may export some of the excess sulfur in organic form. Cleaves, Godfrey, and Bricker (1970) found a similar excess of sulfur input in rain over river output in Pond Branch, Maryland. They attributed the difference to biomass sulfur storage. Pačes (1985) found for four basins in central Europe that the net accumulation of sulfur in the basin, which was defined as the excess of total input in precipitation, dry deposition, and fertilizer, over river output plus removal of forest or crop material, was about the *same* for all basins despite the large differences in dry deposition and fertilizer inputs between the basins. The sulfur accumulation was assumed to be in the regolith, the soil material above the bedrock including organic material, bedrock-derived material and water.

Lowland Amazon Basin rivers (Stallard and Edmond 1981) also have more rain input than river output of sulfate. The release of biogenic sulfur gases (such as H_2S, DMS, etc.) was suggested as a possible way sulfur is being lost in this area. Since Pond Branch has bogs and the Satilla is very swampy, the conditions might also be favorable for biogenic sulfur release in those areas as well. However, sulfur released to the atmosphere would be recycled to the land in rain. In summary, in terms of the overall sulfur balance on land, (1) it is possible that additional sulfur is carried by rivers as organic sulfur in addition to dissolved sulfate, and (2) sulfur may be retained as organic sulfur on land, perhaps up to some limit.

Sulfate Pollution and Acidic Rivers

Sulfuric acid rain is a source of sulfate pollution in rivers, particularly in areas such as the northeastern United States, Canada, and northern Europe. Hubbard Brook, New Hampshire (Likens et al. 1977), an example of such a river, has an unusually high sulfate concentration relative to its TDS (see Table 5.14) and is, in fact, acidic (pH = 4.9). It is a very dilute river in an area of glaciated crystalline bedrock with thin glacial till soil cover. There is a lack of carbonate rocks, and silicate

TABLE 5.14 Composition of Some Low-pH Sulfate-Rich and Organic-Rich Rivers Mentioned in Text

River	pH	Ca^{++}	Mg^{++}	Na$^+$	K$^+$	Cl$^-$	SO$_4^-$	SiO$_2$	HCO$_3^-$	TDS	DOC	Rain pH	Comments	Reference
Hubbard Brook, N.H.	4.9	1.7	0.4	0.9	0.3	0.55	6.3	4.5	0.9	19	1.0	4.1	Sulfuric acid rain weathering	Likens et al. 1977
Pond Br., Maryland	6.7	1.4	0.8	1.7	0.9	2.1	1.3	9.3	7.7	25	—	4.6	Sulfuric acid rain weathering	Cleaves, Godfrey, and Bricker 1970
X-14, Elbe Basin, Czech.	4.9	17.2	6.0	5.2	1.6	3.7	66.8	16.4	0	108.		4.2 (3.2)[b]	Sulfuric acid rain weathering	Pačes 1985
Ichilo R., Amazon Basin	5.28	4.4	2.0	2.4	0.9	0.2	23.7	8.3	0.6	44	—	4.8–5.0	Natural pyrite weathering	Stallard 1980
Moshannon R., Pa.	2.9	44	21	4	1	10	300	7	0	387	—	acid	Mine drainage pyrite weathering	Lewis 1976
Satilla R., Ga.	4.3	1.3	0.7	4.1	0.8	5.4	1.2	6.85	—	20	24	acid	Organic-dominated	Beck, Reuter, and Perdue 1974
Pine Barrens, N.J.	4.5	1.1	0.6	2.7	0.6	4.7	6.4	4.3	—	20	2.2	4.4	Organic + sulfuric acid	Yuretich et al. 1981
U. Negro (above Branco), Amazon Basin	4.64	0.4	0.06	0.3	0.26	0.25	0.19	3.4	—	5	12	4.8–5.0	Organic-dominated	Stallard 1980; DOC-Leenheer 1980
Negro R. (above Manaus), Amazon Basin	5.36	0.17	0.16	0.4	0.24	0.24	0.15	4.3	0.55	6	10	4.8–5.0	Organic-dominated	Stallard 1980; DOC-Leenheer 1980
Matari R., Amazon Basin	4.7	0.14	0.13	0.67	0.15	1.14	0.125	2.4	—	5	(5)[a]	4.8–5.0	Organic and marine rain	Stallard 1980

Concentration (mg/ℓ)

[a] (DOC) is rough estimate based on color measurements.

[b] Effective pH including dry deposition.

weathering is insufficient to neutralize the sulfuric acid rain (pH 4.1). However, sulfuric acid silicate weathering does result in, in addition to sulfate, a relatively high concentration of dissolved silica and Ca^{++} in the river water (see Table 5.14) and also dissolved Al. (We shall discuss this type of water further in the section on acid lakes in Chapter 6.) Pond Branch, Maryland (Cleaves, Godfrey, and Bricker 1970), another dilute river in a crystalline basin, also receives sulfuric acid rain (pH 4.6), but here there is a thick soil cover and sufficient silicate weathering in the basin to neutralize the rain, and the river, as a result, has a pH of 6.7. This river has a high silica concentration and considerable Mg^{++}, K^+ and SO_4^{--} over sea-salt contributions as a result of weathering by the acid rain (see Table 5.14).

On a larger scale, trends in concentration of streamwater sulfate, alkalinity (HCO_3^-), and pH for the period of 1965 to 1980 have been studied at 47 U.S.G.S. Hydrologic Benchmark stations in undeveloped stream basins all over the United States (Smith and Alexander, 1983). In the northeastern United States as a whole, small decreases in stream sulfate (and increases in alkalinity) occurred at streams in areas where SO_2 emissions have been reduced over this period, and small increases in sulfate (and decreases in alkalinity) occurred at southeastern and western sites where SO_2 emissions have increased. This tends to confirm the suspicion that fossil fuel SO_2 and the resulting sulfuric acid rain contribute to stream sulfate and acidity. Stream sulfate and bicarbonate concentrations tend to be inversely correlated if sulfuric acid acts to neutralize stream HCO_3^- rather than reacting with minerals in the stream basin; this relationship is strongest at low alkalinity stations. The ratio of HCO_3^- to major cation concentration (sum of Na, K, Ca, and Mg) is a sensitive indicator of acidification and declines with increased acidity in the stream basin because it is a measure of the relative importance of weathering by H_2CO_3 to that by H_2SO_4 (and HNO_3). The ratio of HCO_3^- to major cation concentration has declined at most streams west of the Mississippi River, and increased at most northestern U.S. stations approximately inversely to stream sulfate changes. Changes in the pH of these streams do not always follow sulfate changes, however, because in some cases acidification is due to HNO_3 rather than H_2SO_4 (Lewis and Grant 1979).

Sulfur in European rivers is also strongly influenced by sulfuric acid rain and dry fallout. For example, Odén and Ahl (1978) in a study of Swedish rivers, estimate that 65% of the sulfate-sulfur in Swedish rivers comes from anthropogenic sulfur in rain and dry fallout with minor pollution from fertilizer sulfur (7%). In the central European Elbe River Basin, Pačes (1985) studied an acidic river (pH 4.9) draining a forested mountain basin where part of the conifer forest has died and which faces an industrial area. The mountain basin receives acid rain (pH 4.2), but more importantly, because the air concentration of SO_2 is very high, about ten times as much acid comes from SO_2 dry deposition as from the rain. Thus, the combined effects of dry deposition and acid rain would be equal to pH 3.2 rain. The river has very high sulfate concentrations and high Ca and silica in addition to being acidic.

Natural weathering of pyrite-rich black shales can produce high sulfate concentrations and quite acid streamwater because the pyrite is oxidized to sulfuric acid (see Chapter 4). An extreme example is the Ichilo River in the Madeira River drainage of the Amazon Basin (Stallard 1980) (see Table 5.14 and Figure 5.5) which has a very high sulfate concentration (on a mole basis more than double the Ca concentration), and a pH of 5.28. In coal-mining areas, such natural pyrite weathering is greatly accelerated by the exposure of relatively large amounts of pyrite in the coal and increased water circulation due to mining. If the groundwater circulating through the pyrite-rich coal beds does not contain sufficient HCO_3^- (from previous carbonate weathering) to neutralize the sulfuric acid being produced, extremely acid water (pH < 3) results, which can seep into nearby streams. This acid production via pyrite oxidation is greatly accelerated by bacteria (e.g., Kleinmann and Crerar 1979).

Moshannon Creek (see Table 5.14), a tributary of the West Branch of the Susquehanna River in the coal-mining area of Pennsylvania, is an example of a mine drainage river with a very high sulfate concentration (300 ppm) and low pH (2.9) (Lewis 1976). The concentrations of Ca (44 ppm) and Mg (21 ppm) are also very high, presumably because of the H_2SO_4 weathering of any available Ca and Mg carbonate or silicate minerals. (The silica concentration is also somewhat higher than usual for rivers in the area.) Because of the addition of these mine drainage waters, the main river (West Branch of the Susquehanna) also develops a low pH (4.0–4.5) for considerable distances where it flows through sedimentary silicate rocks, which are low in carbonate minerals and provide little HCO_3^- to neutralize the sulfuric acid. The West Branch of the Susquehanna River achieves neutrality (pH ≃ 7.0) only after tributaries which drain carbonates join the main river (Lewis 1976).

Organic Matter

Organic matter in rivers is present both in the dissolved and particulate forms. The concentration of dissolved organic matter in rivers is usually expressed in terms of dissolved organic carbon (DOC). The average amount of dissolved organic carbon in rivers is about 6 mg/ℓ but there is considerable variation dependent upon climatic conditions (Meybeck 1982). High median concentrations of DOC are found in subarctic rivers (such as the Mackenzie), which average 10 mg/ℓ DOC with some values up to 20 mg/ℓ, and in humid tropical rivers (such as the Amazon), which have about 6 mg/ℓ DOC. Low concentrations are found in temperate rivers (3 mg/ℓ DOC) and in arctic rivers (2 mg/ℓ DOC) with the lowest concentrations (<1 mg/ℓ DOC) in mountainous alpine rivers (such as those in the French Alps or New Zealand). Rivers draining swampy areas (such as the Satilla River in Georgia, which we shall discuss below) have the highest concentrations of all (~25 mg/ℓ DOC).

In addition to dissolved organic carbon, there is also considerable transport of organic matter in the form of particulate organic matter (expressed as POC or particulate organic carbon). Generally, dissolved organic carbon makes up about

60% of the total load of organic carbon (TOC) (Meybeck 1982), but as the river suspended load increases, the particulate organic carbon concentration becomes greater relative to dissolved organic carbon.

Because the average river ratio of dissolved organic carbon to total (inorganic) dissolved solids is low (DOC:TDS = 1:18), reactions between organic matter and major inorganic ions do not have a large relative effect on the overall chemistry of most rivers. (By contrast, the behavior of trace metal ions is strongly affected.) However, in rivers with low total dissolved solids and large DOC concentrations, organic matter can dominate the river chemistry (see Table 5.14). An example is the Coastal Plain rivers of southeastern Georgia, particularly the Satilla River, which drains a swampy area, has a low pH (4.3), and a ratio of dissolved organic carbon (24 mg/ℓ) to total dissolved (inorganic) solids (20 mg/ℓ) of roughly 1:1 (Beck, Reuter, and Perdue 1974). As Beck, Reuter, and Perdue point out, it is not the absolute amount of dissolved organic carbon in the Satilla (which, however, is very high) that is so important, but rather its very high ratio to dissolved inorganic solids.

The dissolved organic matter, which dominates the chemistry of organic-rich rivers such as the Satilla, consists of humic material, that is, humic and fulvic acids. Both substances are mixtures of complex (and poorly understood) high-molecular-weight organic polymers, which contain carboxyl groups (COOH) and phenolic groups (OH). (Humic acids are defined as being insoluble in strong acid while fulvic acids are acid-soluble.) The low pH of the Satilla River (3.8–5.0) results from the dissociation of the humic and fulvic acidic carboxyl groups (COOH). The organic acids (R—COOH) lose a hydrogen ion and become (R—COO)$^-$ with a net negative charge:

$$\underset{\substack{\text{organic} \\ \text{acid}}}{(\text{R---COOH})} \;\rightarrow\; \underset{\substack{\text{organic} \\ \text{anion}}}{(\text{R---COO})^-} \;+\; \text{H}^+$$

The river has little or no bicarbonate $(\text{HCO}_3)^-$ because any bicarbonate is used up in neutralizing the H$^+$ from the organic acid dissociation:

$$\text{HCO}_3^- + \text{H}^+ \rightarrow \text{H}_2\text{O} + \text{CO}_2 \uparrow$$

Thus, the overall reaction of HCO_3^- with organic acids is

$$(\text{R---COOH}) + \text{HCO}_3^- \rightarrow (\text{R---COO})^- + \text{H}_2\text{O} + \text{CO}_2 \uparrow$$

As more HCO_3^- ions are added by carbonate-draining tributaries downstream, the hydrogen ions produced by organic acid dissociation are neutralized and the river pH goes up, producing large concentrations of organic anions.

This leads to the other major characteristic of rivers such as the Satilla: the sum of the charge on the major inorganic cations (microequivalents per liter of Na$^+$, Mg^{++}, K$^+$, Ca^{++}, H$^+$) is greater than the sum of the charge on the major inorganic anions (microequivalents of Cl^-, SO_4^{--}, HCO_3^-, NO_3^-), leading to an apparent deficiency of inorganic anion charge. However, since there must be electrical balance in the river, the excess inorganic cation charge is balanced by the organic anions, which result, as discussed above, from the dissociation of organic acid carboxyl groups. Thus, the overall charge balance in an organic-dominated river is

Sum of inorganic cations = Sum of inorganic anions + Sum of organic anions

$(Ca^{++}, Mg^{++}, Na^+, K^+, H^+$ in $\mu Eq/\ell)$

$\quad = (Cl^-, SO_4^{--}, HCO_3^-, NO_3^-, PO_4^{---}$ + Sum of organic anions in $\mu Eq/\ell)$

Another characteristic of organic-rich rivers is their large concentration of "dissolved" iron and aluminum relative to other rivers. Fe and Al form dissolved organic complexes or colloidal oxyhydroxides mixed with organic matter, and this leads to the mobilization and transport of Fe and Al, which are otherwise immobile and insoluble (Beck, Reuter, and Perdue 1974).

The major characteristics of the organic-dominated Satilla River (Beck, Reuter, and Perdue 1974), which might be used as criteria in identifying other rivers whose chemistry is dominated by organic matter, are as follows: (1) the ratio of the concentrations of dissolved organic matter (DOC) to dissolved inorganic matter (TDS) is high (here about 1:1); (2) there is an excess of total inorganic cation charge relative to total inorganic anion charge that is presumably balanced by organic anion charge; and (3) the river tends to be acid (although acid rivers are not necessarily organic).

Beck, Reuter, and Perdue suggest that some of the Amazon tributaries are chemically dominated by organic matter. The Negro River in the Amazon Basin, named for its typically organic-rich black color, is a likely candidate. Stallard (1980) gives the concentrations of major dissolved inorganic ions for the Upper Negro River (pH 4.6–4.8) and the Negro River (pH 4.95–5.4) (see Table 5.14). Both of these rivers have considerably greater inorganic cation charge than inorganic anion charge. The DOC concentration in the Upper Negro (Leenheer 1980) is 12 mg/ℓ and TDS is ~5 mg/ℓ. Thus, the ratio of DOC:TDS in the Upper Negro River is 2.4:1, fitting the criteria for an organic-dominated river. Similarly, the DOC:TDS ratio for the Negro River is 1.7:1. The Matari River, another Amazon tributory, with low TDS (4.8) and low pH (4.7) also has considerably greater inorganic cation charge than inorganic anion charge, and its color (measured by Stallard 1980) suggests a large concentration of organic matter.

There has been considerable discussion about whether some of the effects in rivers and lakes attributed to pollutive sulfuric (and nitric) acid rain might rather be natural due to high concentrations of organic matter (e.g., Krug and Frink 1983). The criteria set forward above for recognizing organic rivers can be used to test whether or not a given acidic river or lake is naturally acidic due to high DOC content. For example, Hubbard Brook (which we discussed above as being affected by sulfuric acid rain) has a DOC:TDS ratio of 1:20 (less than world average river water) and very close to a balance between inorganic cation charge and inorganic anion charge, and thus would not seem to be organic-controlled. This agrees with the general observation that in subalpine northeastern forests, 75% of the charge balance in soil and groundwater is comprised of sulfate (from acid rain) and not organic anions (Cronan et al. 1978).

The Pine Barrens rivers in New Jersey (Yuretich et al. 1981; Crerar et al. 1981) have a low pH (4.5), brown water, a fairly high concentration of organic matter (2.2 mg/ℓ), a low TDS (20 mg/ℓ), and a high concentration of SO_4 (6.4 mg/ℓ). Using the criteria established above for control of river chemistry by organic matter: (1) the ratio of DOC:TDS is 1:10, which is about twice as organic-rich as world average river water but nowhere near the 1:1 ratio of organic-rich rivers discussed above, and (2) the sum of the inorganic cations is nearly balanced by the sum of the inorganic anions, suggesting that organic anions are not important in charge balance. Based on these criteria, one would suspect that, at present, the low pH of the Pine Barrens rivers may be due more to very high sulfate concentrations than to the presence of natural dissolved organic matter. This agrees with the work of Johnson (1978), who finds an increase in the acidity of several Pine Barrens streams from 1963 to 1978 which he attributes to acid rain. However, there is reason to suspect that the groundwater, and possibly the river, were already moderately acid, due to organic acids, before the advent of acid rain, as attested to by the presence of bog iron ores. (The iron most likely was transported in association with organic matter.)

CHEMICAL AND TOTAL DENUDATION OF THE CONTINENTS AS DEDUCED FROM RIVER-WATER COMPOSITION

Using the total concentration of dissolved ions in river water one can calculate the chemical denudation rate of the continents, expressed as the mass of dissolved material removed per unit area per unit time, by multiplying the total concentration of the ions times the water discharge and dividing this by the drainage area. This *chemical* denudation rate tends to increase with increasing runoff (Meybeck 1980; Holland 1978). This is shown in Figure 5.7 (after Meybeck 1980). Although the total concentration of dissolved ions decreases with increasing runoff due to dilution, this effect on the total load is more than compensated by the greater volume of water being carried.

The geology (rock composition and weathering history) of the river basin has an important and probably dominant effect on the chemical denudation rate. As shown in Figure 5.7, the total dissolved river load per unit area in sedimentary (carbonate and evaporite) basins is five times greater than that from crystalline (igneous and metamorphic) rocks and 2.5 times greater than that from recent volcanic rocks (Meybeck 1980). Mixed-source rivers (crystalline plus sedimentary) plot in between rivers from the single rock types, but generally show more sedimentary rock influence. The influence of sedimentary rocks is due mainly to the

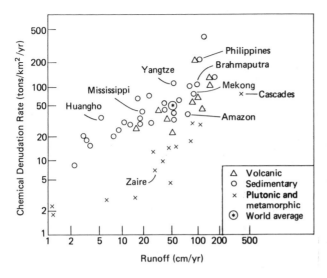

Figure 5.7 Influence of rock composition on total dissolved load per unit area (chemical denudation rate) versus runoff per unit area for major world rivers and some small basins. (Adapted from Meybeck 1980.) Certain major rivers discussed in the text are also included. (Additional data from Ming-hui, Stallard, and Edmond 1982 and Reynolds and Johnson 1972.)

presence of carbonates and evaporites. This is particularly true since HCO_3^- from carbonates comprises one-third of the dissolved load.

Relief does not correlate with the chemical denudation rate on a continental scale as it does with the suspended sediment load (Garrels and Mackenzie 1971; Hay and Southam 1977; Holland 1978). However, on a smaller scale, the chemical denudation rate tends to be higher in mountainous basins, but this may partly be an artifact. As pointed out by Meybeck (1980), there are two problems: (1) Often sedimentary rocks occur in mountainous areas and crystalline rocks in lowlands; some examples are the Amazon Basin, Africa, and the Himalayan drainage area. In addition, volcanic rocks, which weather easily, occur in mountainous Pacific islands. (2) Mountainous areas tend to have high runoff, which results in a large dissolved transport rate. In summary, the dominant influences on the chemical denudation rate are probably geology and runoff rather than relief.

Ming-hui, Stallard, and Edmond (1982) and Meybeck (1980) have generalized on chemical denudation rates for various river types combining several influences. From their data, one can see the important influence of lithology. The highest chemical denudation rates are for rivers with high runoff and in areas of high relief but draining a dominantly carbonate and evaporite terrain (such as the Yangtze and Brahmaputra rivers, whose load is around 100 tons/km²/yr) or rivers whose drainage includes recent volcanic rocks (such as the Philippine rivers and New Guinea rivers with denudation rates of 250 tons/km²/yr and Japanese rivers with 185 tons/km²/yr). By contrast, the Cascades rivers (Reynolds and Johnson 1972) in Washington State, which have high relief, extremely high runoff (409 cm/yr) and glaciation, whose mechanical effects should expedite chemical weathering, have a lower chemical denudation rate (66 tons/km²) because they drain a plutonic-metamorphic terrain. Rivers with a chemical denudation rate close to the world average value (37 tons/km²/yr) are of two types: (1) those draining arid (low runoff) mountainous areas with sedimentary rocks such as the Hwangho, which has a very large dissolved load for its runoff; and (2) those with high runoff but draining extensive lowland areas such as the Mississippi River and the Amazon. (The lower basin of the Amazon contains large shield and sedimentary silicate areas while the upper Amazon drains carbonates and evaporites in the Andes.) Rivers with a very low chemical denudation rate are those draining predominantly crystalline shield terrains such as the Zaire. These different rivers are plotted on Figure 5.7 and tend to follow the trend one might expect for their rock type and runoff.

We have noted that the world average chemical denudation rate is 37 tons/ km^2/yr. This is based on the average "natural" concentration of total dissolved solids (TDS) in river water (100 mg/ℓ) multiplied by the total world water discharge (37,400 km^3/yr) and divided by the area of *external* drainage (100 $\times$ 10^6 km^2) (i.e., drainage to the oceans) (after Meybeck, 1979). However, this is not really the rate at which dissolved ions are being removed from the continents because the value includes ions derived from rainfall and the atmosphere (Holland 1978; Meybeck 1979). Specifically, 64% of the total HCO_3 load in rivers comes from atmospheric CO_2, not carbonate rocks (see Table 5.13). This results in a reduction of 34% in the total dissolved load (or 12.7 tons/km^2/yr). Also correction must be made for the ions in river water that come from atmospheric precipitation—about 4.5% of the total (1.7 tons/km^2). Thus, the corrected chemical denudation rate is about *23 tons/km^2/yr* or about two-thirds of that derived from the total dissolved load. This is the rate at which continental ions are being removed to the oceans by river-dissolved transport.

The suspended sediment or mechanical denudation rate is about 152 tons/ km^2/yr (see previous discussion). Thus, the combined mechanical and chemical denudation rate for the continents amounts to 175 tons/km^2/yr (or 15,850 Tg/yr), of which chemical denudation is only about 15%. The total denudation rate can be translated to a reduction of continental elevation of about 6.5 cm per 1000 years (assuming an average rock density of 2.7). This is less than the actual erosion rate as we noted earlier.

The present dominance of mechanical denudation over chemical denudation may not always have been the case in the geological past. First of all, as we have pointed out earlier in this chapter, the prehuman mechanical erosion rate was probably half that at present. Also, the suspended load is highly dependent upon relief while the dissolved load is not. Thus, at present, when the continents are fairly high, mechanical erosion dominates. With lower continental elevation (as may have been true at various times in the past), chemical erosion could have been more important (Holland 1978). A simple comparison between the percent of the total load that is dissolved for the Congo and Ganges-Brahmaputra rivers illustrates variations in the importance of the dissolved load. The combined Ganges and Brahmaputra rivers are in an area of high relief containing easily dissolved carbonate and evaporite rocks, but they have a dissolved load which is less than 10% of its total load. On the other hand, the Zaire River, draining an area of low relief and weathered crystalline rocks, has a dissolved load which is nearly 45% of its total load.

NUTRIENTS IN RIVER WATER

The two major nutrients in river water that will be discussed here are nitrogen and phosphorus. Since several aspects of the discussion of these elements overlap other water types, the reader is referred to further treatment of nitrogen and phosphorus under the subjects of rainwater and the atmosphere (Chapter 3), lakes (Chapter 6), and estuaries (Chapter 7).

Nitrogen in Rivers: The Terrestrial Nitrogen Cycle

The origin of nitrogen in river water is considerably more complex than it is for most other elements because nitrogen exists in solution in several different forms, is a major constituent of the atmosphere, and is intimately involved in biogeochemical cycling as an essential component of living tissue, both plant and animal. The subject of river-water nitrogen involves study of a wide variety of nitrogen sources which logically leads to study of the *terrestrial nitrogen cycle*. A qualitative diagrammatic representation of this cycle is shown in Figure 5.8. We have discussed the atmosphere-soil part of the terrestrial nitrogen cycle in detail in Chapter 3 so we include only a brief review of the relevant parts here. Nitrogen gas, N_2, which is a dominant part of the atmosphere, is not normally biologically available and must be "fixed" or combined with hydrogen, oxygen, and carbon, in order to be used by plants and organisms on land. Looking at the terrestrial nitrogen cycle in more detail (Table 5.15), we see that there are three major land inputs of fixed nitrogen in forms such as NO_3^- and NH_4^+. These include biological fixation, precipitation and dry deposition of previously fixed nitrogen, and the application of fertilizers (industrially fixed nitrogen).

Biological fixation is the dominant source of fixed nitrogen on land ($\sim60\%$). This is accomplished by microorganisms living symbiotically in higher plants (particularly legumes) and lichens in trees. Overall, humans, by planting crops such as legumes and rice, are responsible for about 44 Tg N/yr out of the total of 139 Tg N/yr which is fixed biologically (Burns and Hardy 1975:55).

The second most important source of terrestrial fixed nitrogen is precipitation and dry deposition (24%). Nitrogen in precipitation is in the forms NO_3^-, NH_4^+, and DON (dissolved organic nitrogen). We estimate, of the total fixed nitrogen delivered to land in precipitation and dry deposition (57 Tg N), on the order of 38 Tg N is due to human influences (90% of NH_4^+-N and 76% of NO_3^--N; see Chapter 3).

The third major terrestrial nitrogen source (17%) is the use of fertilizer, which represents N_2 industrially fixed as NO_3^- and NH_4^+. Humans produce about 40

TERRESTRIAL N CYCLE

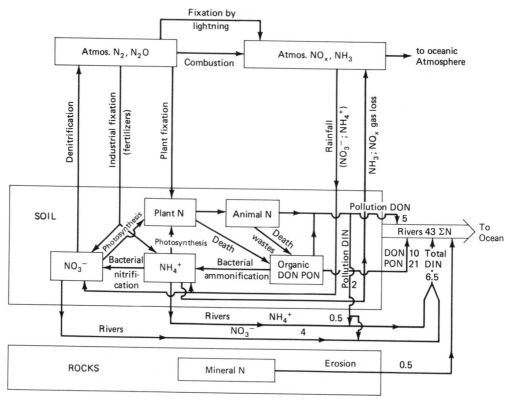

Figure 5.8 The terrestrial nitrogen cycle. Numbers adjacent to arrows represent fluxes in Tg N/yr; (data from Table 5.16). DON = dissolved organic nitrogen; PON = particulate organic nitrogen; DIN = dissolved inorganic nitrogen; Σ N = total nitrogen.

Tg N/yr as fertilizer, and since fertilizer use is growing by 10% a year (Simpson 1977), the human contribution to nitrogen fixation will increase in the future. Overall (Table 5.15), human influences are probably responsible for about 50% of the total fixed nitrogen input to land (polluted rain plus fertilizers).

Once it reaches the land, nitrogen is involved in various transformations in the terrestrial nitrogen cycle (see Figure 5.8). As we have noted, some nitrogen is incorporated in organic matter directly from the atmosphere by fixation in plants. Plants also convert dissolved NO_3^-, and NH_4^+ (from fertilizer, rain, or recycling of organic matter) into plant organic matter, some of which is eaten by animals and becomes animal organic nitrogen. The total amount of nitrogen converted into organic matter by photosynthesis (460 Tg N/yr) is much larger than the yearly input of nitrogen (~236 Tg N/yr); thus, a large amount of nitrogen is recycled within the biosphere.

TABLE 5.15 Terrestrial Nitrogen Cycle

Process	Total Flux (Tg N/yr)	Percent of Total Input or Output	Anthropogenic Flux (Tg N/yr)	Reference
Land Input				
Biological fixation	139	59	44	Burns and Hardy 1975
Fertilizers	40	17	40	Crutzen 1976
Precipitation and dry deposition	57	24	38	(See Chapter 3, Table 3.15)
Total input	236	100	122	
Land Output				
River N	43	18	7	Meybeck 1982
Denitrification to N_2 and N_2O	153	65	?	To balance (see text)
NH_3 gas loss	23	10	?	(See Chapter 3)
NO_x: soil gas loss and forest burning	17	7	?	(See Chapter 3)
Total output	236	100	?	

Note: Tg $= 10^6$ metric tons $= 10^{12}$ g.

When plants and animals die, their organic matter is broken down by bacteria into ammonia, some of which is dissolved in soil water as NH_4^+, and some of which escapes from the soil as NH_3 gas. Bacteria can also convert NH_4^+ into NO_2^- and NO_3^- in the soil (*nitrification*). Most of the NH_4^+ and NO_3^- from organic matter decay is recycled by plants. However, nitrogen can be lost from the land either directly to the atmosphere in the form of nitrogen containing gases (as we have seen in Chapter 3) or in river water (our primary concern here). Some 82% of the total nitrogen output from land is gaseous (Table 5.15); denitrification breaks down soil nitrate to release N_2 (the major gaseous output) and N_2O, and lesser amounts of NH_3 and NO_x gas are also released at various stages of the nitrogen cycle.

River output of nitrogen has been recently estimated by Meybeck (1982) as 43 Tg N/yr (see Table 5.16). This amounts to only about 18% of the total nitrogen loss from the land (see Table 5.15). There are three types of river nitrogen: (1) dissolved inorganic nitrogen (ammonium, nitrate, and lesser amounts of nitrite) (6.5 Tg N/yr); (2) organic nitrogen, as dissolved organic nitrogen (15 Tg N/yr) and particulate organic nitrogen (21 Tg N/yr) from soil organic matter that has not been completely broken down; and (3) direct land erosion of old minerals containing nitrogen (0.5 Tg N/yr) (Meybeck 1982).

The importance of biological processes to river-borne nitrogen is demonstrated by the fact that 85% of river nitrogen is organic N and most of the rest

TABLE 5.16 River Nitrogen Transport (in Tg N/yr)

Source of River Nitrogen	Dissolved		Particulate	Total
	Natural	Pollutional		
Erosion of previous minerals	—	—	0.5	0.5
Dissolved Inorganic N				
$\quad$ NO$_3$-N	4.0 $\Big\}$	2.0[a]	—	6.5
$\quad$ NH$_4$-N	0.5			
$\quad\quad$ Total dissolved inorganic N	4.5	2.0	—	6.5
Organic N				
Dissolved organic N (DON)	10	5.0[a]	—	15.0
Particulate organic N (PON)	—	—	21	21.0
$\quad\quad$ Total organic N	10	5.0	21.0	36.0
$\quad\quad$ Total river nitrogen	14.5	7.0	21.5	43.0

Source: After M. Meybeck. "Carbon, Nitrogen, and Phosphorus Transport by World Rivers,"
American Journal of Science, 282, p. 425. © 1982 by the Amer. J. of Science, reprinted by permission
of the publisher.

[a] Total dissolved pollutional nitrogen of 7 Tg N (Meybeck, 1982) has been divided between
dissolved inorganic N and DON in the same ratio as dissolved natural nitrogen.

(inorganic NO$_3^-$ and NH$_4^+$) is derived from organic matter decomposition (see
Table 5.16).   However, the total river output of organic nitrogen amounts to only
8% of the amount of nitrogen assimilated annually by the terrestrial biosphere
through photosynthesis (460 Tg N/yr).   This shows that biological recycling is very
efficient.

Human activities have influenced the river nitrogen load considerably. Meybeck (1982) estimates that, of the total *dissolved* nitrogen (both organic and inorganic) in river water, some 7 Tg N/yr, or about one-third is pollutional in origin
(see Table 5.16).   (An undetermined part of the particulate nitrogen is also due
to pollution.)   Since we estimate above that the total pollutional input of fixed
nitrogen to the land is about 122 Tg N/yr, the value of 7 Tg N/yr shows that only
5% of this pollutional input is lost as dissolved nitrogen in river water.

The nitrogen cycle shown in Table 5.15 simply assumes that the yearly river
and gaseous output of nitrogen from land equals the input to the land.   In other
words, we balance the terrestrial cycle by assuming that *denitrification* (release of
N$_2$ and N$_2$O gas), makes up the difference between input and the total known
output in rivers plus NH$_3$ and NO$_x$ gas loss.   The flux of N$_2$ to the atmosphere is
poorly known, and any imbalance due to changes in the amount of N$_2$ added to
or subtracted from the atmosphere would not be apparent, because of the large
mass and turnover time of the N$_2$.   It is possible, therefore, that the terrestrial
nitrogen cycle is *not balanced* and that fixed nitrogen is building up in soils, ground-

water, rivers, and lakes because of a large pollutional input (Delwiche 1970). The National Research Council (1972), based on an input-output budget, crudely estimates that nitrogen is being stored in U.S. soil and water (the amount being 20% of the U.S. fertilizer input at that time). However, the evidence is not good enough to say with any certainty whether the terrestrial nitrogen cycle is balanced by gas release (making up for pollutional nitrogen increases), or whether nitrogen is actually being stored in terrestrial soils and waters.

The type of nitrogen found in polluted waters varies. Nitrate is more important than ammonium in most polluted waters (see Table 5.16), but in poorly oxygenated rivers resulting from excessive organic matter loading, ammonium may reach 80% of the total dissolved inorganic nitrogen. (Nitrite, NO_2^-, is much less important.) Urban wastes are higher in ammonium, and agricultural runoff is higher in nitrate as are combustion products in rain. Europe and the United States are the biggest polluters, contributing 70% of the total pollutive nitrogen in world river water (Meybeck 1982).

The anthropogenic sources of nitrogen in rivers include (1) so-called point sources, which are discharged directly into surface waters, such as municipal and industrial sewage, septic tanks, refuse dumps, and animal feedlots; (2) diffuse sources, which result from runoff and leaching from rural and urban land; and (3) precipitation directly to lakes and streams (National Research Council, 1972).

Municipal and industrial wastes, because they are discharged directly into the rivers, can produce large local increases of river nitrogen particularly in urban areas. However, since sewage can be treated to remove around 40% of the nitrogen, this nitrogen source is easier to control than diffuse sources. Diffuse sources include natural leaching of nitrate from soil (a process which humans have accelerated by deforestation and cultivation), rainfall (see Chapter 3), and runoff from agricultural land. Nitrogen from agricultural land arises from application of nitrogen fertilizers, from animal wastes, and from the cultivation of plants such as legumes which are nitrogen fixers. An idea of the relative importance of various sources of pollutional nitrogen in European rivers is given (at the end of the next section) in Table 5.18.

Phosphorus in Rivers: The Terrestrial Phosphorus Cycle

Phosphorus, unlike nitrogen (or sulfur), has no stable gaseous phases in the atmosphere. For this reason, in contrast to nitrogen, most phosphorus is lost from land by way of river runoff, and a considerably smaller proportion, on the average, of land input is provided by precipitation. The dominant feature of the terrestrial phosphorus cycle is the fact that phosphorus is an important and often limiting nutrient. The amount of phosphorus incorporated into organic matter each year is much greater than its production by weathering or its loss by rivers. The deficit is made up by biologically recycled material. Phosphorus tends to be strongly conserved by biological systems so that most phosphorus released by organic decay is rapidly converted into organic matter.

The major ultimate source of phosphorus is weathering; that is, the removal of phosphorus from rocks. However, phosphorus in rocks and sediments is in a relatively insoluble form as the calcium phosphate mineral, apatite. Even when released as soluble phosphate (PO_4^{---}) by weathering, phosphorus is usually quickly tied up in the soil as iron, aluminum, and calcium phosphates or by clay minerals to produce insoluble forms not accessible to plants. Because of the relative insolubility of phosphorus, it is often a limiting nutrient in biological systems; that is, it is in short supply. Humans have intervened in the phosphorus cycle by deforestation, which increases erosion; by the use of phosphorus fertilizers; and through the production of industrial wastes, sewage, and detergents. (For more details on the anthropogenic phosphorus cycle see Stumm 1972 and Pierrou 1976.) Particularly in lakes, the introduction of greatly increased phosphorus has stimulated productivity, and led to eutrophication (see Chapter 6). Here, however, we shall focus on river runoff from the land as it relates to the overall terrestrial phosphorus cycle.

Phosphorus concentrations in rainfall are small ($0.01-0.04$ mg/ℓ total P) and difficult to measure due to contamination. The amount of phosphorus delivered to land in precipitation has been estimated at 1 Tg P/yr (Meybeck 1982) with total P deposition (including dry deposition) being 3.2 Tg P/yr (Graham and Duce 1979). This is of the same order of magnitude as the *dissolved* river transport (2 Tg P/yr) (see Table 5.17). In fact, in some remote areas, such as Hubbard Brook, New Hampshire (Likens et al. 1977), precipitation input appears to be greater than dissolved river output.

The main source of phosphorus deposited on the continents, via precipitation and dry deposition, is soil dust (3.0 Tg P/yr or 94% of total deposition; Graham and Duce 1979). (Other, minor phosphorus sources in rain include industry and combustion, sea salt, and biogenic aerosols.) However, since atmospheric soil dust is derived from weathering on land (and is thus recycled), we have not considered rain and dry deposition of soil dust phosphorus as primary inputs to the terrestrial system (Table 5.17).

The total amounts of suspended particulate phosphorus (inorganic and organic) transported by rivers at present is approximately 20 Tg P/yr (Meybeck 1982). Part of this is from the erosion of phosphate minerals in rocks, primarily apatite, and much of the rest is from human effects—increased soil erosion and transport due to deforestation and agriculture, and increased organic matter removal due to agriculture and fertilization. Froelich et al. (1982) estimate that the preagricultural river phosphorus flux due to weathering was about 10 Tg P/yr (based on a preagricultural continental denudation rate of 10^4 Tg/yr [Judson 1968; Gregor 1970]; and 0.1% P in the earth's crust). Using our preagricultural denudation rate of 7.1 $\times 10^3$ Tg/yr (see earlier in this chapter) would give a value of 7.1 Tg P/yr for the preagricultural mechanical weathering flux. The present P suspended load from weathering likewise is estimated as 14.0 Tg/yr, based on a total suspended load of

TABLE 5.17 Terrestrial Phosphorus Fluxes (in Tg P/yr)

Source	Total Flux	Anthropogenic Part	Reference
P Inputs to Land			
Rock weathering	14.0	6.9	Calculated from total suspended load (see text)
Mining of phosphate rock (1974) for fertilizers and industry	12.6	12.6	Pierrou 1976 (see text)
Rain and dry deposition (nonsoil dust)	0.2	0.2	(See below)
Total input	26.8	19.7	
P Outputs from Land (River Runoff)			
Dissolved PO_4-P	0.8	0.4	Meybeck 1982
Dissolved organic-P[a]	1.2	0.6	Meybeck 1982
Total dissolved P	2.0	1.0	Meybeck 1982
Particulate organic-P	8.0	?	Meybeck 1982
Particulate inorganic-P[a]	12.	?	
Total particulate P	20.	?	Meybeck 1982
Total output	22	>1.0	
P Recycling			
Primary productivity	61	—	Likens, Bormann, and Johnson 1981
P in Rain + Dry Deposition			
Soil particle origin	3.0	0.2	Graham and Duce 1979
Industry, combustion	0.21	0.21	Graham and Duce 1979
Sea salt	0.03	—	Graham and Duce 1979
Total rain and dry deposition	3.2	0.41	
Rain only	1.0	—	Meybeck 1982

[a]Calculated by difference from total; no data.

14.0×10^3 Tg/yr with 0.1% P.   Thus, roughly half (6.9 Tg/yr) of present-day river suspended phosphorus transport may be due to human activities, including deforestation and agriculture.   Since phosphorus is relatively insoluble in water, much of the phosphorus weathered from rocks is not involved in the biologic cycle and is transported directly by rivers as suspended rock debris.   In addition, there are insoluble Fe, Al, and Ca phosphates and phosphate adsorbed on clay minerals.

Land plant material contains considerable phosphorus (C:P ratio of 800:1) so that plants convert 61 Tg P/yr into organic matter through photosynthesis (Likens, Bormann, and Johnson, 1981).   A large proportion of this phosphorus is from recycled organic matter but some is also new phosphorus from weathering.   As a result of the biologic involvement of phosphorus, about half of the dissolved river phosphorus and 40% of the suspended load is in an organic form (Table 5.17). All of the dissolved phosphorus (organic and inorganic or 2 Tg P/yr) and perhaps 10% of particulate phosphorus (another 2 Tg P/yr) is *reactive*, that is, biologically active.   (However, the part of the river particulate phosphorus that is reactive is not well known.)   Thus, the river "reactive phosphorus" flux, which can be considered as P lost by terrestrial organisms, is only about 7% of the phosphorus involved in the terrestrial biological cycle (61 Tg P/yr; see "primary productivity" in Table 5.17).   This further demonstrates the efficiency of biological recycling.

About half of the 2 Tg/yr of dissolved phosphorus transported by rivers is present as inorganic phosphate, generally orthophosphate anions (PO_4^{---}, HPO_4^{--}, $H_2PO_4^-$), which are the best known, and most commonly measured, type of phosphorus in rivers.   Dissolved river phosphate can come from natural weathering and solution of phosphate minerals, accelerated dissolution due to human-induced soil erosion and transport, natural and artificially enhanced (agricultural) release of phosphate from organic P, release of P from fertilizers, and soluble phosphate from detergents and domestic and industrial wastes (Stumm 1972).   Meybeck estimates that half of the total dissolved river phosphate (or 1 Tg P/yr) is anthropogenic.   Because the natural dissolved phosphorus levels are low in rivers (0.025 mg/ℓ P), the addition of pollutive phosphorus can result in large increases.   Although the *world* river dissolved phosphorus load is estimated to have been doubled by human activities, in the polluted rivers of the United States and North America, the phosphorus concentration in many places is ten times natural levels (Meybeck 1982).

The contributions of the various pollutive phosphorus (and nitrogen) sources to dissolved phosphorus (and nitrogen) in river runoff in a typical European situation are shown in Table 5.18 (after Stumm 1972).   Municipal sewage (predominantly human waste) makes a much greater contribution to phosphorus pollution (70%) than agricultural sources (animal wastes and fertilizer, each 15%).   Even if phosphorus detergents were eliminated (they have been considerably reduced), sewage would still be a somewhat greater source than agricultural runoff.   This is in contrast to nitrogen pollution, (Table 5.18) which is dominated by animal wastes and fertilizers.

Phosphate rock mined worldwide in 1974 yielded 12.6 Tg P of which 85%

TABLE 5.18   Sources of Dissolved P and N in
Runoff in Europe as Percent of Total Runoff

Source	P	N
Agricultural Runoff		
Animal wastes	13.5	40
Fertilizer	15	20
Rain	≤1	3
Total agricultural	30	63
Municipal (Sewage)		
Human wastes	30	29
Detergents	30	—
Industrial wastes	5	4
Highways	5	4
Total municipal	70	37

Source: After W. Stumm. "The Acceleration of the
Hydrogeochemical Cycling of Phosphorus," In *The
Changing Chemistry of the Oceans*, ed. D. Dryssen
and D. Jagner, Nobel Sympos. 20, Almqvist and
Wiksell, Stockholm, p. 336.   Copyright © 1972.
Reprinted by permission of the publisher.

was used to make fertilizer and the rest processed by other industries (Pierrou
1976).   Figure 5.9 shows the exponential increases in the mining of phosphate rock
from 1955 to 1970, along with the total dissolved phosphorus in the Rhine and
Thames rivers which also increased greatly over the same period (Stumm 1972).
On a worldwide basis, if one assumes that of the 10.7 Tg P/yr used to make fertilizer
3% is lost as dissolved phosphorus in runoff (Stumm 1972), then fertilizer pollution
in streams would amount to 0.3 Tg P/yr or a third of the dissolved pollutive P in
river water.   Pierrou (1976) estimates that the total production of P from industry
and household wastes is on the order of 2 Tg P/yr (we assume this is soluble).   If
Meybeck (1982) is correct in estimating that total dissolved pollutive phosphorus
in streams amounts to only 1 Tg P/yr, then the excess pollutive phosphorus must
be building up in lakes, soils, and elsewhere, as suggested by Stumm (1972).   This
may well be true since several of the Great Lakes in the United States and a number
of European lakes exhibit recent buildups in dissolved P (see later discussion of
lakes in Chapter 6).

Further perusal of the total input and output data of Table 5.17 suggests that
the terrestrial phosphorus cycle may be out of balance and that there may be an
ongoing loading of P on the land, presumably due to human activities.   This would
be analogous to the situation for nitrogen discussed earlier.   However, the data
for inputs and outputs are not well established (for example, the value given in
Table 5.17 for the rate of rock weathering is simply that for physical erosion of

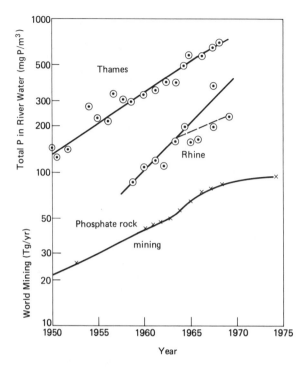

Figure 5.9 Increases in world mining of phosphate rock (Tg/yr) from 1950–1970 along with increases in concentration of total P in river water (mg P/m³) for the Rhine River (border Netherlands/Germany) and the Thames River (Laleham) in the same period. (After W. Stumm. "The Acceleration of the Hydrogeochemical Cycling of Phosphorus," In *The Changing Chemistry of the Oceans*, ed. D. Dryssen and D. Jagner, Nobel Sympos. 20, *Almqvist and Wiksell*, Stockholm, p. 330. Copyright © 1972. Reprinted by permission the publisher.) The additional increase in phosphate mining to 1974 (Pierrou 1976) is also shown.

primary rock) and, as a result, sufficient errors may be present that the input and output values may actually be equivalent. Further measurements are needed to check to see if the land surface is in fact in steady state or not with respect to phosphorus.

REFERENCES

BECK, K. C., J. H. REUTER, and E. M. PERDUE. 1974. Organic and inorganic geochemistry of some coastal plain rivers of the southeastern U.S., *Geochim. Cosmochim. Acta* 38: 341–364.

BERNER, R. A. 1971. Worldwide sulfur pollution of rivers, *J. Geophys. Res.* 76: 6597–6600.

BERNER, R. A., A. C. LASAGA, and R. M. GARRELS. 1983. The carbonate-silicate geochemical cycle and its effect on atmospheric carbon dioxide over the past 100 million years, *Amer. J. Sci.* 283: 641–683.

BERNER, R. A., and R. RAISWELL. 1983. Burial of organic carbon and pyrite sulfur in sediments over Phanerozoic time: A new theory, *Geochim. Cosmochim. Acta* 47 (5): 855–862.

BURNS, R. C., and R. W. F. HARDY. 1975. *Nitrogen fixation in bacteria and higher plants.* Berlin, Heidelberg, and New York: Springer-Verlag.

CLEAVES, E. T., A. E. GODFREY, and O. P. BRICKER. 1970. Geochemical balance of a small watershed and its geomorphic implications, *Geol. Soc. Amer. Bull.* 81: 3015–3032.

CRERAR, D. A., J. L. MEANS, R. F. YURETICH, M. P. BORCSIK, J. L. AMSTER, D. W. HASTINGS, G. W. KNOX, K. E. LYON, and R. F. QUIETT. 1981. Hydrochemistry of the New Jersey coastal plain, 2. Transport and deposition of iron, aluminum, dissolved organic matter, and selected trace elements in stream, ground-, and estuary water, *Chem. Geol.* 33: 23–44.

CRONAN, C. S., W. A. REINERS, R. C. REYNOLDS, and G. E. LANG. 1978. Forest floor leaching: Contributions from mineral, organic and carbonic acids in New Hampshire subalpine forests, *Science* 200: 309–311.

CRUTZEN, P. J. 1976. Upper limits on atmospheric ozone reductions following increased application of fixed nitrogen to the soil, *Geophys. Res. Letters*, 3 (3): 169–172.

CURTIS, W. F., J. K. CULBERTSON, and E. B. CHASE. 1973. *Fluvial-sediment discharge to the oceans from the conterminous United States*, U.S.G.S. Circular No. 670. 17 pp.

DELWICHE, C. C. 1970. The nitrogen cycle, *Scientific American* 223: 137–146.

DION, E. P. 1983. Trace elements and radionuclides in the Connecticut River and Amazon River estuary. Ph.D. dissertation, Dept. of Geology and Geophysics, Yale University, New Haven, Conn. 233 pp.

DREVER, J. I. 1982. *The geochemistry of natural water*. Englewood Cliffs, N.J.: Prentice-Hall. 388 pp.

FAIRBRIDGE, R. W. 1968. *Encyclopedia of geomorphology*, pp. 177–186. New York: Reinhold.

FETH, J. H. 1971. Mechanisms controlling world water chemistry: Evaporation-crystallization process, *Science* 172: 870–871.

———. 1981. Chloride in natural continental water—A review, U.S.G.S. Water Supply Paper No. 2176. 30 pp.

FRIEND, J. P. 1973. The global sulfur cycle. In *Chemistry of the lower atmosphere*, ed. S. I. Rasool, pp. 177–201. New York: Plenum Press.

FROEHLICH, P. N., M. L. BENDER, N. A. LUEDTKE, G. R. HEATH, and T. DEVRIES. 1982. The marine phosphorus cycle, *Amer. J. Sci.* 282: 474–511.

GARRELS, R. M., and F. T. MACKENZIE. 1971. *Evolution of sedimentary rocks*. New York: W. W. Norton, 397 pp.

GARRELS, R. M., F. T. MACKENZIE, and C. HUNT. 1975. *Chemical cycles and the global environment*. Los Altos, Calif.: Wm. Kaufman, Inc., 206 pp.

GIBBS, R. J. 1967. Amazon River: Environmental factors that control its dissolved and suspended load, *Science* 156 (3783): 1734–1737.

———. 1970. Mechanisms controlling world water chemistry, *Science* 170: 1088–1090.

———. 1971. Mechanisms controlling world water chemistry: Evaporation-crystallization process, *Science* 172: 871–872.

GRAHAM, W. F., and R. A. DUCE. 1979. Atmospheric pathways of the phosphorus cycle, *Geochim. Cosmochim. Acta* 43: 1195–1208.

GREGOR, B. 1970. Denudation of the continents, *Nature* 228: 273–275.

HAY, W. W., and J. R. SOUTHAM. 1977. Modulation of marine sedimentation by the continental shelves.   In *The fate of fossil fuel CO₂ in the oceans*, ed. N. R. Andersen, and A. Malahoff, pp. 569–604.   New York: Plenum Press.

HOLEMAN, J. N.   1968.   The sediment yield of major rivers of the world, *Water Res.* 4 (4): 737–747.

————.   1980.   Erosion rates in the U.S. estimated by the Soil Conservation Services inventory, *EOS* 61 (46): 954.

HOLLAND, H. D.   1978.   *The chemistry of the atmosphere and oceans*.   New York: John Wiley.   351 pp.

HUSAR, R. B., and J. D. HUSAR.   1985.   Regional sulfur runoff, *J. Geophys. Res.* 90 (C1): 1115–1125.

JOHNSON, A. H.   1979.   Evidence of acidification of headwater streams in the New Jersey pinelands, *Science* 206 (16): 834–836.

JUDSON, S.   1968.   Erosion of the land, *Amer. Scientist* 56 (4): 356–374.

KENNEDY, V. C., and R. L. MALCOLM.   1977.   Geochemistry of the Mattole River of Northern California.   U.S.G.S. Open-File Report no. 78-205.   324 pp.

KLEINMANN, R. L. P., and D. A. CRERAR.   1979.   *Thiobacillus ferrooxidans* and the formation of acidity in simulated coal mine environments, *Geomicrobiol. J.* 1: 373–388.

KRUG, E. C., and FRINK, C. R.   1983.   Acid rain on acid soil: A new perspective, *Science* 221: 520–525.

LEENHEER, J. A.   1980.   Origin and nature of humic substances in the waters of the Amazon River Basin, *Acta Amazonica* 10 (3): 513–526.

LERMAN, A.   1980.   Controls on river water composition and the mass balance of river systems.   In *River inputs to ocean systems*, ed. J.-M. Martin, J. D. Burton, and D. Eisma, pp. 1–4.   SCOR/UNEP/UNESCO Review and Workshop, FAO, Rome.

LEWIS, D. M.   1976.   The geochemistry of manganese, iron, uranium, lead-210, and major ions in the Susquehanna River.   Ph.D. dissertation, Dept. of Geology and Geophysics, Yale University, New Haven, Conn.

LEWIS, W. M., and M. C. GRANT.   1979.   Changes in the output of ions from a watershed as a result of the acidification of precipitation, *Ecology* 60 (6): 1093–1097.

LIKENS, G. E., F. H. BORMANN, N. M. JOHNSON.   1981.   Interaction between major biochemical cycles in terrestrial ecosystems.   In *Some perspectives of the major biogeochemical cycles*, ed. G. E. Likens, pp. 93–112.   SCOPE 4th Gen. Ass., Stockholm.   New York: John Wiley.

LIKENS, G. E., F. H. BORMANN, R. S. PIERCE, J. S. EATON, and N. M. JOHNSON.   1977.   *Biogeochemistry of a forested ecosystem*.   New York: Springer-Verlag.   146 pp.

LIVINGSTONE, D. A.   1963.   *Chemical composition of rivers and lakes*.   U.S.G.S. Prof. Paper no. 440G. 64 pp.

LOGAN, J. A.   1983.   Nitrogen oxides in the troposphere: Global and regional budgets, *J. Geophys. Res.* 88(C15): 10785–10807.

MARTIN, J.-M., and M. MEYBECK.   1979.   Elemental mass-balance of material carried by major world rivers, *Marine Chem.* 7: 173–206.

MARTIN, J.-M., and M. WHITFIELD.   1981.   The significance of river input of chemical elements to the ocean.   In *Trace metals in sea water*, eds. C. S. Wong, E. Boyle, K. W. Bruland, J. D. Burton and E. D. Goldberg, pp. 265–296.   New York: Plenum Press.

MATTOX, R. B. (ed.). 1968. Saline deposits, a Symposium based on papers from the International Conference on Saline Deposits, Houston, Texas, 1962. *Geological Soc. of Amer. Spec. Paper* 88, 701 pp.

MEYBECK, M. 1979. Concentrations des eaux fluviales en éléments majeurs et apports en solution aux océans, *Rev. Géol. Dyn. Géogr. Phys.* 21(3): 215–246.

———. 1980. Pathways of major elements from land to ocean through rivers In *Proceedings of the review and workshop on river inputs to ocean-systems*, ed. J.-M. Martin, J. D. Burton, and D. Eisma, pp. 18–30. Rome: FAO.

———. 1982. Carbon, nitrogen and phosphorus transport by world rivers, *Amer. J. Sci.* 282: 401–450.

———. 1983. Atmospheric inputs and river transport of dissolved substances. In *Dissolved loads of rivers and surface water quantity/quality relationships*. Proceedings of the Hamburg Symposium, August 1983. IAHS Publ. no. 141.

———. 1984. Les fleuves et le cycle géochimique des éléments. Thèse d'état (no. 84-35), École Normal Supérieure, Laboratoire de Géologie, Univ. Pierre et Marie Curie, Paris 6, France.

———. 1986. Origin of riverborne elements derived from continental weathering, *Amer. Journal. Sci.* (in press).

MILLER, W. R., and J. I. DREVER. 1977. Water chemistry of a stream following a storm, Absaroka Mountains, Wyoming, *Geol. Soc. Amer. Bull.* 88: 286–290.

MILLER, A., and J. C. THOMPSON, R. E. PETERSON, and D. R. HARAGAN. 1983. *Elements of meterology*, 4th ed. Columbus, Ohio: Chas. E. Merrill. 362 pp.

MILLIMAN, J. D. 1980. Transfer of river-borne particulate material to the oceans. In *River inputs to ocean systems*, ed. J.-M. Martin, J. D. Burton, and D. Eisma, pp. 5–12. SCOR/UNEP/UNESCO, Review and Workshop, FAO, Rome.

MILLIMAN, J. D., and MEADE, R. H. 1983. World-wide delivery of river sediment to the oceans, *J. Geology* 91 (1): 1–21.

MING-HUI, H., R. F. STALLARD, and J. M. EDMOND, 1982. Major ion chemistry of some large Chinese rivers, *Nature* 289 (5): 550–553.

MÖLLER, D. 1984. Estimation of the global man-made sulfur emission, *Atmos. Environ.* 18 (1): 19–27.

NATIONAL RESEARCH COUNCIL, 1972. *Accumulation of nitrate*. Publication of Committee on Nitrate Accumulation. Washington, D.C.: National Academy of Sciences, National Research Council. 106 pp.

NRIAGU, J. O. 1978. Production and uses of sulfur. In *Sulfur in the environment, Part 1*, ed. J. O. Nriagu, pp. 1–21. New York: Wiley-Interscience.

ODÉN, S., and T. AHL. 1978. The sulfur budget of Sweden. In *Effects of acid precipitation on terrestrial ecosystems*, ed. T. C. Hutchinson and M. Havas, pp. 111–122. New York: Plenum Press.

PAČES, T. 1985. Sources of acidification in Central Europe estimated from elemental budgets in small basins, *Nature* 315 (6014): 31–36.

PIERROU, U. 1976. The global phosphorus cycle. In *Nitrogen, phosphorus and sulfur—Global cycles*, ed. B. H. Svenson and R. Söderlund. pp. 75–90. SCOPE Report. no. 7. Stockholm: Ecol. Bull. 22.

REEDER, S. W., B. HITCHON, and A. A. LEVINSON. 1972. Hydrogeochemistry of the surface waters of the Mackenzie River drainage basin, Canada: 1. Factors controlling inorganic compositions, *Geochim. Cosmochim. Acta* 26: 825–865.

REYNOLDS, R. C., Jr., and N. M. JOHNSON. 1972. Chemical weathering in the temperature glacial environment of the Northern Cascade Mountains, *Geochim. Cosmochim. Acta* 36: 537–554.

SIMPSON, H. J. 1977. Man and the global nitrogen cycle. In *Global chemical cycles and their alterations by man*, ed. W. Stumm, pp. 253–274. Berlin: Dahlem Konferenzen.

SKINNER, B. J. 1969. *Earth resources*. Englewood Cliffs, N.J.: Prentice-Hall. 150 pp.

SMITH, R. A., and R. B. ALEXANDER. 1983. *Evidence for acid-precipitation induced trends in stream chemistry at Hydrologic Bench-Mark Stations*. U.S.G.S. Circular no. 910. 12 pp.

STALLARD, R. F. 1980. Major element geochemistry of the Amazon River system. Ph.D. dissertation, MIT/Woods Hole Oceanographic Inst., WHOI-80-29. 366 pp.

———. 1985. River chemistry, geology, geomorphology and soils in the Amazon and Orinoco Basins. In *The chemistry of weathering*, ed. J. I. Drever, pp. 293–316. Boston: D. Reidel Publish. Co.

STALLARD, R. F. and J. M. EDMOND. 1981. Geochemistry of the Amazon 1: Precipitation chemistry and the marine contribution to the dissolved load, *J. Geophys. Res.* 86 (C10): 9844–9858.

———. 1983. Geochemistry of the Amazon 2: The influence of the geology and weathering environment on the dissolved load, *J. Geophys. Res.* 88: 9671–9688.

STUMM, W. 1972. The acceleration of the hydrogeochemical cycling of phosphorus. In *The changing chemistry of the oceans*, ed. D. Dyrssen and D. Jagner, pp. 329–346. Nobel Sympos. 20. Stockholm: Almqvist and Wiksell.

SUGAWARA, K. 1967. Migration of elements through phases of the hydrosphere and atmosphere. In *Chemistry of the Earth's Crust*, v.2, ed. A.P. Vinogradov, pp. 501–510. Israel Program for Scientific Translation Ltd., Jerusalem. Reprinted in *Geochemistry of Water*, ed. Y. Kitano, pp. 227–237. New York: Halsted Press, 1975.

TRIMBLE, S. W. 1975. Denudation studies: Can we assume stream steady state? *Science* 188: 1207–1208.

———. 1977. The fallacy of stream equilibrium in contemporary denudation studies, *Amer. J. Sci.* 277: 876–887.

YURETICH, R. F., D. A. CRERAR, D. J. J. KINSMAN, J. L. MEANS, and M. P. BORCSIK. 1981. Hydrogeochemistry of the New Jersey Coastal Plain, 1: Major element cycles in precipitation and river water, *Chem. Geol.* 33: 1–21.

ZEHNDER, A. J. B., and S. H., ZINDER. 1980. The sulfur cycle. In *The handbook of environmental chemistry*, ed. O. Hutzinger, vol. 1, pt. A., pp. 105–145. New York: Springer-Verlag.

ZOBRIST, J., and W. STUMM. 1980. Chemical dynamics of the Rhine catchment area in Switzerland: Extrapolation to the "pristine" Rhine river input into the ocean. In *River inputs to ocean systems*, ed. J.-M. Martin, J. D. Burton, and D. Eisma, pp. 52–63. SCOR/UNEP/UNESCO Review and Workshop, FAO, Rome.

ZVEREV, V. P. and V. Z. RUBEIKIN. 1973. The role of an atmospheric precipitation in circulation of chemical elements between atmosphere, hydrosphere and lithosphere, *Hydrogeochemistry* 1: 613–620.

6
Lakes

INTRODUCTION

Although lakes constitute only about 0.01% of the total water at the earth's surface, they have received proportionately greater attention because of their importance to humans. Lakes are used as a source of drinking water, as a receptacle for sewage and agricultural runoff, for recreation, and for industrial purposes. Because of their generally small size, they can be severely altered by these activities so that any discussion of lake-water chemistry must include the effects of humans. Also, lakes have received so much attention that a whole field devoted to their study, *limnology*, exists, and a large variety of problems have consequently been discovered. Our goal in this chapter will be to discuss some fundamental aspects of lakes and to show how lake-water composition is affected by physical, biological, geological, and anthropomorphic processes. In the process, we hope again to demonstrate the necessity of using a multidisciplinary approach to the study of natural waters. For a more extensive discussion of limnology the reader should refer to books devoted to the subject such as Hutchinson 1957, Wetzel 1975, and Lerman 1978.

PHYSICAL PROCESSES IN LAKES

Water Balance

Lakes, in some respects, can be considered as "little oceans." Like oceans they receive inflow from rivers, exhibit vertical stratification, undergo biological cycling and sedimentation, lose water through evaporation, and so forth. However, most lakes differ from oceans in one important aspect: they have outlets. Water is lost from the oceans only by evaporation (see Chapter 8) but in freshwater lakes the water also leaves via a surface or subsurface outlet. In this way a lake (with a surface outlet) is a sort of "fat" and "slow" portion of a river or, in other words, a portion of a drainage system where water is retained for considerably larger periods than in normal river channels. In some lakes, there is no outlet because of high aridity and interior drainage, and in this case the ocean simile is better. Waters entering such lakes leave only by evaporation and, as in the ocean, high salinities can result. Nevertheless, most lakes consist of fresh water and this water is fresh because of the presence of outlets.

   Table 6.1 and Figure 6.1 summarize the various inputs and outputs of water to and from lakes. The relative importance of each process depends upon the lake. For example, as mentioned above, in arid regions lakes often have no outlet and water is lost only by evaporation. Inputs can also vary considerably. Lake Victoria in Africa is a shallow but areally extensive lake with minor stream input and is located in a region of high rainfall. As a result, roughly three-fourths of the water input to the lake is provided by rainfall on the lake surface (Hutchinson 1957). This is unusually high and for most larger lakes the major input is rivers and streams with rainfall accounting for only a small percent of the total. By contrast, many very small lakes (*spring-fed lakes*) are totally supplied by underground springs and some, known as *karst lakes*, also lose water by underground seepage. Karst lakes develop in karst terrains, or regions of extensive underground dissolution of limestone (see Chapter 4), and because of the permeable nature of the underlying partly dissolved limestone bedrock, water is able to readily seep

TABLE 6.1   Processes of Water Input and Output in Lakes

Inputs	Outputs
1. Rainfall (and snowfall) on lake surface	1. Evaporation
2. Stream and river flow	2. Outflow at surface via natural outlets such as a stream, waterfall, etc. (usually only one outlet)
3. Spring discharge along lake margin	3. Outflow at surface via manmade conduits, irrigation channels, dams, etc.
4. Groundwater seepage through lake floor	4. Seepage out through lake floor
5. Artificial conduits	

Source: After Hutchinson 1957.

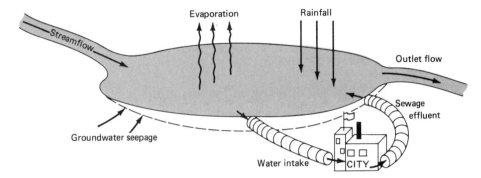

Figure 6.1   Schematic representation of lake water inputs and outputs.

into and out of the lake floor.   Most lakes, however, because of the presence of relatively impermeable clay-sediment on their bottoms, do not undergo major underground seepage.

The volume of water in a lake, and consequently the level of its surface, depends on how the inputs and outputs of water balance.   If there is a good balance, the water level fluctuates very little.   However, because of year-to-year changes in rainfall and other climatic factors, some fluctuation is expected.   The degree to which fluctuation affects the lake level depends on the size of the lake and the rate of water addition and removal.   For lakes receiving most of their water from streamflow, this can be represented in terms of the time it would take to replace the water in the lake; that is, the time necessary, at the present rate of inflow, to fill the lake to its present volume.   Mathematically the *replacement*, or filling, time for water is defined as

$$\tau_w = V/F_i \tag{6.1}$$

where

τ_w = replacement time for water.

V = volume of water in the lake.

F_i = rate of streamflow addition.

Low values of τ_w mean that short-term fluctuations in input are rapidly witnessed by lake level changes whereas high τ_w values mean that the lake level (volume) is relatively impervious to rapid input changes.   Not surprisingly, low τ_w values are characteristic of small lakes and ponds and high τ_w values of large lakes.   Large lakes are able to adjust their level to long-term averages of inputs and outputs. Values of τ_w for most lakes lie between 1 and 100 years.

If a lake maintains a constant volume, because inputs and outputs are equal, it is said to be in a *steady state* with respect to its water content.   In this case the replacement time can also be viewed as a *residence time* of water in the lake.   This is the average time a water molecule spends in the lake before being removed

through the outlet.   The concept of residence time is used extensively in the
literature on both limnology and oceanography, but it is sometimes forgotten that
it only makes sense when there is a steady state.   Otherwise, more general concepts,
such as that of replacement time, are preferable.

Thermal Regimes and Lake Classification

The chemistry of freshwater lakes depends to a large extent upon a circulation
driven by temperature changes.   The density of water in lakes is primarily a function
of temperature.   As temperature changes, density changes, and if less dense water
becomes overlain by more dense water, convection, or lake-water overturn, takes
place.   The effect of temperature on water density has been shown already in
Figure 1.3.   Recall that an unusual situation occurs between 0° and 4° C.   With
heating, instead of decreasing in density like most liquids, dilute (fresh) water
actually increases in density over this temperature range.   At 4° C the water is at
a maximum density and above this temperature the density then continues to
decrease with temperature like a normal liquid.   (The causes for this maximum
density are discussed in detail in Chapter 1.)   This unusual temperature-density
behavior of fresh water provides the fundamental basis for classifying lakes and
for describing their overall circulation behavior.

   In order to provide a better understanding of the effects of temperature on
lake circulation and classification, we shall discuss the effect of seasonal temperature
changes on profiles of temperature versus depth for a typical medium-sized lake
of temperate region (for further details consult Hutchinson 1957).   This is shown
in Figures 6.2 and 6.3.   Let us start the discussion with late summer (Figure 6.2).

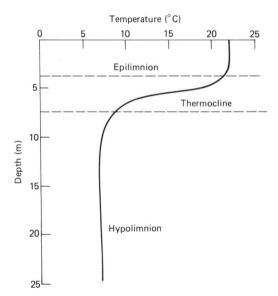

Figure 6.2   Temperature profile of a
typical temperate freshwater lake in
summer.

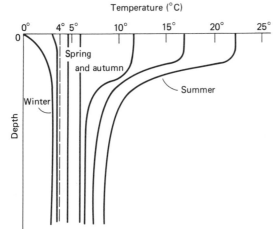

Figure 6.3 Schematic evolution of temperature-versus-depth profiles over the year for a typical dimictic lake of temperate climates. Dashed line represents the maximum density at 4° C. (See also Hutchinson 1957, and Wetzel 1975.)

At this time air temperature is at a maximum and so is the surface water temperature. The depth of heating of the water depends on wind stirring and extends down to where wind effects die out. The shallow, wind-stirred portion of the lake is called the *epilimnion*. Below the epilimnion we encounter a region where temperature falls rather rapidly with depth. This is the *thermocline* or *mesolimnion*. Below the thermocline is the deep, cold portion of the lake known as the *hypolimnion*. Because the temperature of all the lake water is above 4° C, the coldest water is most dense and is located at the bottom. The situation depicted in Figure 6.2 represents stable density stratification, which means that there is little vertical circulation of water and the hypolimnion is effectively isolated from the atmosphere. This situation is characteristic of most of the oceans but only occurs, in temperate freshwater lakes, during the summer.

Let us examine (Figure 6.3) what happens as the air temperature drops during the autumn. As the air cools, the epilimnetic water also cools until its temperature matches that of the hypolimnion. In this case the thermocline disappears and there is a constant temperature from top to bottom. Further cooling of the surface water causes the density distribution to become unstable with heavier water on top and, as a result, vertical convection arises. The lake mixes or *overturns* from top to bottom and this overturn is aided by winds, which are generally strong in the autumn. As fall cooling continues, the constant top-to-bottom temperature drops with time (Figure 6.3) until 4° C is reached, at which point new behavior takes over.

Further cooling of surface water below 4° C causes it to become less dense than the deeper water. Stable density stratification develops and the top-to-bottom overturn ceases. Eventually the surface water reaches 0° C and freezes. During the winter, the cold surface water plus ice cap prevent any wind stirring of deeper water and we have a situation known as *winter stratification* (as opposed to summer stratification) where the deep water again becomes isolated from the atmosphere.

As spring approaches, the ice melts and the surface water begins to warm up. When the surface water temperature reaches that of the deep water, we again have thermal instability and, as a result, there is a *spring overturn*. (The temperature of the bottom water can be somewhat less than 4° C due to excessive wind stirring prior to winter stratification and to conductive cooling during the winter.) The water then warms up from top to bottom during further spring warming until stable density stratification ensues and vertical mixing is inhibited. Further heating on into the summer increases this stable summer-type stratification until our original situation of late summer is reattained and the seasonal cycle is completed.

The situation depicted above and in Figure 6.3, where there are two overturns per year, is typical of temperate lakes, and where this occurs the lakes are referred to as being *dimictic* (mixing twice a year). Other situations can arise in different climates. In warmer (e.g., Mediterranean-type climates) the mean monthly air temperature does not drop below 4° C in the winter and there is no winter stratification or freezing. As a result, the water overturns throughout the winter until stratification sets in during the spring. The stratification is then maintained through the summer into the fall. Temperature-versus-depth profiles resemble only those to the right of the 4° C dashed line of Figure 6.3, and under these conditions (warm) *monomictic* lakes arise, which mix only once a year. Typical examples are those of the Italian Lake District. Another kind of monomictic lake, the cold monomictic, occurs rarely in alpine or high-latitude cold climates. Here the water temperature never exceeds 4° C and, as a result, there is continuous mixing during the summer and ice cover plus winter stratification throughout the rest of the year. Temperature-versus-depth profiles resemble only those to the left of the 4° C dashed line of Figure 6.3.

In tropical regions the air temperature changes little over the year and, as a result, well-defined, cyclic temperature-versus-depth regions in lakes are not present. Overturn in this case depends on a variety of unpredictable factors, and stratification is variable from place to place. Such lakes are referred to as being *oligomictic*.

All density stratification in lakes is not brought about by seasonal temperature changes. In certain lakes that are fed by saline springs or rivers, a situation known as *meromixis* sets in. Meromictic lakes never undergo overturn due to the presence of dense, saline water at depth and these lakes stand in fundamental contrast to all of the lakes discussed above, which overturn at least once a year and are thereby termed *holomictic*.

A summary of the types of lakes described above is shown in terms of the classification presented in Table 6.2. Added to these are two types of lakes that make up special cases because of their unusual morphology. One type is very shallow lakes. Some lakes are so shallow that they are always stirred by the wind and never develop stratification or a hypolimnion. An opposite situation is provided by very deep lakes like those of the East African rift system. Here, because of the large volume of deep water, little or no heating or cooling of the hypolimnion

TABLE 6.2 Classification of Freshwater Lakes

I. Holomictic (mixing between epilimnion and hypolimnion)
    A. Dimictic (mixes twice a year)
    B. Monomictic (mixes once a year)
        1. Warm monomictic
        2. Cold monomictic
    C. Oligomictic (mixes irregularly)
    D. Shallow lakes (continuous mixing)
    E. Very deep lakes (mixing in upper portion of hypolimnion only)
II. Meromictic (no mixing between epilimnion and hypolimnion)

Note: See Hutchinson 1957 for details.

occurs during the spring and fall. In other words, the deep water acts as a sort of thermal buffer. Below a given depth there is a large reservoir of maximum-density deep water at 4° C that is not affected by wind stirring. This water, thus, does not overturn and remains permanently isolated from the atmosphere.

Throughout this discussion of lakes we have mentioned that, during stratification, deep water is isolated from the atmosphere. This is especially important because isolation from the atmosphere, accompanied by biological oxygen consumption in the deep water, results in a lowering of dissolved O_2. If the isolation is prolonged, all O_2 may be consumed and, as a result, the deep water becomes anoxic resulting in dramatic changes in water quality. These changes are important and will be dealt with later in the present chapter. However, it is important to remember that it is the physical process of stratification that initiates these chemical changes.

Lake Models

Like water, other substances are added to and removed from lakes. Rivers and groundwater carry dissolved constituents into a lake, where they may undergo chemical reactions, and this is followed by removal via water flow through an outlet. One way of quantitatively treating rates of addition and removal is by means of box modelling. In box modelling one assumes that a portion of a lake or the whole lake is so well stirred that it is homogeneous in composition and can be treated as a uniform "box." Rates of addition (or removal) to each box are slow enough, relative to mixing, that high concentrations of added substances do not build up around each source. (This situation is not always obeyed; see Imboden and Lerman 1978). The concentration of a given substance in a given "box" is controlled by the relative magnitude of inputs and outputs. If inputs balance outputs there is a steady state and concentrations do not change with time. This is analogous to (but not necessarily connected with) the situation of steady-state water content discussed earlier.

The simplest kind of box model is that of a single box representing a whole lake. In this case we have input of a dissolved substance from streams (groundwater and rainwater inputs are neglected), output by a surface outlet, precipitation and removal to bottom sediments, and addition via dissolution, or bacterial regeneration, of suspended and sedimented solids (see Figure 6.4). Rates of these processes can be represented as

F_i = rate of water inflow from streams (volume per unit time).

F_o = rate of water outflow through outlet.

M = total mass of the dissolved substance in the lake.

R_p = rate of removal via precipitation and sedimentation to the bottom (mass per unit time).

R_d = rate of addition via dissolution of solids (mass per unit time).

C_i = concentration of the dissolved substance in streamwater (mass per unit volume).

C = concentration in lake water.

t = time.

Assuming steady state with respect to water (constant volume lake) the rate of change of mass with time in the lake $\Delta M / \Delta t$ is

$$\frac{\Delta M}{\Delta t} = C_i F_i - C F_o + R_d - R_p \tag{6.2}$$

If there is a steady state also with respect to the dissolved substance, $\Delta M / \Delta t = 0$, then

$$C_i F_i - C F_o + R_d - R_p = 0 \tag{6.3}$$

Finally, if dissolution represents redissolution of the same material that was previously precipitated and sedimented to the bottom, then

$$R_s = R_p - R_d \tag{6.4}$$

where R_s = rate of burial in sediments (mass per unit time). Using Equation (6.3) one can calculate, for example, from a knowledge of measured water flow,

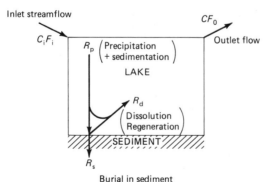

Figure 6.4   One-box model for lakes. (For explanation of symbols, see text.)

rainfall, and sedimentation rates, the maximum allowable input concentration (C_i) of a pollutant, P, to a lake, if the lake concentration of the pollutant (C) is not to exceed a certain level.   Suppose the water inflow rate (F_i) to the lake is equal to 100 m³/sec (appropriate for a small river) and rainfall (minus evaporation) averaged over a year is 50 m³/sec.   Then the outflow rate (F_o) should be 150 m³/sec in order to maintain constant lake volume.   Let the sedimentation rate (R_p) be 250 mg P/sec, and assume that the lake concentration (C) may not exceed 5 µg P/ℓ (which equals 5 mg P/m³ and is sufficiently low that it should inhibit eutrophication; see discussion below).   Then upon substituting in Equation (6.3) and letting $R_d = 0$

$$C_i = (CF_o + R_p)/F_i$$

$$= \frac{(5 \text{ mg } P/\text{m}^3) \cdot (150 \text{ m}^3/\text{sec}) + (250 \text{ mg } P/\text{sec})}{100 \text{ m}^3/\text{sec}}$$

$$C_i = 10 \text{ mg } P/\text{m}^3$$

or

$$C_i = 10 \text{ µg } P/\ell$$

Two useful concepts, already applied to water itself, are those of replacement time and residence time.   Replacement time is the time necessary to replace the mass of a dissolved substance, via the present rate of stream addition, if all of the substance were suddenly removed.   It gives a measure of the sensitivity of lake concentration C to changes in input concentration, C_i or water inflow, F_i.   The replacement time of a dissolved substance is defined as

$$\tau_r = \frac{\text{Mass in lake}}{\text{Rate of stream input to lake}} = \frac{M}{C_iF_i} \tag{6.5}$$

Recalling from Equation (6.1) that the replacement time for water is

$$\tau_w = V/F_i$$

(V = volume of lake) and that $M = CV$, then Equation (6.5) can be rewritten as

$$\tau_r = (C/C_i)\tau_w \tag{6.6}$$

If the lake is at a steady state with respect to the dissolved substance of interest (and water), the τ_r (and τ_w) can be viewed as residence times as well as replacement times. In other words, for steady state the value τ_r represents the average time spent by a dissolved species in the lake prior to removal either via sedimentation or through the outlet.

If, on the other hand, the lake is not at steady state, and an attempt is being made to lower the lake concentration, C, of a pollutant by reducing the input concentration, C_i (see calculation above), one can see from Equation 6.6 that lakes with short water-replacement times are more responsive to efforts to reduce pollution.   In other words, the replacement time of a dissolved substance, τ_r, is directly proportional to the water replacement time, τ_w.

An additional concept, that of relative residence time, (Stumm and Morgan 1981) is very useful. Relative residence time is the residence time of a given dissolved substance relative to that of water:

$$\tau_{rel} = \tau_r/\tau_w \tag{6.7}$$

or, from Equation (6.6):

$$\tau_{rel} = C/C_i \tag{6.8}$$

Relative residence time is an indication of the type of behavior to be expected for a given substance. A relative residence time of 1 indicates that the substance does not react chemically in the lake ($R_d = 0$; $R_p = 0$) and it simply accompanies water as it passes through the lake. Dissolved Cl^- or Na^+ are examples. In this case the substance acts as a tracer of water motion. If τ_{rel} is less than 1, the substance tends to undergo removal via sedimentation in the lake ($R_p > 0$) indicating its chemical reactivity. (An example is dissolved Al.) If τ_{rel} is greater than 1, the substance tends to be trapped in the lake while the water that brought it in is removed. This can take place if the substance is cycled within the lake, that is, it is precipitated and sedimented to the bottom ($R_p > 0$), then redissolved ($R_d > 0$), then reprecipitated, and so on. This is characteristic of elements involved in biological processes, for example, P, N, Si, and Ca, and such biological cycling within the lake can result in a relative residence time of each of these elements appreciably greater than 1.

Simple one-box models, although applicable to all lakes to express their *average* properties, are most accurate as representations of shallow lakes that do not undergo stratification. For the more usual case of stratified lakes, a two-box model is more appropriate (e.g., Imboden and Lerman 1978; Stumm and Morgan 1981). One box is used to represent the epilimnion and the other box the hypolimnion. This is shown in Figure 6.5. In the two-box model we have *fluxes* between the reservoirs (boxes) as well as inputs and outputs for the whole lake. Figure 6.5 represents the situation expected for a biological element such as phosphorus or nitrogen. There is input of dissolved material by streams to the epilimnion and output via an outlet. There is exchange of water containing the dissolved substance of interest between hypolimnion and epilimnion which is represented by up and down (short) arrows. (Actual exchange occurs sporadically during seasonal overturn, but for modelling this is averaged over a year.) Finally there are chemical reactions; in this case these include removal of the substances from the epilimnion via precipitation and transfer downward by sedimentation, injection of a portion to solution in the hypolimnion, and burial of the remainder.

Mathematical representation of the rates in a two-box lake model is similar to that presented above for a one-box lake. Besides the parameters defined for the one-box model we also have the following:

F_U = rate of water transfer from hypolimnion to epiliminion.
F_D = rate of water transfer from epilimnion to hypolimnion.

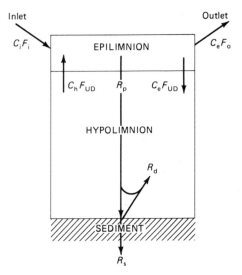

Figure 6.5 Two-box model for lakes (see text).

R_p = rate of removal, by precipitation and sedimentation, from epilimnion.
R_d = rate of addition, via dissolution, to hypolimnion.
M_e = mass dissolved in epilimnion.
C_e = concentration in epilimnion (mass per unit volume).
M_h = mass dissolved in hypolimnion.
C_h = concentration in hypolimnion (mass per unit volume).

If we have steady state with respect to water for both boxes (constant volumes of the epilimnion and hypolimnion, V_e and V_h, with time), then

$$F_U = F_D$$

and we can refer to either rate as F_{UD} (see Figure 6.5).

Using the above definition, and referring to Figure 6.5, we obtain the rates of change of mass in solution for each box $\Delta M_e/\Delta t$ and $\Delta M_h/\Delta t$, by summing inputs and outputs:

$$\frac{\Delta M_e}{\Delta t} = C_iF_i - C_eF_o + (C_h - C_e)F_{UD} - R_p \tag{6.9}$$

$$\frac{\Delta M_h}{\Delta t} = R_d - (C_h - C_e)F_{UD} \tag{6.10}$$

If, in addition we have steady state, both with respect to the dissolved substance and its solid precipitated form, and dissolution represents redissolution of sedimenting solids, we have

$$C_iF_i - C_eF_o + (C_h - C_e)F_{UD} - R_p = 0 \tag{6.11}$$

$$R_d - (C_h - C_e)F_{UD} = 0 \qquad (6.12)$$

$$R_p = R_d + R_s \qquad (6.13)$$

From these equations we can, thus, gain information on rates of processes from measurements of other rates and concentrations and, in this way, the equations are most useful.

BIOLOGICAL PROCESSES IN LAKES AS THEY AFFECT WATER COMPOSITION

Photosynthesis, Respiration, and Biological Cycling

One of the main reasons why hypolimnia and epilimnia differ in composition is that each is greatly, but differently, affected by biological processes. The starting point is photosynthesis. In photosynthesis CO_2 and H_2O are converted to organic matter by green plants, which use the energy of sunlight for this process. Organic compounds synthesized by the plants contain not only carbon, hydrogen, and oxygen but also other essential nutrient elements, chief of which are nitrogen and phosphorus (see Table 4.1). The overall process of photosynthesis in lakes (and the ocean) can be represented by the following reaction (e.g., Stumm and Morgan 1981):

$$106CO_2 + 16NO_3^- + HPO_4^{--} + 122H_2O + 18H^+$$

$$(+ \text{ trace elements and energy}) \qquad (6.14)$$

$$\rightarrow C_{106}H_{263}O_{110}N_{16}P_1 + 138O_2$$

(Here the stoichiometric elemental composition of average marine plankton is used simply for the purposes of illustration.) Photosynthesis thus involves the uptake of dissolved CO_2, phosphate, and nitrogen, and the production of O_2. (Nitrogen can be taken up from NH_4^+ or N_2 as well as NO_3^-.)

Hand in hand with photosynthesis is the process of aerobic respiration. Respiration occurs in all organisms including plants, animals, and bacteria, and, in the presence of O_2, can be considered as the above reaction written backwards. In other words, aerobic respiration is essentially the reverse of photosynthesis, and involves the breakdown of organic matter while photosynthesis involves its formation. Although in natural waters the rates of photosynthesis and respiration are closely matched, there is almost always a slight excess of photosynthesis. In lakes this excess in surface waters manifests itself as a downward flux of dead organic matter to the bottom. Photosynthesis can occur only in water sufficiently shallow that light, necessary for the process, can penetrate it. At greater depths it is absent. In sufficiently deep (or murky) lakes, respiration dominates over photosynthesis at depth and this helps to destroy much cf the dead organic matter sedimenting from above. In other words, in sufficiently deep lakes the processes of photosynthesis and respiration are in part separated from one another with net photosynthesis at the surface and net respiration at depth. If the lake is also stratified, this separation can lead to dramatic changes in water composition.

Consider a stratified lake with photosynthesis confined to the epilimnion (reference here to the two-box model diagram, Figure 6.5, is helpful). The nutrients phosphorus and nitrogen are removed from the epilimnetic water to form organic matter; some of this organic matter is eaten and respired by zooplankton, fish, and other organisms; some of it is decomposed by bacteria; and the remainder falls into the hypolimnion. Here it is further decomposed, especially by bacteria living in the water and on the lake bottom, and phosphorus and nitrogen are liberated to solution. Some remaining organic matter that escapes decomposition is buried in bottom sediments. The P and N liberated to solution in the hypolimnion is not reused by plants there because the water is too deep for photosynthesis. However, the P and N are eventually, but slowly on the average, returned to the epilimnion by occasional overturn. By these processes, concentrations of P and N in the hypolimnion, where there is net respiration, build up to much higher concentrations than exist in the epilimnion, where there is net photosynthesis, and continued biological cycling maintains these concentration differences. Only during overturn is there homogenization of concentrations.

Besides P and N, other elements undergo a differentiation of concentration due to biological cycling. Dissolved O_2 neither builds up nor is depleted in the epilimnion because it readily exchanges with the atmosphere. By contrast, in the hypolimnion there is no contact with the atmosphere and any O_2 lost by respiration is not immediately replaced. As a result, the concentration of O_2 in the hypolimnion is lowered and this lowering, depending upon the rate of addition of dead organic matter, can be all the way to zero. Once all dissolved oxygen is removed, new anoxic, bacterially mediated chemical reactions can take place. This includes the reduction of collodial ferric and Mn^{++++} oxides to Fe^{++} and Mn^{++} in solution, the reduction of SO_4^{--} to H_2S, and the production of methane, all of which accompany the anoxic decomposition of organic matter. Some examples of chemical changes brought about by anoxia are shown in Table 6.3. (For a further discussion of anoxic bacterial processes see Chapter 8 and consult Claypool and Kaplan 1974; Fenchel and Blackburn 1979; and Berner 1980.) Also, the production of anoxia can have catastrophic effects on higher organisms dwelling in the deep water or on the bottom, as will be discussed later under eutrophication.

Accompanying changes brought about by organic matter itself are changes involving mineral matter synthesized by photosynthetic organisms. Opaline silica (not quartz, but a different, amorphous form of SiO_2 found in skeletal material) is secreted by microscopic floating organisms, chiefly diatoms. Calcium carbonate is secreted by some algae. Together these substances fall into deep water where they ordinarily undergo dissolution. This dissolution, along with secretion in the epilimnion, leads to concentration differences in dissolved H_4SiO_4 and Ca^{++} and HCO_3^-, between epilimnion and hypolimnion. In addition, the nonbiogenic precipitation of $CaCO_3$ can also occur in the epilimnion. Dissolution of $CaCO_3$ in

TABLE 6.3 Some Chemical Changes Brought about by
Anoxic Conditions in Natural Waters

Bacterial nitrate reduction (denitrification)
$$5CH_2O + 4NO_3^- \rightarrow 2N_2 + 4HCO_3^- + CO_2 + 3H_2O$$
Bacterial sulfate reduction
$$2CH_2O + SO_4^{--} \rightarrow H_2S + 2HCO_3^-$$
Bacterial methane formation
$$2CH_2O \rightarrow CO_2 + CH_4$$
Iron reduction
$$CH_2O + 7CO_2 + 4Fe(OH)_3 \rightarrow 4Fe^{++} + 8HCO_3^- + 3H_2O$$
Manganese reduction
$$CH_2O + 3CO_2 + H_2O + 2MnO_2 \rightarrow 2Mn^{++} + 4HCO_3^-$$
Ferrous sulfide precipitation
$$Fe^{++} + H_2S \rightarrow FeS + 2H^+$$
Manganese and ferrous carbonate precipitation
$$Mn^{++} + 2HCO_3^- \rightarrow MnCO_3 + CO_2 + H_2O$$
$$Fe^{++} + 2HCO_3^- \rightarrow FeCO_3 + CO_2 + H_2O$$
Ferrous phosphate precipitation
$$8H_2O + 3Fe^{++} + 2PO_4^{---} \rightarrow Fe_3(PO_4)_2 \cdot 8H_2O$$

Note: CH_2O represents decomposing organic matter.   (See also Table 8.4.)

the hypolimnion is due primarily to a higher acidity there, arising from production of excess carbonic acid from respiratory CO_2.   The reactions are

$$C_{organic} + O_2 \rightarrow CO_2$$

$$CO_2 + H_2O \rightarrow H_2CO_3$$

$$H_2CO_3 + CaCO_3 \rightarrow Ca^{++} + 2HCO_3^-$$

In the case of opaline silica, dissolution occurs because the water is undersaturated with silica throughout the lake.   (Formation of opaline silica in shallow water occurs only because of the input of photosynthetic solar energy.)

Eutrophication

Historically, lakes have been classified as *oligotrophic* or *eutrophic* based on either their concentrations of plant nutrients or their productivity of organic matter (Hutchinson 1973; Vallentyne 1974; Rodhe 1969).   *Trophic* means nutrition, and *oligotrophic* lakes are poorly fed, that is, have a low concentration of nutrient elements such as nitrogen and phosphorus.   The lack of nutrients results in few plants and, thus, a low rate of organic matter production by photosynthesis.   Oligotrophic lakes are usually deep and have relatively plankton-free clear water, which is well oxygenated at depth.   On the other hand, *eutrophic* or "well-fed" lakes have high concentrations of plant nutrients and large concentrations of plankton due to high organic productivity.   Their waters are murky with suspended plankton and often depleted in oxygen at depth. (The term *mesotrophic* has also been applied to lakes with properties intermediate between those that are termed as eutrophic or oligotrophic.)

Recently, the process of *eutrophication* has been more broadly defined as high biological productivity resulting from increased input of either nutrients or of organic matter, with the ultimate development of a decreased volume of the lake (Likens 1972). Therefore, this definition also includes certain eutrophic lakes that result from a large input of organic matter from the surrounding area as opposed to internally produced organic matter. Natural eutrophication occurs as lakes gradually fill in with organic-rich sediments over a long period of time and eventually become swamps and then disappear. However, humans have greatly accelerated the process by artificially enriching lakes with too many nutrients and/or with excess organic matter; this has been called *cultural eutrophication* (Hasler 1947). The characteristics of eutrophic and oliogotrophic lakes are summarized in Table 6.4.

The process of natural eutrophication (see Figure 6.6) can be described in terms of a typical lake in northern North America that was formed after the glacial retreat 10,000–15,000 years ago (Hutchinson 1973; Vallentyne 1974). The original oligotrophic lake is clear, with a clean bottom, and has a small population of phytoplankton (minute aquatic plants), zooplankton (small aquatic animals), and

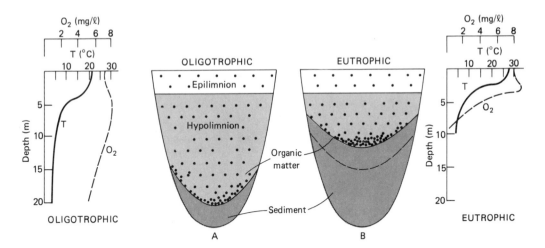

Figure 6.6 Comparison (in late summer) between an oligotrophic lake (A) and a naturally eutrophic lake (B) with moderate nutrient supply, showing organic matter falling through the epilimnion and hypolimnion, and organic matter and sediment accumulation at depth. (A) The oligotrophic lake has a large oxygenated hypolimnion and little drop in oxygen with depth (see graph). (B) The eutrophic lake is either a sediment-filled late stage of a deep lake (solid-line lake bottom) or a shallow contemporary lake (broken-line lake bottom). Breakdown of organic matter in the small hypolimnion has depleted the oxygen supply (see graph). (After G. E. Hutchinson. "Eutrophication," *American Scientist*, 61, p. 270. © 1973 by American Scientist, reprinted by permission of the publisher.)

TABLE 6.4 Oligotrophic versus Eutrophic Lakes

Criteria	Oligotrophic $\longrightarrow$	(Mesotrophic) $\longrightarrow$	Eutrophic $\longrightarrow$
1. Concentration of plant nutrients (N, P)	"Poorly fed"; low concentration of plant nutrients; P<10 μg/l[a]		"Well fed"; high concentration of plant nutrients; P>20 μg/l[a]
2. Organic matter enrichment	Low concentration of organic matter		High concentration of organic matter, either authochtonous (produced in lake) or allochthonous (transported from environment)
3. Biological productivity (organic matter productivity) in g org. C/m²/yr; generally phytoplankton productivity via photosynthesis	Low productivity—few plants (phytoplankton) because of low nutrient concs. Primary productivity <150 g C/m²/yr; concentration chlorophyll A (phytoplankton) <3 μg/l[a]		High productivity—large concentration of plants, particularly plankton (green & blue algae) and/or macrophytes (rooted plants); primary productivity >250 g C/m²/yr; concentration chlorophyll A (phytoplankton) >6 μg/l[a]
4. Depth (volume) of lake	Deep, large volume (>15–25 m depth)		Shallow, small volume (≤10–15 m depth)
5. Water	Clear, blue; Secchi depth >5m[a]		Murky, dark, turbid; Secchi depth <3 m[a]

6. O_2 in hypolimnion in summer	Well-oxygenated hypolimnion (see Figure 6.6); >50% O_2 saturation after stratification[a]	Oxygen depleted in hypolimnion by high biological oxygen demand to decompose organic matter (see Figure 6.6); <10% O_2 saturation ($\sim$<1 mg/ℓ O_2) after stratification[a]
7. Bottom fauna	Diversified bottom fauna, deep-water fish: char, whitefish, trout	Bottom fauna tolerant of low oxygen conditions, no deep-water fish; carp
8. Bottom sediment	Sandy, inorganic, low N	Organic rich muck, high in nitrogen
9. Examples	L. Superior (U.S.), L. Huron (U.S.), L. Geneva (Switzerland), Great Bear Lake (N.W. Canada), Great Slave Lake (N.W. Canada)	Western L. Erie (U.S.), L. Lugano (Switzerland-Italy)

Sources: Chapra and Dobson 1981, where indicated; for other sources see text.

[a] Rounded values from Chapra and Dobson 1981, primarily for offshore Great Lakes.

perhaps fish.   When the organisms die, their remains, along with organic debris from the area surrounding the lake, settle to the bottom into the hypolimnion. Initially there is considerable erosion of glacial till into the lake that rapidly buries the sedimenting organic matter and prevents return of its nutrients to the lake water.   Later, after this first phase, a rather steady supply of nutrients from soil and organic remains is carried into the lake, supporting a rather constant biological productivity over the year.   For a typical dimictic lake, when the lake overturns in the spring and fall, nutrients are carried to the surface, where they stimulate phytoplankton growth.   The lake water becomes stratified in summer and dead plankton and organic debris accumulate and decompose on the bottom, consuming oxygen from the hypolimnion in the process and releasing nutrients.   Gradually a layer of organic-rich sediment builds up (10–15 m since glacial time; Vallentyne 1974).   If the lake was originally deep, the hypolimnion has a large volume and can supply adequate oxygen for organic decomposition.   In this case the lake tends to remain oligotrophic.

   If, however, the lake was originally shallow, as it fills up the volume of the hypolimnion becomes too small to supply sufficient oxygen to counteract organic matter decomposition and, as a result, the bottom becomes depleted in oxygen in the summer.   The rate at which nutrients can reach the surface in such a lake is increased because of the greater bottom area relative to lake volume and better water circulation (the hypolimnion disappears in very shallow lakes).   Also, anoxic conditions tend to make phosphate, which is an important nutrient, more soluble, thus accelerating eutrophication.

   A shallow lake is likely to become moderately eutrophic but it may remain this way for thousands of years in the typical case of undisturbed natural eutrophication (see Figure 6.6; Hutchinson 1973).   The biological productivity in such a lake is perhaps ten times that in an oligotrophic lake but not nearly as excessive as that produced by cultural eutrophication.   Ultimately, a very shallow, naturally eutrophic lake will fill up with enough sediments so that rooted plants can grow on the bottom, and the depletion of bottom oxygen will be so great that most fish cannot live there.   The lake finally becomes a marsh or bog.   Humans have often speeded up this "natural" moderate eutrophication by deforestation and cultivation of the land, which produces greater flow of water into the lake carrying more sediment, nutrients, and organic matter.   This accelerates both sedimentation and biological productivity in the lake.

   Cultural or human-induced eutrophication of a lake is vastly speeded-up eutrophication due to the accelerated input to the lake of nutrients and organic matter from sewage, agriculture, and industries.   This results in greatly increased biological productivity.   A naturally eutrophic lake might only produce 75–250 g $C/m^2/yr$ while a culturally eutrophic lake can support as much as 700 g $C/m^2/yr$ (Rodhe 1969).   Mats of blue-green algae are typical of eutrophic lakes.   Although many naturally eutrophic lakes support fish, in a culturally eutrophied lake the summer oxygen supply is often inadequate for most fish and bottom fauna, except those very tolerant of oxygen deprivation.   The time scale of cultural eutrophication is greatly shortened and, in this way, the lake "dies." However, the process is *reversible* if the nutrient supply is reduced, whereas natural eutrophication is slow and irreversible since it involves filling of the lake basin with sediments.

Attempts have been made to quantify the symptoms of cultural eutrophication in terms of surface water quality. The bounding values shown in Table 6.4 are rounded numbers given by Chapra and Dobson (1981) for the offshore waters of the Great Lakes. The values used by other workers for the Great Lakes (Vollenweider, Muniwarily, and Stadelmann 1974) and other lakes (Wetzel 1975) are somewhat different, so rigid boundaries do not exist for all lakes. The main characters for classifying of surface waters according to Chapra and Dobson (1981) are (1) *primary production* of organic matter via photosynthesis by the phytoplankton: < 150 g organic $C/m^2/yr$ = oligotrophic; > 250 g organic $C/m^2/yr$ = eutrophic; (2) *phytoplankton biomass* (in the surface water), which is measured by the total concentration of chlorophyll A: <3 $\mu g/\ell$ = oligotrophic; >6 $\mu g/\ell$ = eutrophic; (3) *phosphorus concentration* <10 μg P/ℓ = oligotropic; >20 μg P/ℓ = eutrophic; and (4) *water transparency* (or clarity) measured by the depth at which a white (Secchi) disk lowered in the water is no longer visible: >5 m = oligotrophic; <3 m = eutrophic. (Intermediate values for each characteristic are considered to be *mesotrophic*.)

Chapra and Dobson (1981) point out that the degree of depletion of dissolved oxygen in the hypolimnion during (summer) stratification should also be considered when characterizing the trophic state of a lake. Because 10% oxygen saturation (≈ 1 mg/ℓ dissolved O_2) is a threshhold for various biological and chemical processes, a lake is eutrophic when its hypolimnetic O_2 falls below this value during summer stratification. Oligotrophy begins at or above 50% O_2 saturation ($\sim \geqslant$ 6 mg/ℓ dissolved O_2). Dissolved O_2 tends to be a problem in lakes with a thin hypolimnion, that is, shallow lakes which become stratified and have limited oxygen reserves. An example is shallow central lake Erie (18 m) whose hypolimnion has a severe oxygen depletion problem during summer stratification (eutrophic on the above O_2 scale) but whose surface water exhibits other characteristics (productivity, etc.) that place this portion of the lake in the mesotrophic category. Thus, the same area of a lake can be placed in different classifications depending upon which criteria for trophic state are used.

Limiting Nutrients

Organic compounds produced by algal photosynthesis in lakes can be represented as consisting of carbon, hydrogen, oxygen, nitrogen, and phosphorus (see preceding section). Of the nutrient elements needed for photosynthesis, hydrogen and oxygen are readily available. Also carbon is generally available from atmospheric carbon dioxide. The major elements that are not always available are nitrogen and phosphorus. Relatively small amounts of N and P can produce relatively large amounts of organic matter. For instance, to synthesize 100 g (dry weight) of algae, only 7 g of N and 1 g of P (see Equation 6.14) are required.   The total amount of organic matter produced will be determined by the availability of the nutrient element that is the least abundant; this is the so-called *limiting nutrient* and in lakes it is usually phosphorus (less commonly nitrogen) that is limiting.   Vallentyne (1974) has calculated the ratio of *demand* (amount required) by freshwater plants (algae, diatoms, rooted plants) to *supply* from average river water for various nutrients in a world average lake (in late winter before the spring algal growth bloom).   This is shown in Table 6.5.   Those elements whose ratio of demand to supply is greater than 1500 are listed below.

	Demand/Supply	
Element	Late winter	Midsummer
P	80,000	Up to 800,000
N	30,000	Up to 300,000
C	5000	Up to 6000
Si	2000	

This explains why phosphorus might be expected to most often be the limiting nutrient in lakes, followed by nitrogen.   This is also why the addition of these nutrients could lead to greatly expanded biological productivity and eutrophication.

Schindler (1974, 1977) has studied the effects of adding various nutrients (phosphorus, nitrogen, and carbon) to experimental lakes in a remote area of Canada.   He found that the standing crop of phytoplankton is proportional to the concentration of total phosphorus in most lakes and that phosphorus is the limiting nutrient even when the nutrient ratios of lake inputs (streams, rainfall, etc.) might be expected to favor nitrogen or carbon limitation.   This is because biological mechanisms exist in lakes which can correct for carbon deficiencies and, in some cases, nitrogen deficiencies.   Carbon dioxide from the atmosphere makes up for deficiencies in other carbon inputs to the lake.   Nitrogen deficiencies (low N/P ratios) can be made up by development in the lake of blue-green algae which are capable of fixing N_2 from the atmosphere.   In fact, 20%–40% of the total nitrogen input in the experimental lakes is from nitrogen fixation.   The importance of blue-green algae varies.   In a study of a large number of lakes, Smith (1983) found that blue-green algae are only abundant when the ratio of total N/total P is less than 2.9.   (Blue-green algae also make objectionable floating mats, but in lakes with adequate nitrogen they are supplanted by green algae, which do not form such mats.)

Since phosphorus does not occur as a gas in the atmosphere, a lake has no way of compensating for phosphorus deficiencies, and phosphorus thus becomes the limiting nutrient.   Schindler (1977) suggests that only when sudden or very large increases of phosphorus input occur during cultural eutrophication do lakes show temporary carbon or nitrogen limitation, which will ultimately be corrected by biological and environmental mechanisms.   This results in phytoplankton growth

TABLE 6.5 Concentrations of Essential Elements for Plant Growth in Living Tissues of Freshwater Plants (Demand), in Mean World River Water (Supply), and the Plant/Water (Demand/Supply) Ratio of Concentrations

Element	Symbol	Demanded by Plants (%)	Supplied by Water (%)	Demand/Supply (Plant/Water) Ratio (approx.)
Oxygen	O	80.5	89	1
Hydrogen	H	9.7	11	1
Carbon[a]	C	6.5	0.0012	5,000
Silicon	Si	1.3	0.00065	2,000
Nitrogen[a]	N	0.7	0.000023	30,000
Calcium	Ca	0.4	0.0015	<1,000
Potassium	K	0.3	0.00023	1,300
Phosphorus[a]	P	0.08	0.000001	80,000
Magnesium	Mg	0.07	0.0004	<1,000
Sulfur	S	0.06	0.0004	<1,000
Chlorine	Cl	0.06	0.0008	<1,000
Sodium	Na	0.04	0.0006	<1,000
Iron	Fe	0.02	0.00007	<1,000
Boron	B	0.001	0.00001	<1,000
Manganese	Mn	0.0007	0.0000015	<1,000
Zinc	Zn	0.0003	0.000001	<1,000
Copper	Cu	0.0001	0.000001	<1,000
Molybdenum	Mo	0.00005	0.0000003	<1,000
Cobalt	Co	0.000002	0.000000005	<1,000

Source: J. R. Vallentyne, *The Algal Bowl: Lakes and Man.* Copyright © 1974. Environment Canada. Reprinted by permission of the publisher.

[a]Concentrations in water for inorganic forms only.

being proportional to phosphorus concentration. In the water of the lakes he studied, the *weight ratios* of C/P (174) and N/P (31) were more than four times that required for the formation of average (marine) plankton (C/P = 41 and N/P = 7; see reaction 6.14), showing that carbon and nitrogen were present to excess. Also, plots of N versus P for many other lakes indicate that P is exhausted before N.

The algal bloom in a Canadian lake resulting from experimental phosphorus additions (Schindler 1974; see Figure 6.7) is rapid (several weeks) and spectacular, showing the role of phosphorus in cultural eutrophication. If phosphorus enrichment ceases, these small lakes revert to their noneutrophic state rapidly (within a year). Apparently in these lakes phosphorus is quickly sedimented to the lake bottom and stored there and not released again to the water, even under anoxic conditions. (Rapid release of phosphorus from bottom sediments prevents quick recovery from eutrophication and this often occurs in other lakes.)

Figure 6.7 Dramatic illustration of the role of phosphorus in eutrophication. The far lake basin, with murky, opaque water was artificially fertilized with phosphorus, nitrogen, and carbon, whereas the near basin, with clear water, was fertilized only with nitrogen and carbon. Murkiness has been caused by an algal bloom in the far basin. (After D. W. Schindler. "Eutrophication and Recovery in Experimental Lakes: Implications for Lake Management," *Science*, 184, p. 897. Copyright © May 24, 1974 by the Amer. Assoc. for the Advancement of Science, reprinted by permission of the publisher.)

Phosphorus Cycle in Lakes

Since phosphorus is normally the limiting nutrient, we shall consider the phosphorus cycle in a typical lake, as summarized by Stumm and Baccini (1978) (see Figure 6.8). They assume that, on the average, of every 1 mg of phosphorus delivered to a stratified lake, 0.2 mg of phosphorus is insoluble and unreactive, such as mineral apatite from rock weathering, Fe and Al phosphate, and that adsorbed on clays. This unreactive material will not enter into lake reactions, but will be only deposited on the lake bottom. The other 0.8 mg of P is present as dissolved phosphate (from rain, sewage, or agricultural runoff), which can be used as a nutrient by lake algae in photosynthesis, producing organic matter. This 0.8 mg of P combines with 0.2 mg of P from upwelling to produce 100 mg (dry weight) of algal organic matter in the epilimnion of the lake. This algal organic matter (containing 1 mg of organic P) then falls into the hypolimnion of the lake where the phosphorus is liberated to solution by bacterial organic matter decomposition (respiration).

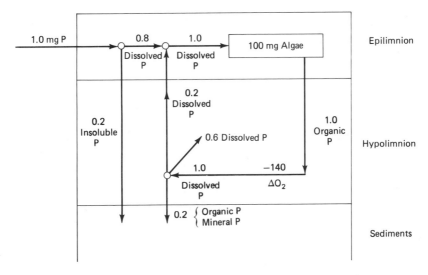

Figure 6.8 Phosphorus cycle in a stratified lake. Units are in mg per arbitrary unit of time. (Adapted from W. Stumm and P. Baccini. "Man-made Chemical Perturbation of Lakes," In *Lakes: Chemistry, Geology, Physics*, ed. A. Lerman, Springer-Verlag, New York, p. 106. Copyright © 1978. Reprinted by permission of the publisher.)

According to the photosynthesis-respiration reaction (6.14), for every atom of P in organic matter which is released, 138 O_2 molecules are consumed or, in weight ratios, approximately 140 mg of O_2 per 1 mg of P. (Molar and weight ratios are almost the same because 1 mole of O_2 has almost the same weight as 1 mole of P.) Thus, oxygen in the hypolimnion of the lake is being used up in the process of oxidizing organic matter. Since the lake is stratified, the hypolimnion is cut off from the atmosphere and little new atmospheric O_2 is thereby available. This leads to the oxygen depletion mentioned previously in eutrophic lakes (where large amounts of organic matter are being produced and falling into the hypolimnion). Figure 6.9 shows that the ratio between the measured concentrations of O_2 and dissolved P in the deep water of an actual lake are in fact close to the ratio 138 O_2:1 P, thus confirming control by respiration.

Of the 1 mg of phosphorus in organic matter introduced into the hypolimnion of our idealized lake (Figure 6.8), 20% or 0.2 mg of soluble P is liberated to solution and returned to the epilimnion where it serves as an algal nutrient again. This occurs by eddy diffusion, even during stratification. Another 0.6 mg of soluble P (60%) accumulates in the lake hypolimnion to be returned to the surface water during lake overturn. This addition to the epilimnion of relatively large quantities of phosphorus in spring and fall overturns often results in algal blooms. The remaining 20% of the phosphorus in the hypolimnion (0.2 mg P) is sedimented onto the lake bottom and stored in lake sediments as residual organic phosphorus

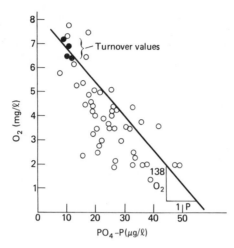

Figure 6.9 The correlation between measured concentrations of dissolved O_2 and PO_4-P in deep lake water (Lake Gersau, Switzerland). The line is for a ratio of 138 O_2:1 P, which is that expected for control by respiration (see text). (Molar and weight ratios of O_2:P are the same because 1 mole of O_2 has nearly the same weight as 1 mole of P). (From Stumm and Baccini, 1978, after H. Ambühl, 1975. *Swiss Journal of Hydrology*, v. 37, pp. 35–52, published by Birkhäuser Verlag AG, Basel, Switzerland.)

or newly formed iron and calcium phosphate or as that adsorbed on clays. Thus, it is effectively removed from the lake cycle.

Sources of Phosphorus in Lakes

The primary sources of phosphorus (and nitrogen) in lakes are direct rainfall and snowfall on the lake and runoff from the surrounding drainage area. In oligotropic lakes most phosphorus in runoff comes from rock weathering and soil transport. However, in areas influenced by humans there are additional sources of phosphorus, including agricultural runoff (containing phosphorus from fertilizers and animal wastes) and sewage (containing phosphorus from human wastes, detergents, and industrial wastes), which are discharged directly into the lake or its inlet tributaries. (For a more detailed discussion of phosphorus on land, sources of phosphorus pollution, and the phosphorus cycle, see Chapter 5). Atmospheric precipitation may be a very important source of phosphorus (and nitrogen) for oligotrophic lakes, particularly those in areas of granitic terrain with low contributions of nutrients from weathering and those lakes whose area is large compared to the drainage area (for example, Lake Superior). Likens, Eaton, and Galloway (1974) found that 50% of the phosphorus in some oligotrophic lakes comes from precipitation. As anthropogenic influences increase, runoff becomes more important, and for eutrophic lakes the average precipitation contribution is only 7% of the phosphorus and 12% of the nitrogen. Table 6.6 summarizes nutrient sources for various lakes.

Chapra (1977) has modelled changes in the sources of phosphorus over the last 150 years for the U.S.–Canada Great Lakes. Those lakes most influenced by human activities, Lakes Michigan, Erie, and Ontario, are depicted in Figure 6.10. There are two phases of increased phosphorus input. The first began around 1850 with much greater land runoff of phosphorus resulting from the conversion of forests to agricultural land. The second phase, which began about 1945, is a

TABLE 6.6 Phosphorus and Nitrogen Sources for Selected Lakes as a
Percent of Total Annual Input

	Phosphorus				Nitrogen			
Lake	Precipitation	Urban Runoff & Waste	Rural Runoff & Waste	Total Runoff	Precipitation	Urban Runoff & Waste	Rural Runoff & Waste	Total Runoff
Disturbed Lakes								
Lake Erie	4			96	18			82
Lake Ontario	4			96	28			72
Lake Ontario[a]	10			90				
European lakes[b]	1	70	29	99	3	37	60	97
Lake Mendota	6	35	59	94	17	11	66	77
Lake Canadaigua	2	46	52	98	3	6	91	97
Eutrophic lakes[c]	7			93	12			88
Undisturbed Lakes								
Lake Superior	46			54	47			53
Lake Huron	27			73	62			38
Oligotrophic lakes[c]	50			50	56			44

Source: Likens, Eaton, and Galloway 1974, except as noted.

[a] From Robertson and Jenkins 1978.

[b] From Vollenweider 1968; see also Stumm 1972.

[c] Summary of 18 lakes.

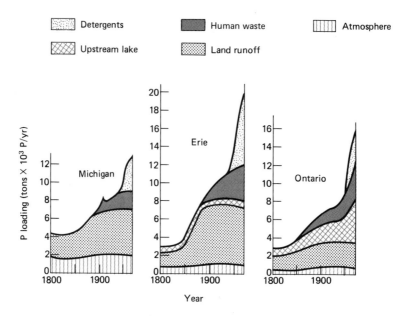

Figure 6.10 Historical loading of total phosphorus (in 10^3 tons/yr) from 1800 to 1970 for three Great Lakes—Michigan, Ontario, and Erie—based on model calculation. (From Stumm and Morgan, 1981, after S. C. Chapra. "Total Phosphorus Model for the Great Lakes," *Journal of Div. of Environmental Engineering*, 103 (EE2), page 153. Copyright © 1977. American Society Civil Engineering, reprinted by permission of the publisher.)

result of population growth, accompanied by the use of sewers, which has introduced human wastes and phosphate detergents directly into the lakes. Lake Erie has the most phosphorus loading and is eutrophic in its western basin, while mesotrophic in the central and eastern basins (Chapra and Dobson 1981). Since it is upstream from Lake Ontario, it greatly increases the phosphorus load in the latter, which is mesotrophic offshore (Chapra and Dobson 1981).

Vollenweider (1968) has plotted annual phosphorus inputs to lakes versus mean depth (Figure 6.11) as a means of predicting which lakes should become eutrophic. He also includes predictions of how much the phosphorus load of Lake Erie and Lake Ontario could be reduced between 1968 and 1986 with phosphate input control, and the growth of the phosphorus load without phosphate control. His diagrams and predictions called attention in the United States and Canada to the importance of human activities as a control on lake-water composition and the dominant (but reversible) role played by phosphorus pollution.

Since Vollenweider made his predictions there have been considerable changes in the composition of the Great Lakes. The peak phosphorus loading in the Great Lakes Basin occurred in 1972 (see Figure 6.11). In 1972, the United States and Canada established a phosphorus control program for municipal sewage water entering the Great Lakes (Lee, Rast, and Jones 1978); the phosphorus concentration in sewage was reduced from 6 mg P/ℓ in the late 1960s to 2 mg P/ℓ from 1975 to 1980 and 1 mg P/ℓ after 1980 (Chapra 1980). In addition, in 1973 the P content in detergents in the Great Lakes drainage basin was greatly reduced.

The results of the Great Lakes phosphorus control program can be seen in Lake Ontario which has a seven-year water residence time and deep well-oxygenated bottom waters (avoiding anaerobic P feedback from bottom sediments) (Chapra 1980). Phosphorus concentrations in Lake Ontario dropped significantly from about 25 µg P/ℓ (eutrophic) in 1973 to about 16 µg P/ℓ (upper level mesotrophic) in 1980 (Kwiatkowski 1982). This 35% drop in P was accompanied by a 15% reduction in open-lake algal biomass concentrations. Lake Ontario would be maintained at upper level mesotrophy (15–20 µg P/ℓ) until 2000 by the phosphorus controls in effect in 1980. But in order to restore oligotrophy in Lake Ontario (i.e., P concentrations < 10 µg P/ℓ), it would be necessary to both reduce diffuse land P runoff (a more difficult task) and reduce the P concentrations in Lake Erie, which supplies 40% of Lake Ontario's phosphorus (Chapra 1980).

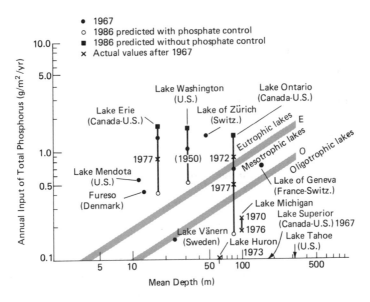

Figure 6.11 Annual phosphorus input versus depth for some North American and European lakes. Oligotrophic lakes lie below line *O*, eutrophic lakes above line *E* with mesotrophic (intermediate) lakes in between. (Diagram, 1967 data, and 1986 predictions from Vollenweider 1968; diagram modified by Vallentyne 1974; see also Stumm and Baccini 1978. Great Lakes data after 1967: Lake Huron in 1973 from Vollenweider, Muniwarily, and Stadelmann 1974; Lake Michigan in 1970 from Chapra 1977, and in 1976 from Eisenreich, Emmling, and Beeton 1977; Lake Ontario in 1972 from Chapra 1980, and in 1977 from Fraser 1980; Lake Erie in 1977 from Vallentyne and Thomas 1978, see also Charlton 1980.)

Lake Erie, the shallowest of the Great Lakes, is still considered eutrophic as a whole, although there have been improvements in lake quality between 1974 and 1980 with P concentrations dropping 33% and algal biomass decreasing 18% (Barica 1982). Lake Erie is often considered as three separate basins. Western Lake Erie (8 m deep) was still eutrophic (30–35 μg P/ℓ) in 1977–1978 despite a 50% drop in P concentration from 1970, and a dramatic reduction in green and blue-green algae. This is largely due to a reduction in P loading from the Detroit River (Nicholls, Standen, and Hopkins 1980). Because of its shallowness, western Lake Erie does not have thermal stratification and thus has no problems with O_2 depletion. As we noted earlier, the central and largest basin of Lake Erie has meso-trophic surface waters but a severe oxygen depletion problem occurs in the bottom waters during summer stratification (often leading to anoxia) (Chapra and Dobson 1981). The goal of restoring year-round oxic conditions to the central Lake Erie bottom waters may be difficult to achieve by reducing P loading (and reduced organic production) because of the thinness of the hypolimnion and its consequently small O_2 reserves (Barica 1982; Charlton 1980). Eastern Lake Erie, which is much deeper, has much less of a problem with O_2 depletion, and, as a result, exhibits both mesotrophic bottom and surface waters (Chapra and Dobson 1981).

POLLUTIVE CHANGES IN MAJOR LAKES: POTENTIAL LOADING

The predicted input of pollutional substances to a lake can be expressed in terms of *potential loading* (Stumm and Morgan 1981). Potential loading is equated to human energy consumption per unit volume of lake water (for example, watts/m³ as in Table 6.7). The expression for potential loading is

$$\text{potential loading} = \frac{\text{drainage area}}{\text{lake area}} \times \frac{1}{\text{lake depth}}$$

$$\times \frac{\text{inhabitants}}{\text{drainage area}} \times \frac{\text{energy consumption}}{\text{inhabitants}}$$

This expression assumes that waste production or "pollution" will be roughly pro-portional to energy consumption and considers the fact that the energy consumption per capita is much greater in heavily industrialized countries like the United States than it is in underdeveloped countries (see Figure 6.12). For any one country where the energy consumption per capita is roughly constant, then the population density per lake volume will be the major criterion for potential pollutional loading. This concept of potential loading is useful for various pollutional inputs (Cl, SO_4, etc.) in addition to those nutrients which cause eutrophication (P, N). Table 6.7 gives loading parameters for various lakes. The first six lakes (including Lake Erie) are or have been eutrophic (Stumm and Morgan 1981).

The potential pollutional loading for the Great Lakes can be compared to

TABLE 6.7 Loading Parameters for Some Lakes

Lake	Country	Drainage Area/ Lake Area	Mean Depth (m)	Inhabitants per m² of Drainage Area	Energy Consumption (10^3 W/inhab.)[c]	Potential Loading[a] = Energy Consump./ Lake Vol. (W/m³)[c]
Greifensee[b]	Switzerland	15	19	441	5.2	1.81
Washington[b]	United States	~15	18	~50	11.4	0.48
Constance[b]	Switzerland-Germany-Austria	19	90	114	5.0	0.12
Lugano[b]	Switzerland-Italy	11	130	264	4.9	0.11
Erie[b]	United States	1.34	19	293	10.0	0.21
Biwa[b]	Japan	4.5	41	~150	4.4	0.07
Winnipeg	Canada	35	13	~3	8.6	0.07
Ontario[b]	United States-Canada	3.2	85	108	10.0	0.041
Michigan	United States-Canada	2.3	84	42.6	10.0	0.012
Huron	United States-Canada	2.0	59	16.9	10.0	0.0057
Titicaca	South America	14	~100	~40	0.2	0.001
Victoria	Africa	3	40	~70	0.4	0.002
Baikal	USSR	17	730	~5	0.8	0.0005
Tanganjika	Africa	4	572	~50	0.3	0.0001
Superior	United States-Canada	1.5	145	~5	10.0	0.0005

Source: Modified from W. Stumm and J. J. Morgan, *Aquatic Chemistry*, 2nd ed., p. 693. Copyright © 1981. John Wiley & Sons, Inc. Reprinted by permission of John Wiley & Sons, Inc.

[a] Potential loading $= \dfrac{\text{drainage area}}{\text{lake area}} \times \dfrac{1}{\text{lake depth}} \times \dfrac{\text{inhabitants}}{10^6 \text{ m}^2 \text{ drainage area}} \times \dfrac{\text{energy consumption}}{\text{inhabitants}} = \dfrac{\text{energy consumption}}{\text{lake volume}}$

[b] Eutrophic lakes, presently or in past.

[c] W = watts

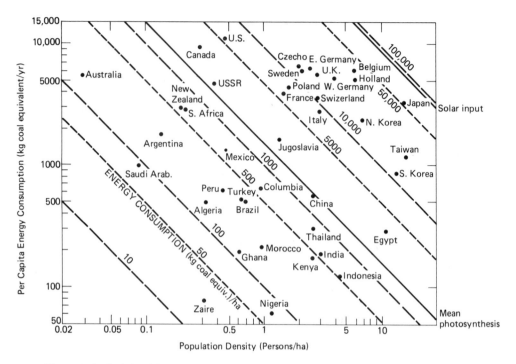

Figure 6.12 Relationship between per capita energy consumption (kg coal equivalent per year) and population density (people per hectare; 1 ha = 10^4 m²). One kg coal equiv./yr/ha ≃ 1.1×10^{-4} watts/m² (Stumm and Morgan 1981). Diagonal lines represent constant values of energy consumption per unit area. (From Stumm and Baccini 1978 after Y. H. Li, 1976, "Population growth and environmental problems in Taiwan (Formosa): A case study," *Environmental Conservation* 3, p.176. Reprinted by permission of the publisher.)

actual changes in their chemistry. For this purpose the nonnutrient ions chloride, sulfate, calcium, and sodium plus potassium (Figure 6.13), which are sensitive to pollution in lakes as they are in rivers (see Chapter 5), can be used. These ions are especially useful because there is an abundance of concentration data on them covering the past 120 years. Lake Superior, which is the least affected by human activities, shows little change in chemistry with time and the concentrations of Lake Superior water resemble incident precipitation (Beeton 1969). Lake Michigan shows considerable increases in Cl^- and particularly SO_4^{--} over the past 90 years. These would undoubtedly be greater except for the fact that sewage from the city of Chicago is diverted into the Chicago River away from the lake. Lake Huron, which is fed by both Lake Superior and Lake Michigan, tends to have concentrations of all ions intermediate between them. However, increased Cl^- and SO_4^{--} concentrations in Lake Huron are due partly to effects from its own drainage area in addition to input from Lake Michigan. Lake Erie, as noted previously, is heavily polluted and this is reflected not only in large increases in sulfate and chloride but for other ions as well. Lake Ontario, as mentioned previously for phosphorus,

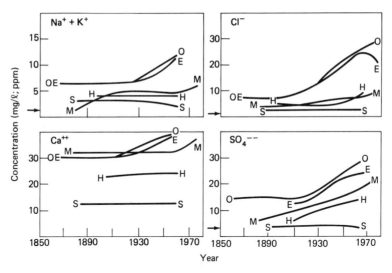

Figure 6.13 Changes in the chemical composition of the Great Lakes from 1850 to 1967. S is Lake Superior; M, Lake Michigan; H, Lake Huron; E, Lake Erie; O, Lake Ontario. Arrows represent concentration of ions in Lake Superior precipitation (Beeton 1969). (Data from Beeton 1969. Diagram modified from J. R. Vallentyne, *The Algal Bowl*, p. 176. Copyright © 1974. Environment Canada. Reprinted by permission of the publisher. Later data: Lake Michigan in 1976 from Bartone and Schelske 1982; Lake Ontario Cl in 1978 from Fraser 1980; northeastern Lake Erie Cl in 1970 and 1978 from Heathcote, Weiler and Tanner 1981.)

shows smaller increases in all ion concentrations than Lake Erie, which feeds it, but the changes mirror Lake Erie changes. Robertson and Jenkins (1978) attribute 94% of the Cl^- input to Lake Ontario as coming from Lake Erie. Thus, cleaning up Lake Erie would clean up Lake Ontario as well.

In summary, it can be seen that, in addition to influences of the surrounding drainage area, shown by potential loading, the chemistry of the Great Lakes (particularly Lakes Huron and Ontario) is complicated by pollution from upstream lakes.

ACID LAKES

During the past several decades many lakes have been found to exhibit large decreases in pH and this has proven to be a serious environmental problem. The acidification of dilute freshwater lakes and streams in large areas of southern Scandinavia, southeastern Canada, and the northeastern United States is linked to the increasingly acid precipitation (pH 4.0–4.6) received in these areas (Wright and Gjessing 1976; for further discussion of acid rain consult Chapter 3). Figure 6.14 shows the change in pH of higher elevation lakes in the Adirondack Mountains

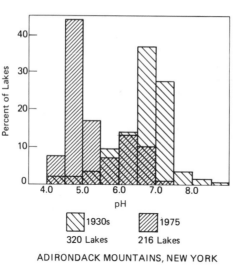

Figure 6.14 Frequency distribution of pH in lakes in the Adirondack Mountains, New York, in the 1930s and in 1975. (From R. F. Wright and E. T. Gjessing. "Changes in the Chemical Composition of Lakes," *Ambio*, 5, (5–6), p. 221. Copyright © 1976 by the Royal Swedish Academy of Sciences, reprinted by permission of the publisher. Based on data of Schofield 1976a.)

of New York from the 1930s to 1975. Fifty-one percent of the lakes had pH values less than 5 in 1975 as compared to only 4% in the 1930s (Schofield 1976a). The vast majority of the Adirondack lakes in the 1930s (80%) were in the pH range of 6.9 to 7.5, which is what one might expect in a noncalcareous granitic area of this type. The average yearly change in pH was −0.05 pH units, which is also typical of changes in pH in Scandinavian and Ontario lakes which have received acid precipitation (Wright and Gjessing 1976).

The most convincing evidence for the acidification of lakes, even where no long-term records of pH exist, has been their declining fish populations (many species of fish do not survive in acid lakes). For example, 90% of the Adirondack lakes that had a pH less than 5 in 1975 had no fish (Schofield 1976a). A group of remote lakes about 65 km from the metal smelters in Sudbury, Ontario, experienced very rapid acidification (change in pH from 6.3 to 4.9) in the 1960s (Beamish *et al.* 1975) and this was accompanied by a simultaneous loss of fish in these lakes. (This change of lake pH occurred *after* the building of a higher smokestack at Sudbury in order to spread SO_2 over a larger downwind area.)

Acidification of lakes occurs in areas that are unusually sensitive to acid precipitation because of a characteristic bedrock geology and soil. They are underlain by weathering-resistant igneous and metamorphic rocks or noncalcareous sandstones and have thin, patchy acid soils, neither of which are conducive to acid neutralization. Also calcium carbonate, which has the buffering ability to provide carbonate and bicarbonate ions to solution and thus neutralize acid precipitation, is generally lacking. In contrast, other areas that are also receiving acid precipitation but have calcareous sedimentary rocks (limestone and calcareous sandstone) contain lakes whose pH values are essentially unaffected by acid precipitation (Norton 1980). Figure 6.15 shows areas in North America that would be sensitive to acid precipitation based on bedrock geology. Areas with acid lakes in Scandinavia have similar geology. Pristine lakes in these sensitive noncalcareous areas are dilute because of slow chemical weathering and, as a result, they have low bicarbonate concentrations and pH values, before acidification, between 6 and 7. In contrast, lakes developed in areas with calcareous sedimentary rocks can be expected to have a natural (pre-acidification) pH of around 8.0 and higher bicarbonate concentrations. Thus, initial lake composition can be used as a guide to susceptibility to acidification. Consistently acid lakes (pH < 5), as opposed to those that have episodes of acidity, seem to develop in areas with susceptible bedrock geology that are also receiving precipitation more acid than about pH 4.6 (Henriksen 1979).

The effect of acid rain on lakes varies depending upon what part of it falls directly on the lake and what part is runoff from the land. Soils generally are better able to neutralize acid precipitation than lake water. In Norway, runoff averages around 0.2–0.9 pH units higher than precipitation, and lakes are consequently 0.3–1.0 units higher than precipitation, (Wright and Gjessing, 1976). Thus, even in susceptible areas receiving acid precipitation, lakes will vary in acidity depending upon how much contact their input water has had with the soil and what the local characteristics of the soil and bedrock are. (This can be seen in Figure 6.14 in the distribution of pH of Adirondack lakes in 1975 where, although most lakes are acid, 10% of the lakes had a pH > 6.5.) The local characteristics include the geometry of the lake basin and the thickness and type of soil. A lake, with a small area and volume, whose input comes from a large watershed will receive

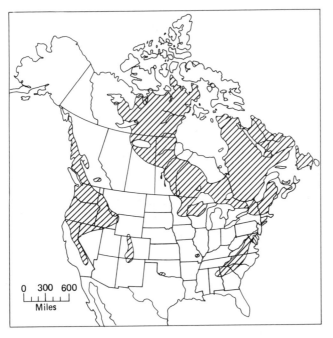

Figure 6.15 Regions in North America containing lakes that would be sensitive to potential acidification by acid precipitation (shaded areas). These areas have igneous or metamorphic bedrock geology which results in dilute lakes with low HCO_3^- concentrations (<0.5 mEq HCO_3^-/ℓ). Unshaded areas have calcareous or sedimentary bedrock geology. (From J. N. Galloway and E. B. Cowling. "The Effects of Precipitation on Aquatic and Terrestrial Ecosystems, A Proposed Precipitation Chemistry Network," *Journal of the Air Pollution Control Assoc.*, 28(3), p. 233. Copyright © 1978. J. of the Air Pollution Control Assoc. Reprinted by permission of the publisher.)

0 300 600
Miles

runoff that has had considerable soil contact. Such a lake is less likely to become acid. In contrast, acid lakes are favored by steep slopes and exposed bedrock, with little vegetation and soil development; in this way they receive precipitation virtually unaffected by the soil (Galloway and Cowling 1978). Also lakes fed by headwater streams, which are more acid than larger streams, will tend to become more acid (Johnson 1979).

The soil cover in susceptible areas is frequently absent or thin and the soil varies in grain size and composition. Carbonate minerals in rock outcrops, in soils, or in calcareous glacial till will prevent lake acidity by buffering acid precipitation. In areas of noncalcareous soils, the presence of unconsolidated sediment, which has a greater surface area for minerals to contact acid precipitation, aids acid neutralization. For example, Northern Ontario lakes in a quartz sandstone area receiving acid precipitation have a pH of 3.8–4.4 when on bedrock, while lakes with gravel- or sand-sized sediment have a pH > 5 (Kramer 1978). In addition to grain size, the presence of clays and organic matter in soils also helps to provide acid neutralizing capacity (see below); (see also Gorham and McFee 1978 for further details on neutralization.)

As pointed out by Krug and Frink (1983) and others, in the eastern North American and northern European areas that are receiving acid rain, many soils and lakes may be naturally acid due to the presence of organic acids, particularly humic acid. Forest regrowth occurring in these areas tends to produce acid soils with thick organic layers. Thus, how can one decide whether or not any given lake owes its acidity to natural processes as opposed to pollution? The criteria, mentioned previously, that indicate the presence of organic acidity in rivers, should also be true for lake acidity partially or totally due to (natural) organic acids. If acidity is organic in origin, there should be a high concentration of dissolved organic carbon relative to dissolved inorganic matter. Also, the sum of the major inorganic cations should be greater than the sum of the major inorganic anions because organic anions supply part of the negative charge balance.

In addition to chronically acidified lakes, there are also episodic pH drops in poorly buffered, but usually nonacidic, lakes caused by acid snowmelt runoff in the spring. This comes about because acid precipitation accumulates in the snow over the winter, leading to large concentrations of sulfuric acid at the time of spring runoff. One Adirondack lake dropped from pH 7 to pH 5.9 over a period of a week or two and temporarily had greatly increased Al concentrations resulting in fish mortality (Schofield 1980).

Acid lakes have a distinctive chemical composition. In general they have hydrogen-calcium-magnesium sulfate waters as opposed to unaffected lakes, which have calcium-magnesium bicarbonate waters (Wright and Gjessing 1976). Table 6.8 gives the chemical composition of two sets of lakes, both located in similar areas, one in Norway and the other in Ontario, Canada. The acid lakes in each case receive very acidic precipitation (pH < 4.5) while the unaffected lakes receive less acidic precipitation (pH > 4.8). The sulfate concentration in the acid lakes is three to five times that in unaffected lakes. All the bicarbonate (HCO_3^-) has

TABLE 6.8 Mean Chemical Composition of Acid Lakes in Areas Receiving Highly Acidic Precipitation and of Otherwise Similar Nonacidic Lakes in Areas Not Receiving Highly Acidic Precipitation

Area	Number of Lakes	Lake pH	Rain pH	Concentration (mg/ℓ)						
				Na^+	K^+	Ca^{++}	Mg^{++}	HCO_3^-	SO_4^-	NO_3^-
Norway:										
Southern	26	4.76	<4.5	9	4	50	25	11	92	4
West Central	23	5.2	>4.8	9	3	16	7	13	30	5
Ontario, Canada:										
La Cloche Mtns. S.E. Ontario	4	4.7	<4.5	9	10	150	65	0	290	—
Experimental Lakes Area, N.W. Ontario	40	5.6–6.7	>4.8	4	10	80	65	60	55	<1.5

Source: R. F. Wright and E. T. Gjessing. "Changes in the Chemical Composition of Lakes," *Ambio*, 5, (5-6), pp. 220–221. Copyright © 1976 by the Royal Swedish Academy of Sciences, reprinted by permission of the publisher.

Note: Data corrected for sea salt on the basis of Cl^-.

been lost in the acidic Ontario lakes. While the acidic Norwegian lakes have a slightly lower HCO_3^- concentration than the "nonacidic" Norwegian lakes, both concentrations are low and the nonacidic lakes may already have been somewhat affected by acid precipitation. The increased Ca^{++} and Mg^{++} concentrations in acid lakes are often accompanied by increased dissolved Al and heavy metal concentrations (Wright and Gjessing 1976). The cation increase reflects rock and soil effects in neutralizing acid precipitation. Sulfate concentrations are higher in acid lakes because of the nature of acid precipitation, which is predominantly sulfuric acid (H_2SO_4) (see the section on acid rain in Chapter 3).

Bicarbonate loss in acid lakes reflects bicarbonate dependence upon the pH of lake water and represents a loss of *buffering* in the lake. Buffering in lake water is the ability of the water to neutralize input of acid (or base). If a lake is buffered, its pH is not greatly changed by the addition of moderate quantities of acid (or base). Most lakes are buffered by carbonate species (bicarbonate in particular) and the most effective bicarbonate buffering occurs in the pH range (6.0–8.5) of most lakes (Hem 1970). Bicarbonate (HCO_3^-) buffering of acid comes about via the reaction of H^+ with HCO_3^- to produce neutral carbonic acid (H_2CO_3):

$$HCO_3^- + H^+ \leftrightharpoons H_2CO_3$$

Thus, the buffering ability of a lake depends upon its concentration of HCO_3^-.

Figure 6.16 shows the relative molar concentrations of H_2CO_3 and HCO_3^- at different pH values in a fairly dilute solution (total carbonate concentration of 10^{-3} moles per liter) typical of fresh water. The concentrations of HCO_3^- and H_2CO_3 are equal at pH 6.3 and at pH values greater than that there is little change in the concentration of HCO_3^- for a change in pH (i.e., there is good buffering). At pH values less than 6, the concentration of HCO_3^- drops very rapidly with a decrease in pH. There is no buffering at all below the point where the concentration of H^+ equals the concentration of HCO_3^-. At a total carbonate concentration of 10^{-3}, as in Figure 6.16, this point is at pH 4.65. (For a more dilute solution with total carbonate concentration of 10^{-4} moles/ℓ, represented by the dashed lines in Figure 6.16 and found in some lakes susceptible to acidification, the buffer capacity would be lost at about pH 5.15; for rainwater, which is in equilibrium with atmospheric CO_2 and is even more dilute, this point is at pH 5.65.)

Kramer (1978) shows why the buffering ability of lakes in a calcareous area are much greater than those in a noncalcareous area. A lake in a calcareous regime has calcareous minerals (such as calcite, $CaCO_3$) that can resupply HCO_3^- previously lost by reaction with H^+. A lake in a calcareous area in equilibrium with atmospheric CO_2 would have a pH of 8.4 and a molar concentration of HCO_3^- of about 10^{-3} (Garrels and Christ 1965). This pH and HCO_3^- concentration favor a high buffering capacity (see Figure 6.16). In addition, dissolution of $CaCO_3$

$$H^+ + CaCO_3 \rightarrow Ca^{++} + HCO_3^- \tag{6.15}$$

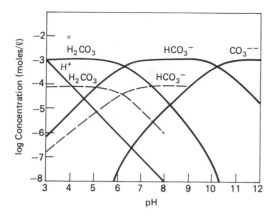

Figure 6.16 Concentration of carbonate species (H_2CO_3, HCO_3^- and CO_3^{--}) in fresh water at 25° C. (Total carbonate concentration = $[H_2CO_3]$ + $[HCO_3^-]$ + $[CO_3^-]$ = 10^{-3} moles/ℓ). Concentration of H^+ is also shown. Dashed lines shows how the curves shift for total carbonate concentration = 10^{-4} moles/ℓ. (Redrawn from W. Stumm and J. J. Morgan, *Aquatic Chemistry*, 2nd ed., p. 176. Copyright © 1981. John Wiley & Sons, Inc. Reprinted by permission of John Wiley & Sons, Inc.)

replaces HCO_3^- previously lost from solution by reaction with acid rain. In this way calcareous lakes may have essentially "infinite" buffering capacity.

By contrast, dilute lakes in noncalcareous regions have a much lower initial pH of 6 to 7, and much lower HCO_3^- concentration (about 10^{-4} moles/ℓ). If the only input were acid rain of pH 4 (H^+ = 10^{-4} M) the HCO_3^- would be used up upon addition of a volume of rain equal to that in the lakes and no HCO_3^- would be resupplied by mineral dissolution. Thus, lakes in noncalcareous area, which are poorly buffered and have little ability to neutralize incoming H^+ ions, tend to become acid. Other terrain and soil factors discussed previously therefore become very important in such areas.

Lakes in susceptible noncalcareous areas with an initial pH of 6 to 7 tend to show a characteristic pattern of acidification in response to acid precipitation (Wright and Gjessing 1976). This is determined by the H_2CO_3 - HCO_3^- buffering curve (Figure 6.16). The concentration of HCO_3^- relative to H_2CO_3 is at a maximum at pH 6.3 (and higher) and it changes very little with pH changes. Thus, in the first stages of lake acidification the lake pH will change slowly toward pH 6 because there is adequate HCO_3^- for buffering. However, as the lake pH drops below 6.0, the concentration of HCO_3^- declines rapidly and further additions of H^+ result in a much more rapid drop in pH (i.e., buffering is poor). The lake is also very sensitive to temporary changes in pH in the pH range between 5.0 and 6.0. Below about pH 5.0, lakes are unbuffered due to loss of HCO_3^- and are chronically acid. Wright and Gjessing (1976) observed that, because of local variations in terrain and soil contact, the frequency distribution of pH of lakes receiving acid precipitation (see Figure 6.14 for Adirondack lakes in 1975) tends to reflect the bicarbonate buffering curve with a number of lakes in the better-buffered pH range above pH 6, few lakes in the poorly buffered range of pH 5.5–6.0, and many acid lakes below pH 5.0.

The reaction of H^+ with $CaCO_3$ above (Equation 6.15) also releases Ca^{++} ions; if the carbonate mineral is dolomite, $CaMg(CO_3)_2$, it also releases Mg^{++}

ions. This is part of the reason for the increased Ca^{++} and Mg^{++} concentration in acidic lakes in comparison to similar nonacidic lakes.

Even in the absence of calcareous minerals, soils can provide some buffering or neutralization of H^+ ions. This is accomplished by cation exchange and by chemical weathering (Norton 1980; Galloway et al. 1981). Acid neutralization via cation exchange involves uptake of H^+ to replace cations associated with clays and organic (humic) substances. It is more rapid than weathering because exchangeable cations are less strongly held by the host phases. Highly acidic weathering involves the uptake of H^+ via the dissolution of aluminous phases (see Chapter 4). For example:

$$Al_2Si_2O_5(OH)_{4_{kaolinite}} + 6H^+ \rightarrow 2Al^{+++}_{aq} + 2H_4SiO_{4_{aq}} + H_2O$$

$$4H_2O + NaAlSi_3O_{8_{plagioclase}} + 4H^+ \rightarrow Na^+_{aq} + Al^{+++}_{aq} + 3H_4SiO_{4_{aq}}$$

$$Al(OH)_{3_{gibbsite}} + 3H^+ \rightarrow Al^{+++}_{aq} + 3H_2O$$

Most of these reactions release Al^{+++} in solution in exchange for the H^+ ions that they neutralize. Since aluminum solubility increases at low pH (below pH 5; Cronan and Schofield 1979), this aluminum can be transported in solution in soil water to lakes and, as a result, the aluminum concentration of acid lakes tends to be higher than similar nonacid lakes. This is shown in Figure 6.17. The excessive Al in acid lakes probably contributes to fish mortality (Cronan and Schofield 1979; Schofield 1980).

Schindler et al. (1985) *artificially* acidified a small experimental lake from pH 6.8 to 5.0 over an eight-year period to determine changes in lake biota in response to increased acidity. These changes occurred in direct response to increased H^+

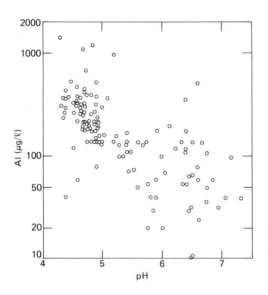

Figure 6.17 Aluminum versus pH for 217 high-altitude lakes in the Adirondack Mountains, New York. (From Galloway et al. 1981, generalized from Schofield 1976b.)

ion since all other ion concentrations were not greatly increased. Decline in the fish population was caused by reproductive failure (below pH 5.4) and loss of key species in the fish food chain. Algal mats appeared in shallow areas where fish normally spawned. However, contrary to expectations, acidification did not cause decreases in primary productivity or rates of decomposition nor did decreases in phosphorus concentration result.

Eutrophic lakes, which have anoxic bottom water during summer stratification, may be protected against becoming acid. In two of the Canadian experimental lakes (see earlier discussion) that were artificially made eutrophic, a model calculation shows that enough persistent alkalinity could be produced by bacterial processes in the hypolimnion to neutralize typical levels of acid deposition. (This is not true for a noneutrophic lake; see Kelly et al. 1982.) Acid deposition results in higher levels of sulfate and nitrate deposition in addition to H^+. In lakes and soils, bacteria reduce SO_4^{--} to H_2S and NO_3^- to N_2 under anoxic, highly organic-rich conditions. These bacterial reduction processes both result in the production of alkalninity (HCO_3^-) (for reactions, see Table 6.3), which is capable of neutralizing H^+ in a lake. However, the alkalinity produced under anoxic ʋottom water conditions during summer stratification may be only temporary unless H_2S is converted to FeS_2 (pyrite) and permanently removed to sediments, and unless N_2 escapes. Otherwise, when aerobic conditions are restored on lake overturn, H_2S is reoxidized to SO_4^{--} and N_2 to NO_3^-, and the HCO_3^- alkalinity is accordingly consumed.

SALINE AND ALKALINE LAKES

In arid to semiarid climates, lakes often are saline. This comes about primarily because there is no outflow of water from such lakes other than evaporation. Waters containing dissolved salts flow in but only pure water is lost by evaporation, thus leaving the salts behind to accumulate in the lake. Arid conditions, however, don't always produce saline lakes. Necessary conditions for saline lake formation and persistence, according to Eugster and Hardie (1978: 237–238) are as follows (our additions in brackets): "(1) outflow of water must be restricted, as it is in a hydrologically closed basin; (2) evaporation must exceed inflow [during initial stages]; and (3) [for persistence] the inflow must be sufficient to sustain a standing body of water." An unusually favorable locale for saline lake formation, according to Eugster and Hardie, is in arid basins located near high mountains, which serve as precipitation traps and sources of groundwater. Some examples are the numerous saline lakes found in intermontane basins of the western United States including the Great Salt Lake.

In general, the volume of water in saline lakes fluctuates considerably both seasonally and from year to year depending upon climatic conditions. Nevertheless, for highly saline lakes one can visualize a (quasi-) steady-state water balance of input by springs and streams and output by evaporation and a steady-state salt balance of input by springs and streams and output by the *precipitation of saline minerals*.   In other words, dissolved salts cannot build up forever in saline lakes, and ultimately saturation is reached with respect to soluble minerals.   One of the characteristic features of saline lakes is the unusualness of the minerals formed from them and found in their sediments.   Some examples are shown in Table 6.9. (Besides soluble phases, some insoluble silicate minerals are listed which form by reaction of saline solutions with silicate detritus; see below).   Pathways of evaporation necessary to form different minerals and to bring about different lake-water compositions are discussed and summarized by Eugster and Hardie (1978), Drever and Smith (1978), and Eugster and Jones (1979).

Saline lakes are often also highly alkaline and exhibit a high pH.   Whereas most freshwater lakes (excluding the special case of acid lakes) have pH values ranging from roughly pH 6 to 8 (Baas Becking, Kaplan, and Moore 1960), the pH of saline lakes can rise to values greater than 10.   The cause for high pH has been studied by several workers (e.g., Garrels and Mackenzie 1967) and can be explained fundamentally in terms of the natural processes of weathering, evaporation, and CO_2 gas equilibration.   The reasoning begins as follows: In areas underlain by acid igneous rocks, the weathering of feldspars and volcanic glass by carbonic acid (see Chapter 4) results in the production of groundwaters containing dissolved HCO_3^- that is balanced by Na^+ and K^+ as well as by Ca^{++} and Mg^{++}.   In other words, the concentration of HCO_3^- is more than twice the concentration of Ca^{++} plus Mg^{++}.   This means that upon evaporation of the groundwater after passing

TABLE 6.9   Some Typical Minerals Formed from Saline Lakes

Mineral	Composition
Halite	NaCl
Gypsum	$CaSO_4 \cdot 2H_2O$
Calcite	$CaCO_3$
Dolomite	$CaMg(CO_3)_2$
Thenardite	Na_2SO_4
Mirabilite	$Na_2SO_4 \cdot 10H_2O$
Glauberite	$CaNa_2(SO_4)_2$
Trona	$Na_2CO_3 \cdot NaHCO_3 \cdot 2H_2O$
Nahcolite	$NaHCO_3$
Pirssonite	$CaNa_2(CO_3)_2 \cdot 2H_2O$
Gaylussite	$CaNa_2(CO_3)_2 \cdot 5H_2O$
Smectite	Mg-aluminosilicate $\cdot nH_2O$
Sepiolite	$Mg_2Si_3O_8 \cdot nH_2O$
Analcime	$NaAlSi_2O_6 \cdot H_2O$
Aphthitalite	$K_3Na(SO_4)_2$

Note: Many other minerals have been identified depending on lake composition (e.g., borates), but, for lack of space, are not listed here.

into a lake, the lake cannot precipitate all HCO_3^- as Ca^{++} or Mg^{++} carbonates, and some HCO_3^- remains behind to be concentrated by evaporation.

Let us follow the course of composition change during the evaporation of a typical igneous-derived groundwater according to the scheme proposed by Garrels and Mackenzie (1967). This is shown in Figure 6.18. As the water undergoes initial evaporative concentration it reaches saturation with calcium carbonate and,

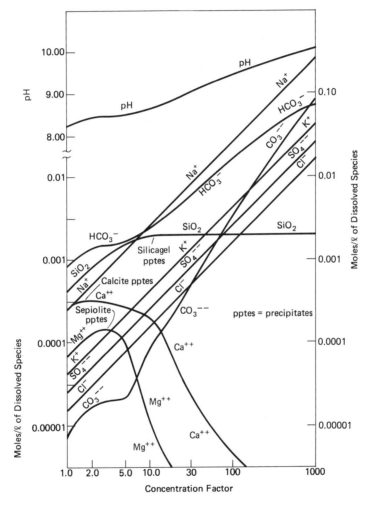

Figure 6.18 Calculated results for the evaporation of typical springwater, from a granitic terrain, which is in equilibrium with atmospheric CO_2. (After R. M. Garrels and F. T. Mackenzie. "Origin of the Chemical Compositions of Some Springs and Lakes," In *Equilibrium Concepts in Natural Water Systems*, Adv. Chem. Series 67, p. 239. Reprinted with permission from the American Chemical Society. Copyright © 1967 by the American Chemical Society.)

at about the same time, also with magnesium silicate (represented in Figure 6.18 by the mineral sepiolite). Further evaporation results in $CaCO_3$ precipitation (see below) and in Mg-silicate precipitation according to the generalized reaction:

$$2Mg^{++} + 4HCO_3^- + 3H_4SiO_4 \rightarrow Mg_2Si_3O_8 \cdot nH_2O + 4CO_2 + 8H_2O$$

Note that Mg-silicate precipitation (along with the removal of silica) results in the removal of two HCO_3^- ions for each Mg^{++} ion. If the concentration of silica exceeds that of Mg^{++}, and the concentration of HCO_3^- exceeds twice the concentration of Mg^{++}, on a molar basis, then eventually evaporation should result in a loss of most all the Mg^{++}. This is shown in Figure 6.18. During and after Mg-silicate precipitation, $CaCO_3$ precipitation results in the removal of Ca^{++} and HCO_3^- according to the reaction:

$$2HCO_3^- + Ca^{++} \rightarrow CaCO_3 + CO_2 + H_2O$$

Note that, analogous to Mg-silicate formation, HCO_3^- is removed in a ratio with Ca^{++} of 2:1. If the remaining concentration of HCO_3^- exceeds twice that of Ca^{++}, as would be the case for our typical igneous-derived water, than upon further evaporation all Ca^{++} is removed and eventually an alkaline lake results.

After Mg-silicate and $CaCO_3$ formation, further evaporation causes the concentration of HCO_3^- to rise at about the same rate as other ions, such as Cl^-, which are not involved in precipitation reactions. While this is going on the pH is also rising (Figure 6.18). This is so because increase in HCO_3^- concentration causes the reaction

$$H^+ + HCO_3^- \rightarrow H_2O + CO_2 \uparrow$$

to be driven to the right resulting in the loss of hydrogen ions. (Carbon dioxide does not build up in solution to back-react, but instead is readily lost to the atmosphere.) In this way an alkaline lake comes about. Because the rise in pH causes some HCO_3^- to be converted to CO_3^{--} (see Figure 6.16), the concentration of CO_3^{--} ions also rises and the rise is faster than that expected for simple evaporative concentration (see Figure 6.18).

If evaporation continues to a very great extent, not shown in Figure 6.18, eventually saturation with alkali carbonates, for example $NaHCO_3$, may be reached and in this way rare soda lakes can arise.

From the above one can see that without weathering of Na-K silicates, especially volcanic glass (which weathers rapidly), the evolution of alkaline lakes would not follow the path shown in Figure 6.18. In the situation where HCO_3^- concentration does not exceed that necessary to precipitate all Ca^{++} and Mg^{++}, no HCO_3^- can build up on continued evaporation and no high pH results. This explains why many salt lakes (e.g., the Great Salt Lake) are not highly alkaline.

The evaporation path shown in Figure 6.18 was merely calculated by Garrels and Mackenzie (1967). However, more recently this "thought experiment" has actually been duplicated using real water. Gac et al. (1978) have evaporated dilute water from the River Chari, in the country of Chad in Africa, and found that it underwent concentration changes during evaporation that more or less agree with those shown in Figure 6.18.   One modification that Gac et al. found was that the Garrels and Mackenzie predictions were better matched if detrital clays, collected with the river water, were left in the water during evaporation.   Removal of the clays resulted in much delayed and subdued precipitation of Mg-silicate.   Apparently the aluminous clays reacted with Mg^{++} and H_4SiO_4 upon evaporation to form a Mg-aluminosilicate, such as smectite, which precipitates more easily than pure Mg-silicates (i.e., sepiolite).

REFERENCES

AMBÜHL, H.   1975.   Versuch der Quantifizierung der Beeinflussing der Oekosystems durch chemische Faktoren: Stehen Gewässer, *Schweiz. Z. Hydrol.* 37: 35–52.

BAAS BECKING, L. G. M., I. R. KAPLAN, and D. MOORE.   1960.   Limits of the natural environment in terms of pH and oxidation-reduction potentials, *J. Geology* 68: 243–284.

BARICA, J.   1982.   Lake Erie oxygen depletion controversy, *J. Great Lakes Res.* 8 (4): 719–722.

BARTONE, C. R., and C. L. SCHELSKE.   1982.   Lake-wide seasonal changes in limnological conditions in Lake Michigan in 1976, *J. Great Lakes Res.* 8 (3): 413–427.

BEAMISH, R. J., W. L. LOCKHART, J. C. VAN LOON, and H. H. HARVEY.   1975.   Long-term acidification of a lake and resulting effects on fishes, *Ambio* 4 (2): 98–104.

BEETON, A. M.   1969.   Changes in the environment and biota of the Great Lakes.   In *Eutrophication: Causes, consequences and correctives*, pp. 150–187.   Natl. Acad. Sci./ Natl. Res. Council Publ. no. 1700.

BERNER, R. A.   1980.   *Early diagenesis: A theoretical approach*.   Princeton, N.J.: Princeton Univ. Press, 241 pp.

CHAPRA, S. C.   1977.   Total phosphorus model for the Great Lakes, *J. Environ Eng. Div., Amer. Soc. Civ. Eng.*, 103 (EE2): 147–161.

———.   1980.   Simulation of recent and projected total phosphorus trends in Lake Ontario, *J. Great Lakes Res.*, 6 (2): 101–112.

CHAPRA, S. C., and H. F. H. DOBSON.   1981.   Quantification of the lake trophic typologies of Naumann (surface quality) and the Thienemann (oxygen) with special reference to the Great Lakes, *J. Great Lakes Res.* 7 (2): 182–193.

CHARLTON, M. N.   1980.   Oxygen depletion in Lake Erie: Has there been any change? *Can. J. Fish. Aquatic Sci.* 37: 72–81.

CLAYPOOL, G., and I. R. KAPLAN.   1974.   The origin and distribution of methane in marine sediments.   In *Natural gases in marine sediments*, ed. I. R. Kaplan, pp. 99–139.   New York: Plenum Press.

CRONAN, C. S., and C. L. SCHOFIELD.   1979.   Aluminum leaching in response to acid precipitation: Effects on high elevation watersheds in the northeast, *Science* 204: 304–306.

DREVER, J. I., and C. L. SMITH. 1978. Cyclic wetting and drying of the soil zone as an influence on the chemistry of ground water in arid terranes, *Amer. J. Sci.* 278: 1448–1454.

EISENREICH, S. J., P. J. EMMLING, and A. M. BEETON. 1977. Atmospheric loading of phosphorus and other chemicals to Lake Michigan, *J. Great Lakes Res.* 3: 291-3-4.

EUGSTER, H. P., and L. A. HARDIE. 1978. Saline lakes. In *Lakes: Chemistry, geology, physics*, ed. A. Lerman pp. 237–293. New York: Springer-Verlag.

EUGSTER, H. P., and B. F. JONES. 1979. Behavior of major solutes during closed-basin brine evolution, *Amer. J. Sci.* 279: 609–631.

FENCHEL, T., and T. H. BLACKBURN. 1979. *Bacteria and mineral cycling.* New York: Academic Press. 225 pp.

FRASER, A. S. 1980. Changes in Lake Ontario total phosphorus concentrations 1976–1978, *J. Great Lakes Res.* 6 (1): 83–87.

GAC, J.-Y., D. BADAUT, A. AL-DROUBI, and Y. TARDY. 1978. Comportement du calcium, du magnésium, et de la silice en solution. Précipitation de calcite magnésienne, de silice amorphe et de silicates magnésiens au cours de l'évaporation des eaux du Chari (Tchad), *Sci. Géol. Bull. Strasbourg*, 31: 185–197.

GALLOWAY, J. N., and E. B. COWLING. 1978. The effects of precipitation on aquatic and terrestrial ecosystems, a proposed precipitation chemistry network. *J. Air Poll. Control Assoc.* 28 (3): 229–235.

GALLOWAY, J., S. A. NORTON, D. W. HANSON, and J. S. WILLIAMS. 1981. Changing pH and metal levels in streams and lakes in the eastern U.S. caused by acid precipitation. In *Proc. EPA Conference on Lake Restoration*, pp. 446–452.

GARRELS, R. M., and C. L. CHRIST. 1965. *Solutions, minerals and equilibria.* New York: Harper. 450 pp.

GARRELS, R. M., and F. T. MACKENZIE. 1967. Origin of the chemical compositions of some springs and lakes. In *Equilibrium concepts in natural water systems*, pp. 222–242. *Amer. Chem. Soc. Adv. Chem.* ser. 67.

GORHAM, E., and W. W. McFEE. 1978. Effects of acid deposition upon outputs from terrestrial to aquatic ecosystems. In *Effects of acid precipitation on terrestrial ecosystems*, pp. 465–480. New York: Plenum Press.

HASLER, A. D. 1947. Eutrophication of lakes by domestic drainage, *Ecology* 28: 383–395.

HEATHCOTE, I. W., R. R. WEILER, and J. W., TANNER. 1981. Lake Erie nearshore water chemistry at Nanticoke, Ontario, 1969–1978, *J. Great Lakes Res.* 7 (2): 130–135.

HEM, J. D. 1970. *Study and interpretation of the chemical characteristics of natural water.* U.S.G.S. Water Supply Paper no. 1473. 363 pp.

HENRIKSEN, A. 1979. A simple approach for identifying and measuring acidification of freshwater, *Nature* 278: 542–545.

HUTCHINSON, G. E. 1957. *A treatise on limnology*, vol. 1. New York: John Wiley. 1015 pp.

———. 1973. Eutrophication, *Amer. Scientist* 61: 269–279.

IMBODEN, D., and A. LERMAN. 1978. Chemical models of lakes. In *Lakes: Chemistry, geology, physics* ed. A. Lerman, pp. 341–356. New York: Springer-Verlag.

JOHNSON, N. M. 1979. Acid rain: Neutralization within the Hubbard Brook ecosystem and regional implications, *Science* 204: 497–499.

KELLY, C. A., J. W. M. RUDD, R. B. COOK, and D. W. SCHINDLER. 1982. The potential importance of bacterial processes in regulating rate of lake acidification, *Limnol. Oceanogr.* 27 (5): 868–882.

KRAMER, J. R. 1978. Acid precipitation. In *Sulfur in the environment, Part 1: The atmospheric cycle,* ed. J. R. Nriagu, pp. 325–369. New York: John Wiley.

KRUG, E. C., and C. R. FRINK. 1983. Acid rain on acid soil: A new perspective, *Science* 221: 520–525.

KWIATKOWSKI, R. E. 1982. Trends in Lake Ontario surveillance parameters, 1974–1980, *J. Great Lakes Res.* 8 (4): 648–659.

LEE, G. F., W. RAST, and R. A. JONES. 1978. Eutrophication of water bodies: Insights for an age-old problem, *Environ. Sci. Technol.* 12 (8): 900–908.

LERMAN, A., ed. 1978. *Lakes: Chemistry, geology, physics.* New York: Springer-Verlag. 363 pp.

LI, Y. H. 1976. Population growth and environmental problems in Taiwan (Formosa): A case study, *Environ. Conserv.* 3: 171–177.

LIKENS, G. E. 1972. Eutrophication and aquatic ecosystems, in *Nutrients and eutrophication.* ed. G. E. Likens, pp. 3–13. Amer. Soc. of Limnology and Oceanography Spec. Sympos., vol. 1.

LIKENS, G. E. J. S. EATON, and J. N., GALLOWAY. 1974. Precipitation as a source of nutrients for terrestrial and aquatic ecosystems. In *Precipitation scavenging,* ed. R. G. Semonen and R. W. Beadle, pp. 552–570. ERDA Sympos. ser. 41.

NICHOLLS, K. H., D. W. STANDEN, and G. J. HOPKINS. 1980. Recent changes in the near–shore phytoplankton of Lake Erie's Western Basin at Kingsville, Ontario, *J. Great Lakes Res.* 6(2): 146–153.

NORTON, S. A. 1980. Geologic factors controlling the sensitivity of ecosystems to acidic precipitation. In *Atmospheric sulfur deposition: Environmental impact and health effects,* pp. 521–531. Ann Arbor, Mich.: Ann Arbor Science.

ROBERTSON, A., and C. F. JENKINS. 1978. The joint Canadian-American study of Lake Ontario, *Ambio* 7 (3): 106–112.

RODHE, W. 1969. Crystallization of eutrophication concepts in northern Europe. In *Eutrophication: Causes, consequences, and corrective,* pp. 50–64. Natl. Acad. Sci./Natl. Res. Council Publ. no. 1700.

SCHINDLER, D. W. 1974. Eutrophication and recovery in experimental lakes: Implications for lake management, *Science* 184: 897–899.

———. 1977. Evolution of phosphorus limitation in lakes, *Science* 195: 260–262.

SCHINDLER, D. W., K. H. MILLS, D. F. MALLEY, D. L. FINDLAY, J. A. SHEARER, I. J. DAVIES, M. A. TURNER, G. A. LINSEY, and D. R. CRUIKSHANK. 1985. Long-term ecosystem stress: The effects of years of experimental acidification on a small lake, *Science* 228: 1395–1401.

SCHOFIELD, C. L. 1976a. Acid precipitation: Effects on fish, *Ambio* 5 (5–6): 228–230.

———. 1976b. *Dynamics and management of Adirondack fish populations.* Final Report, Proj. no. F-28-R, State of New York.

————. 1980. Processes limiting fish populations in acidified lakes. In *Atmospheric sulfur deposition: Environmental impact and health effects*, pp. 345–355. Ann Arbor, Mich.: Ann Arbor Science.

SMITH, V. H. 1983. Low nitrogen to phosphorus ratios favor dominance by blue-green algae in lake phytoplankton, *Science* 221: 669–671.

STUMM, W. 1972. The acceleration of the hydrogeochemical cycling of phosphorus. In *The Changing Chemistry of the Oceans*, eds. D. Dryssen and D. Jagner, *Nobel Symposium* 20, Stockholm: Almqvist and Wicksell, pp. 329–346.

STUMM, W., and P. BACCINI. 1978. Man-made chemical perturbation of lakes. In *Lakes: Chemistry, geology, physics*, ed. A. Lerman, pp. 91–126. New York: Springer-Verlag.

STUMM, W., and J. J. MORGAN. 1981. *Aquatic chemistry*, 2nd ed. New York: John Wiley. 780 pp.

VALLENTYNE, J. R. 1974. *Algal bowl: Lakes and Man*. Ottawa: Environment Canada. 186 pp.

VALLENTYNE, J. R. and N. A. THOMAS. 1978. Fifth year review of Canada–United States Great Lakes water quality agreement. Report of Task Group III, a technical group to review phosphorus loadings. Windsor, Ontario: I. J. C., 86 pp.

VOLLENWEIDER, R. A. 1968. *Scientific fundamentals of the eutrophication of lakes and flowing waters with particular reference to nitrogen and phosphorus as factors in eutrophication*. OECD Report no. DAS/CSI/68.27, Paris, France.

VOLLENWEIDER, R. A., M. MUNAWARILY, and P. STADELMANN. 1974. A comparative review of phytoplankton and primary production in the Laurentian Great Lakes, *J. Fisheries Res. Bd. Can.* 31 (5): 739–762.

WETZEL, R. G. 1975. *Limnology*. Philadelphia: Saunders. 743 pp.

WRIGHT, R. F., and E. T. GJESSING. 1976. Changes in the chemical composition of lakes, *Ambio* 5 (5–6): 219–223.

7

Marginal Marine Environments:

Estuaries

INTRODUCTION

Marginal marine environments encompass all those bodies of seawater that have salinities decidedly different from that of the open ocean. Throughout most of the ocean, seawater is remarkably uniform in salinity (for further details see Chapter 8). However, along many coastlines it undergoes mixing with river water and glacial meltwater to produce subsaline or brackish marginal marine bodies of water. Such bodies of water vary greatly in size, ranging from brackish ponds and small lagoons to such large water masses as Hudson Bay and the Baltic and Black Seas. Of special interest are the class of drowned river mouths known as estuaries. Here river water meets seawater and the resulting processes that take place provide an important control on cycling of the elements. Thus, much attention will be paid in this chapter to the subject of estuarine chemistry.

Marginal marine environments may also be more saline than the open ocean. In regions of limited runoff and high evaporation, seawater can undergo extensive loss of H_2O with the consequent concentration of dissolved salts. Where mixing with the ocean is impeded, this gives rise to supersaline bodies of water and an *antiestuarine* circulation. A large-scale example (of moderate supersalinity) is the Mediterranean Sea. Although studied much less than brackish marginal marine environments, supersaline environments provide a modern-day analogue for ancient evaporite basins (places where, in the geologic past, vast beds of salt and gypsum were formed by evaporation and mineral precipitation) and, therefore, they will also receive some attention here.

ESTUARIES

Circulation and Classification

Estuaries are drowned river valleys filled with brackish (diluted) seawater. There is a large variation in estuarine circulation patterns depending upon the relative magnitude of the river flow and of oceanic tidal currents. (The following discussion is based on reviews by Bowden [1967], Pickard and Emery [1982], and Pritchard and Carter [1973], which should be consulted for further details.) In the simplest case with minimal tidal mixing, the river tends to flow seaward as a lighter fresh-water layer over the denser seawater. However, the tides, although they do not produce net water transport, mix some seawater upward into the fresh water and thus, a portion of the seawater is carried out of the estuary along with the river flow. In order to conserve water, since the estuary is neither filling nor emptying, there is an inward flow of seawater at depth to replace the saline water being lost along with fresh water at the surface. This is the typical *estuarine circulation* with fresh-to-brackish water flowing out at the top and saline water flowing in at the bottom. Salt must also be conserved in an estuary if it is to maintain a constant salinity. Thus, the amount of salt lost by mixing into the outward-flowing upper layer must be replaced by the inflow of saline ocean water in the deeper layer.

In addition to the river flow and tidal currents, estuaries are also affected by the earth's rotation (Coriolis force; see Chapter 2). The effect of the Coriolis force is greatest in wide estuaries where it causes flow variations across the width of the estuary (i.e., lateral variations). Facing in the direction of flow, both the seaward freshwater flow at the surface and the landward saltwater flow at depth are stronger to the right in the Northern Hemisphere (and to the left in the Southern Hemisphere). This causes the interface (between seawater and fresh water) to slope downwards to the right looking towards the sea (in the Northern Hemisphere), as exhibited, for example, by the Mississippi River estuary and Long Island Sound.

Open estuaries can be classified by their circulation pattern and the resulting distribution of salinity within the estuary (Stommel and Farmer 1952; see also Pritchard and Carter 1973; Pickard and Emery 1982.) This is shown in Table 7.1 and Figure 7.1. (The symbol ‰ refers to parts per thousand or grams per kilogram.) These types grade into one another and the type of estuarine circulation can vary considerably throughout the year and with varying river discharge. In *salt wedge estuaries* the saline ocean water enters the estuary as a wedge beneath the lighter fresh river water flowing out at the surface. The Mississippi River is an example of this type of estuary. The river flow, which is very large, dominates and there is very little tidal mixing. A small amount of saline water is mixed upward by entrainment into the fresh outflowing upper layer, which becomes somewhat more saline seaward, but there is no mixing of fresh water downward so the salt wedge retains its original oceanic salinity. Thus, there is a sharp salinity change at depth between layers (Figure 7.1a). In order to compensate for the small amount of salt water being transported outward by the surface layer, there

TABLE 7.1 Classification of Estuaries

Type	Water Circulation	Physical Processes	Examples
A. *Open Estuaries*			
1. Salt wedge	Salt wedge below river flow.	River flow dominant.	Mississippi River estuary
2. Highly stratified	Two-layer flow with entrainment (upward mixing).	River flow modified by tidal currents.	Fjords—with deep sill below upper layer; deep estuaries
3. Slightly stratified	Two-layer flow with vertical mixing (upward and downward).	River flow and tidal mixing.	Thames River estuary; Long Island Sound; Chesapeake Bay; shallow estuaries
4. Well mixed	Vertically homogeneous (a) with lateral variations (b) laterally homogeneous.	Tidal currents dominant (shallow: with Coriolis; deep: no Coriolis)	Severn estuary
B. *Silled Estuaries* (with shallow sill)	Surface layer flow with entrainment (upward mixing); restricted saltwater influx at depth which is rare if sill is shallow. (Stagnant anoxic water may result.)	River flow modified by tidal currents.	Fjords with shallow sill; Black Sea

Source: Modified from Stommel and Farmer 1952, Bowden 1967, and Pickard and Emery 1982.

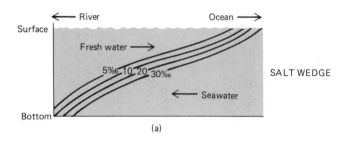

(a)

SALT WEDGE

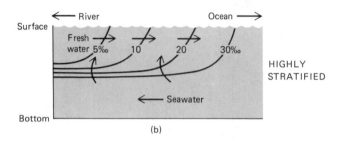

(b)

HIGHLY
STRATIFIED

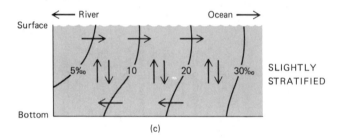

(c)

SLIGHTLY
STRATIFIED

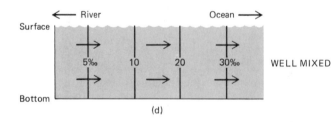

(d)

WELL MIXED

Figure 7.1 Types of estuaries: (a) salt
wedge; (b) highly stratified; (c) slightly
stratified; (d) well mixed.   In all cases
generalized salinity contours (in ‰) are
drawn for an idealized longitudinal cross
section down the estuary.   Arrows represent net water flow, i.e., tidally averaged.   (Adapted from G. L. Pickard
and W. J. Emery, *Descriptive Physical
Oceanography*, 4th ed., p. 220.   Copyright © 1982 by G. L. Pickard, reprinted by permission of the author.)

is weak flow of seawater inward at depth.   In the *highly stratified (entrainment)
estuary* (Figure 7.1b) the river flow still dominates but the tidal currents cause more
mixing of saline water upward (entrainment) into the seaward-flowing surface layer.
The surface layer becomes progressively more saline and its volume increases
seaward as a result, but because there is very little downward mixing of fresh water,
the bottom layer is still nearly of oceanic salinity.   Thus, a strong salinity gradient

still exists between the surface and bottom layers. The volume of water flowing seaward in the surface layer is often 10 to 30 times the river flow itself because of the entrainment of saline water from the bottom layer (Pickard and Emery 1982). Deep narrow estuaries have this type of water circulation (including some fjords with a deep sill).

Slightly stratified estuaries grade into highly stratified estuaries. In slightly stratified estuaries, river flow and tidal mixing are both important. Considerable vertical mixing occurs with fresh river water mixed downward as well as seawater being mixed upward. Although there is still a seaward-flowing surface layer and landward-flowing bottom layer, the salinity gradient between the surface layers and the bottom is not very sharp (see Figure 7.1c) and the surface layer is only a little less saline than the bottom layer. The James River in Chesapeake Bay, Long Island Sound, and the Thames River estuary are of this type. Salinity increases from the head to the mouth of the estuary in both the surface and deep layers.

Well-mixed estuaries are dominated by tidal currents, which overwhelm the river flow. These are shallow estuaries and the Severn River estuary in England is of this type. The result is an estuary well mixed vertically with no salinity gradient from the surface layer to deeper layers. There is a gradual increase in salinity horizontally from the head to the mouth of the estuary. In addition, although a complicated current system may develop, there is net flow seaward at all depths (see Figure 7.1d). The river water must still be discharged seaward and there is also net downstream transport of salt (which is balanced by turbulent diffusion of salt upstream from the sea) (Pritchard and Carter 1973). In wide estuaries of this type, the Coriolis force may produce a strong seaward current on the right side of the estuary and a landward current on the left side (facing downstream in the Northern Hemisphere). In narrow estuaries, however, the lateral current variation is absent because the Coriolis effect is not as strong.

Silled or restricted estuaries are partially closed to the ocean by a shallow near-surface sill and cannot develop a full estuarine circulation (Figure 7.2). The sill or shallow area is generally near the seaward end of the estuary and it restricts flow into the deeper portions of the esturary. Fresher water flows seaward in the surface layer, but if the sill is shallow, inward return flow of saline ocean water is blocked at depth, and occurs only near the surface in the lower part of an entrain-

SILLED ESTUARY

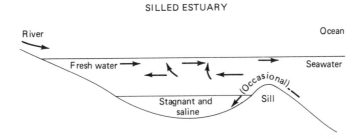

Figure 7.2 Generalized representation of a silled estuary.

ment-type surface layer.   There is a very strong halocline with the bottom water being much more saline and dense than the surface water (silled estuaries are an extreme case of highly stratified estuaries).   The strong density stratification prevents vertical mixing between surface and deep water.   As a result, analogous to stratified lakes, oxygen depletion in bottom waters often takes place.   Only occasionally does outside ocean water flow over the sill into the bottom of the basin.

The periodic flow of outside ocean water into the deep water of silled estuaries is regulated by density differences between the outside ocean water and estuarine deep water.   A small amount of fresh river water becomes mixed downward into the deeper estuarine water; thus, the deep water has less than oceanic salinity. When the outside water becomes heavier than estuarine water at the sill depth and below (due sometimes to upwelling), ocean water flows over the sill into the deep basin layers.   This displaces some of the deep water upward and, thus, the deeper basin water is replaced by new oxygenated ocean water.   In basins with a deeper sill, ocean water flows into the bottom part of the basin annually, but when the sill is very shallow, inflow from the ocean into the bottom of the basin may occur only after long intervals.   The latter case leads to stagnant totally oxygen-depleted or *anoxic* bottom water (Pickard and Emery 1982).

Estuaries with a shallow sill, restricted circulation, and anoxic bottom waters (at least periodically) are typified by fjords.   (Much larger-scale examples are provided by the Black Sea, and parts of the Baltic Sea.)   Fjords are deep, long basins with a U-shaped cross section and a sill between the estuary and the ocean (Pritchard and Carter 1973).   They are characteristic of the coastal regions of Norway, the Canadian West Coast, and Chile, and were formed by the scouring activity of preexisting glaciers.   Those fjords that develop anoxic bottom waters have a shallow sill and a small river runoff (Pickard and Emery 1982).

The Black Sea is a gigantic example of a silled "estuary."   It is fed by a number of major rivers and is connected to the Mediterranean Sea through the Bosporus and Dardanelles passages, which have shallow sills (40–100 m).   Very little saline water from the Mediterranean ever flows over the Bosporus sill to replenish the deeper layers of the Black Sea and, as a result, they are always anoxic below 180 m.   In fact, even if all the flow through the Dardanelles (the outer passage to the Mediterranean) reached the Black Sea it would still take 2500 years to replace the large volume of bottom water (Pickard and Emery 1982).   (We shall discuss the consequences of the formation of anoxic bottom water to chemical cycles later in this chapter.)

Estuarine Chemistry: Conservative versus Nonconservative Mixing

Estuaries and similar marginal marine waters are the principal places where the two major types of earth-surface water meet: fresh land-derived water (dominantly river water) and saline ocean water.   As we have seen, estuaries vary considerably in how these waters mix, depending largely upon the relative influence of river input and tidal mixing combined with basin geometry. Salinity changes are variable both between different estuaries and within any one estuary. In addition, there are temporal variations in salinity due to changes in the amount of river runoff and in the tides. River runoff varies both seasonally and annually depending upon the amount of rainfall and the incidence of floods. Because of time variations, chemical measurements in estuaries have to be made over the whole year and over several different years to be representative (Aston 1978).

Suspended matter, which is chemically equilibrated with fresh water, is carried by rivers and, upon subjection to a series of changing salinities (and pH changes) within an estuary, undergoes chemical reactions. This often causes precipitation of the suspended matter by flocculation and/or aggregation (see section below on suspended sediment deposition). Additionally, there are chemical exchanges between the suspended sediment and estuarine water that change the chemistry of both.

Besides mixing of fresh and saline water in estuaries, there are internal processes within the estuary itself that can change the chemical composition of the water. Exchange of both dissolved and particulate matter occurs between the sediments on the bottom of the estuary and the overlying water. In addition, considerable biological activity occurs in the estuarine water, in the surrounding marsh tidal areas, and in the bottom sediments. Nutrients (C, N, P, Si) are cycled biologically within the estuary and, as a result, dissolved and particulate organic matter is both produced and consumed. Humans cause changes in estuaries both in the amount and type (e.g., sewage sludge) of suspended sediment and of dissolved material reaching estuaries through rivers and land runoff from surrounding urban and rural areas (see Chapter 5). Nutrients are particularly affected by pollution, and estuaries, because they retain water for appreciably long periods, can become eutrophic in a manner similar to lakes (see Chapter 6). There is also concern about trapping of anthropogenic trace metals in estuaries. However, as elsewhere, we shall not deal with trace metals here, but shall confine our discussion to major elements.

The time that a river-borne dissolved constituent or pollutant spends in an estuary obviously affects how available it is for sediment exchange or biological processes. A measure of how long it would take to remove a pollutant that does *not* undergo sediment exchange or biological cycling is given by the *flushing time*. The flushing time is defined as the length of time required to replace the existing volume of fresh water in the whole estuary, or some part of the estuary, at the river discharge rate (Aston 1978). Thus, the flushing time τ is analogous to the replacement time for water in lakes, as given in Chapter 6, and is equal to the total volume of fresh water in the estuary (V_f) divided by the rate of river discharge into the estuary (R):

$$\tau = \frac{V_f}{R}$$

A representative average flushing time for a vertically well-mixed estuary is on the order of days (1–10 days).   This is less than the residence time of most lakes, which are measured in years, but longer than that for many rivers.

It is desirable to know from the point of view of geochemical cycling whether element fluxes calculated from river water concentrations really represent the flux *reaching* the ocean after passing through estuaries, or whether the river flux of an element is reduced and/or added to in passage through the estuary.   In the idealized model for mixing of river water and seawater in an estuary (Boyle et al. 1974; Liss 1976; Officer 1979; Loder and Reichard 1981; Kaul and Froelich 1984), measured concentrations of the dissolved river-water constituent that is being studied are plotted against corresponding measured values of a dissolved estuarine constituent, which is assumed to behave *conservatively* (i.e., to show no loss or gain during mixing).   Measurements are generally made on a series of samples collected along the length of the estuary from the river mouth to the ocean.   The conservative constituent is generally either total salinity or chloride concentration.   In the case of a constituent of interest that is also conservative, its measured concentration plotted against increasing chloride concentration (as a measure of mixing with oceanic water) should lie on a straight line between the concentration of the element in river water (C_R) and its concentration in the oceanic water (C_S) with which it is mixing.   This is the *theoretical dilution line* for a conservative constituent (Liss 1976), which is illustrated in Figure 7.3.

Some elements behave *nonconservatively* in estuaries; that is, they are removed from or added to solution during mixing.   If the dissolved component is being added as the salinity increases, then a concave-down curve above the dilution line will result.   Conversely, if the constituent is being removed during mixing, a concave-up curve below the dilution line will result.   This is also illustrated in Figure 7.3.   To generalize, then, a constituent that is *conservative* in estuaries will show a straight-line plot versus salinity (or chloride concentration) and the plot of a *nonconservative* component will show curvature.

These simple mixing models assume that there are only two well-defined end-member concentrations (river and oceanic).   However, if there is a third end-member such as another tributary to the estuary, then behavior of a dissolved constituent may appear erroneously nonconservative (Boyle et al. 1974).   In addition, the concentrations of the oceanic and riverine end members often are either not well known or variable.   The oceanic end-member in coastal or shelf water may have an intermediate salinity, less than that of the open ocean, and by erroneously assuming ocean salinity (35‰) one can introduce curvature into an otherwise straight-line plot (Boyle et al. 1974).   Also, it is often difficult to determine the exact concentration of the component of interest in the oceanic end-member because of problems in locating and sampling the oceanic end of the salt gradient.   Furthermore, river concentrations may vary with time; regularly in some estuaries, such as in the Florida Ochlocknee estuary where nutrients show a yearly sinusoidal concentration curve (Kaul and Froelich 1984), or irregularly, such as represented by pollutants and silica in the Tamar estuary of England (Morris, Bale, and How-

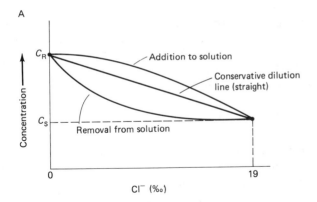

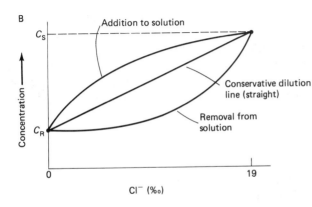

Figure 7.3 Idealized plots for estuarine water of the concentration of dissolved components versus chloride (which serves as a conservative measure of the degree of mixing between fresh water and seawater). C_R = concentration in river water; C_S = concentration in seawater. (A) Component whose concentration in fresh water is greater than it is in seawater (for example P, N, Si). (B) Component whose concentration in fresh water is less than it is in seawater (for example Ca, Mg, K). (Modified from P. S. Liss. "Conservative and Non-Conservative Behavior of Dissolved Constitutents During Estuarine Mixing," In *Estuarine Chemistry*, ed. J. D. Burton and P. S. Liss, p. 95. © 1976 by Academic Press, reprinted by permission of the publisher.)

land 1981). This riverine variation can introduce curvature into mixing curves, particularly in estuaries that are not rapidly flushed (Officer and Lynch 1981; Loder and Reichard 1981). As these problems demonstrate, one should apply caution when using mixing curves of the types shown in Figure 7.3 to elucidate estuarine chemical processes.

Estuarine Chemical Processes

The dissolved constituents of estuarine water can be divided into two groups (Liss 1976):

1. Those which are more abundant in seawater than in river water (e.g., Ca, Mg, Na, K, Cl, and SO_4)
2. Those which are more abundant in river water than seawater (e.g., Fe, Al, P, N, Si, dissolved organic matter)

Since seawater has a much greater salinity than river water, most of the major dissolved elements have a greater concentration in seawater than in river water. However, metals such as Fe, Al, Mn (and trace metals such as Zn, Cu, Co, etc.) as well as nutrients such as P, N, Si, and dissolved organic matter (DOM) generally have a greater concentration in river water than ocean water. Here we shall only briefly discuss the behavior of the elements that are more abundant in seawater because they are discussed in detail in Chapter 8. In this chapter we shall focus instead on those elements that have greater concentrations in river water than seawater.

If the elements in seawater are to maintain a constant concentration ratio, then dissolved constituents that have a greater concentration in river water than in seawater must be removed either in estuaries, where the original mixing of seawater and river water occurs, or later in the oceans (Mackenzie and Garrels 1966). Thus, there is a reason to suspect that removal of elements such as Fe and Al might occur in estuaries.

The elements that are removed in estuaries are removed predominantly by either *inorganic* (nonbiogenic) processes or *biogenic* processes. Removal of some elements (Si, P, C) may occur by both processes in different estuaries or at different times or under different conditions in the same estuary. We shall begin with *inorganic removal* of dissolved constituents in estuaries and confine our discussion to the major rock-forming elements Al, Fe, and Si, to the nutrients N and P, and to dissolved organic matter.

Inorganic Removal in Estuaries

The main questions to be addressed concerning inorganic removal in estuaries are (1) the extent of removal, (2) the dependence of removal upon salinity, (3) the importance of pH changes, (4) the role of organic matter in removal, (5) the role of suspended matter in removal, (6) the importance of the concentration of "dissolved" constituents in river water and in what form each "dissolved" constituent appears, (7) the removal mechanisms and products, and (8) whether the removal in estuaries is permanent or whether there is recycling. Answers exist to some of these questions, but there is debate about others and the answers may vary from one estuary to another. Where possible, mention of these factors will be brought up when discussing each element.

The inorganic removal of "dissolved" (dissolved and colloidal) iron in estuaries is well documented and generally agreed upon (see summary in Boyle, Edmond, and Sholkovitz 1977; Liss 1976; Aston 1978). Removal occurs rapidly upon mixing of river water and ocean water in the low salinity (0‰–5‰) part of the estuary and most removal is complete by the time 15‰ salinity is reached. The evidence for Fe removal comes from plots of Fe concentration versus salinity (or Cl), which show a concave-up curve below the theoretical mixing line (see discussion in the previous section). An example is shown in Figure 7.4. In addition, laboratory studies of experimental mixing of river and ocean water (Sholkovitz 1976;

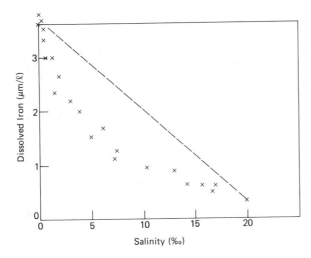

Figure 7.4 Total dissolved iron (μm/ℓ = micromoles per liter) versus salinity in the Merrimack Estuary, Massachusetts. Data points in order of increasing salinity fall on a concave-up (iron removal) curve. Dashed line indicates theoretical conservative mixing between fresh and ocean end-members. (Modified from E. A. Boyle, R. Collier, A. T. Dengler, J. M. Edmond, A. C. Ng and R. F. Stallard. "On the Chemical Mass Balance in Estuaries," *Geochimica et Cosmochimica Acta*, 38, p. 1722. Copyright © 1974 by Pergamon Press, reprinted by permission of the publisher.)

Boyle, Edmond, and Sholkovitz 1977; Crerar et al. 1981) show that "dissolved" iron flocculates or precipitates from solution during mixing. The amount of Fe removal estimated by either of these methods is large (50%–95%) and the higher the Fe concentration in the rivers, the greater the total Fe removal (Boyle, Edmond, and Sholkovitz 1977).

"Dissolved" iron is defined as iron that will pass through a filter with pores of 0.45 μm diameter, that is, having a size less than 0.45 μm. It may consist of any of several forms: (1) *colloidal* (very fine suspended particles) as mixed Fe oxide or oxyhydroxide-organic matter colloids (Crerar et al. 1981; Boyle, Edmond, and Sholkovitz 1977), (2) dissolved inorganic Fe species, and (3) soluble Fe-organic complexes. The same is true of "dissolved" aluminum.

Most recent work (Boyle, Edmond, and Sholkovitz 1977; Crerar et al. 1981) suggests that for iron removal the pH change from most river water (whose pH is generally 6–7) to ocean water (pH 8) is less important than the mixing of seawater electrolytes with fresh water. The marine cations Ca^{++}, Mg^{++}, and Na^+ neutralize the negative charge on river-borne substances and cause flocculation and coagulation.

Sholkovitz (1976) in laboratory experiments on the mixing of Scottish river water with ocean water found flocculation of Fe, Al, Mn, P, and organic substances in a similar low salinity range. He felt that the solubility of the inorganic constituents in river water and their flocculation in seawater was due to their association with organic matter. The mechanism suggested (Boyle, Edmond, and Sholkovitz 1977) is the flocculation of mixed iron-oxide–organic matter colloids (which are stabilized in river water by the organic matter) due to the neutralization by seawater cations of the negative colloid charges. They felt that river-borne "dissolved" Fe in the East Coast estuaries of the United States is almost entirely colloidal (Fe-oxide particles coated with an organic film) and not truly dissolved. In the laboratory experiments, a large percentage (>50%) of the total Fe, Mn, and P was precipitated, as was a considerable part of the Al (10%–70%), but only a small part (3%–11%) of the dissolved organic matter (DOM). The latter was explained as being due to the removal of only the high-molecular-weight (humic acid) part of the DOM (Sholkovitz, Boyle, and Price 1978). It is worth noting that this removal of DOM is too small to be apparent on DOC (dissolved organic carbon) versus salinity plots.

Crerar et al. (1981), as a result of their study of the organic-rich, acid rivers (pH 4–5) in the Pine Barrens area of New Jersey and their estuaries, feel that organic matter in these rivers is less important in *causing* Fe and Al removal. Although they found that Fe, Al, and DOM were all removed in the estuaries, they feel that these substances were not necessarily chemically associated with one another, but rather were removed by a common physical-chemical mechanism— that is, flocculation by seawater electrolytes. For one thing, they found that more than 50% of the Fe and Al in these acid rivers was truly dissolved (dominantly as dissolved inorganic Fe with a small amount of soluble Fe-organic complexes), and the rest was colloidal (mixed Fe oxyhydroxide and organic colloids). They suggest that upon increase of pH during mixing with seawater, the dissolved inorganic Fe and Al become supersaturated and precipitate as Fe and Al oxyhydroxide floccules. The preexisting Fe and Al colloids and high-molecular-weight humics also precipitate and together constitute the primary precipitate. Although a large percentage of the Fe and Al is removed, only about 10% of the DOM is removed (similar to Sholkovitz's [1976] estimate). Crerar et al. feel that the removal of elements varies from estuary to estuary depending upon such effects as the pH and chemistry and DOM content of the river water. (The river water studied by them, being unusually acid, probably has more truly dissolved iron than that of most other rivers.)

In highly polluted estuaries, such as the Belgian-Dutch Scheldt Estuary (Wollast 1981) where *anoxic* bottom waters exist, reactive Fe^{+++} hydroxides carried by the river are reduced to dissolved Fe^{++}. Phosphate, which had been adsorbed by the Fe^{+++} hydroxides, is released as dissolved PO_4^{---}. When the estuarine waters become more oxygenated further downstream, the ferrous iron is reprecipitated as ferric hydroxide, which removes dissolved PO_4^{---} by adsorption, and sediments out. Similar effects may occur at the sediment-water interface where sediments are exposed to alternately anoxic and oxygenated conditions (Krom and Bener 1981; Klump and Martens 1981), with Fe and P being soluble under anoxic conditions.

The role of suspended matter in the estuarine removal of Fe, Al, and other substances has been variously interpreted. Aston and Chester (1973) found in laboratory studies that increased suspended particle concentrations increase the rate of Fe^{+++} removal from solution. The mechanism suggested was the formation of Fe oxides and hydroxides on negatively charged suspended sediment particles that act as nuclei (with further precipitation of Mn, Ni, Co, and Cu). However, the work of others (Sholkovitz 1976; Boyle, Edmond, and Sholkovitz 1977; Crerar et al. 1981) has shown flocculation in estuaries can readily occur without suspended particles. Nevertheless, Sholkovitz, Boyle, and Price (1978) suggest that the *greater extent* of Fe and humic removal at low salinities in the Amazon estuary, as compared to lab experiments, may be due to suspended matter.

Hydes and Liss (1977) found that 30% of "dissolved" riverine Al was removed in the Conway estuary (U.K.) during the early stages of mixing (i.e., < 8‰ salinity). They have suggested that the mechanism for removal of Al is the flocculation of very fine clay particles containing adsorbed Al which are suspended in fresh water but which are irreversibly coagulated upon entering the estuary. Dion (1983) found that, in the Connecticut River and Amazon River, Al seems to be associated with humic acids, probably as organic complexes. This would suggest that Al removal by flocculation of humics (Sholkovitz 1976) is the likely removal mechanism. However, as pointed out by Dion, when humic material (including complexed Al) is adsorbed on clay particles, the two mechanisms become the same.

An alternate model for the flocculation of Al has been suggested by Mackin and Aller (1984). In this case, Al is displaced from river-borne organic complexes and adsorption sites by the increase in cation concentrations and rise in pH in estuaries. The displaced Al reacts in solution with H_4SiO_4 and cations to form authigenic (newly-formed) aluminosilicates which precipitate out. In addition, bottom sediments can act as either a sink or source for Al. Al can diffuse into the sediments, and react with Si in solution to form authigenic aluminosilicates. However, if bottom sediments are resuspended in silica-depleted waters, the clays dissolve and dissolved Al is released into solution.

Because of the documentation of extensive "dissolved" Fe and Al, and lesser DOM removal in estuaries, the river flux of these elements cannot necessarily be considered as the ocean input. However, the question arises as to whether the estuarine removal is permanent and whether material is remobilized and carried out of estuaries (Boyle, Edmond, and Sholkovitz 1977; Bewers and Yeats 1980). Although river Fe flocculates extensively in the Pine Barrens estuaries, it is not found concentrated in the bottom sediments, leading to the asssumption that most flocculated Fe is carried out onto the shelf by currents (Coonley, Baker, and Holland 1971; Crerar et al. 1981). Thus, basically the further removal of Fe from estuaries is controlled by the removal of particulates. In addition, ferric iron that is precipitated onto the bottom sediments or originally deposited as Fe coatings on clay minerals can be remobilized as ferrous iron by reactions in reducing sediments and then released either by wave, current and biological stirring of bottom sediments or by diffusion into bottom water. Thus, there is potentially an Fe source in estuarine bottom sediments.

Thus far we have discussed the well-documented inorganic estuarine removal of dissolved Fe (generally >50% of the river concentration) and to a somewhat lesser extent Al (30%–70% of the river concentration) and dissolved organic matter (3%–10% of the river concentration). Inorganic removal of P (a predominantly biogenic element) is less well documented. Sholkovitz (1976) and Boyle, Edmond, and Sholkovitz (1977) found, associated with humic and Fe removal, the large-scale inorganic removal of phosphate (and varying amounts of Mn and trace elements such as Cu, Ni, Co, and Cd). The removal of phosphate may occur by more than one mechanism. In addition to flocculation of colloidal PO_4 already associated with Fe (the latter discussed above for the Scheldt estuary), adsorption of phosphate on the other colloids probably also occurs (Bale and Morris 1981). The existence of a buffering mechanism (which maintains dissolved PO_4-P at a concentration of 20–40 $\mu g/\ell$), by partially reversible adsorption onto suspended particles in the low-salinity part of estuaries, has been suggested (see review in Liss 1976). Desorption of phosphate from river-derived suspended matter occurs in the low-salinity part in the Pamlico estuary, North Carolina (Upchurch, Edzwald, and O'Melia 1974), and the Zaire estuary (van Bennekom et al. 1978), and results in an increase in dissolved PO_4. (Phosphate desorption and removal is further discussed below in the section on biogenic nutrient removal.)

The behavior of dissolved silica in estuaries is variable. In some estuaries silica is conservative (Boyle et al. 1974; Crerar et al. 1981), and in others it is removed either by inorganic processes or by biological processes (since it is a nutrient). The evidence for inorganic silica removal comes from plots of Si versus salinity (or Cl concentrations) which show concave-up curvature below the mixing line. In laboratory mixing experiments, Sholkovitz (1976) found very little Si removal associated with Fe and humic acid removal. He suggested that this is because Si is truly dissolved and not colloidal. By contrast, Liss (1976) noted that inorganic (nonbiological) silica removal can occur in some estuaries but only during the early stages of fresh water–seawater mixing within the low-salinity (0‰–5‰) part of the upper estuary. Also, the presence of suspended matter as well as a high dissolved silica concentration in the river runoff seems to be required. Liss suggested that some sort of buffering mechanism is involved which requires high silica concentrations (roughly >14 mg/ℓ SiO_2) that are only present before river water has been diluted appreciably by seawater. The total amount of inorganic silica removal is generally 10%–20% where it occurs, and in all cases less than 30%. These suggestions are borne out by observations of inorganic silica removal in the Tamar estuary (Morris, Bale, and Howland 1981).

Faxi (1980) suggests that the mechanism for inorganic silica removal is adsorption on colloidal ferric and aluminum hydroxides formed during early river and ocean water mixing. Since colloidal ferric and Al hydroxides can also adhere to the surface of suspended matter, this would explain the importance of suspended matter in Si removal. Aluminosilicate formation, suggested by Mackin and Aller (1984) as a mechanism for Al removal in estuaries, would also remove silica but only a small percentage of that available because Al concentrations are much lower than Si concentrations.

Estuarine studies of the behavior of Na, K, Mg, Ca, and SO_4, which are much more abundant in seawater than in river water, have shown that they are essentially conservative upon mixing of seawater with river water (see reviews by Liss 1976; or Aston 1978). This does not mean that reactions do not occur, only that, because of their large concentrations in the seawater end-member, small changes in concentration during estuarine mixing are very difficult to detect via the standard mixing models. (Documentation of reactions involving the major ions, upon the addition of river-borne clays to seawater, are given in Chapter 8 based on laboratory experiments and studies of estuarine and marine sediments.)

Biogenic Nutrient Removal in Estuaries: Nitrogen and Phosphorus

Nutrients, specifically N and P (Si will be discussed later as a special case), are generally more abundant in river water than in oceanic water due partly to their removal by organisms from oceanic water and partly to the input to rivers from pollutants and land weathering. As a result, rivers serve as nutrient sources for estuaries and here biogenic removal is often found. Minute phytoplankton (un-icellular floating organisms), particularly diatoms and algae, are the organisms most often responsible for the removal of dissolved nutrients from estuarine surface water. The amount of phytoplankton activity is measured in terms of the *net primary production* in grams of carbon fixed (in organic matter) per square meter of water surface area per year (g $C/m^2/yr$). The less productive parts of the open ocean where there is no upwelling tend to have a primary productivity of around 50 g $C/m^2/yr$ or less. The primary productivity of estuarine and coastal waters is much higher because of greater nutrient supply (from rivers and other sources). Some typical values are 270 g $C/m^2/yr$ for the average coastal environment (Wollast and Billen 1981) and 100–200 g $C/m^2/yr$ for estuaries and coastal waters (Williams 1980). The amount of phytoplankton productivity will generally be higher in polluted areas, which have a greater nutrient supply.

In order for phytoplankton "blooms" or large concentrations of the organisms to occur in estuaries, the surface water must be fairly clear so that the light necessary for photosynthesis can pass through. Thus, large amounts of suspended sediment tend to inhibit organic growth, and in estuaries of rivers such as the Yangtze and Amazon, which have large suspended loads, diatom blooms tend to occur further down the estuary or on the continental shelf, after the suspended load has been deposited and the water has cleared (Milliman and Boyle 1975). Because of the resuspension of bottom sediment, turbulent water is also less favorable for phytoplankton growth than relatively placid water where the sediment can settle out. In temperate estuaries, phytoplankton blooms are seasonal, occurring in the spring and summer when there is more light and warmer water temperatures. This occurs, for example, in the Connecticut River estuary and in Long Island Sound (Dion 1983).

Another factor that influences the quantity of phytoplankton (i.e., primary productivity) in an estuary is the flushing time (see definition above) for fresh water in the estuary. A longer flushing time favors greater phytoplankton growth and therefore greater biogenic nutrient removal. The flushing time for fresh, nutrient-rich water is longer in a well-mixed estuary than in a stratified estuary where the river water rapidly flows out in the surface layer. Flushing time also affects the amount of recycling of nutrients (by breakdown of phytoplankton organic matter and siliceous tests) that is possible. If the estuary is rapidly flushed with fresh water, then there is not time enough for nutrient regeneration to be important in the nutrient supply, and phytoplankton organic debris will be carried out of the estuary. Also, in a highly stratified estuary, nutrients remineralized and returned to the bottom water will accumulate there in the absence of rapid replacement of the bottom water, so nutrient recycling is less efficient than in a well-mixed estuary. The bottom water acts as a so-called nutrient trap (Redfield, Ketchum, and Richards 1963).

In addition to dissolved nutrients carried in freshwater river runoff into estuaries, there are a number of other nutrient sources for phytoplankton in the surface waters of estuaries and coastal waters. These sources, which vary from estuary to estuary, include (1) regeneration (return to a dissolved state) of nutrients from *particulate* organic and inorganic material carried by rivers, (2) regeneration of nutrients from the breakdown of internally produced biogenic debris as it passes downward through the water column, (3) *benthic* nutrient regeneration—regeneration of internal biogenic debris from estuarine bottom sediments, (4) fixation of atmospheric N_2 as plant N, (5) coastal upwelling of nutrient-enriched oceanic bottom water which then enters the estuary, and (6) lateral advection from deep offshore ocean waters.

Edmond et al. (1981) found that the dissolved NO_3^- flux and PO_4 flux of the Amazon River is increased by 75% by the decomposition of riverine particulate organic matter in the shoal area between the main channel and the estuary. This involves breakdown of half of the riverine particulate organic N and P. In the polluted Belgian-Dutch Scheldt River estuary, Wollast (1983) found that some 40% of the river organic load (mainly dissolved organic matter) is respired to form dissolved inorganic N and P within the estuary. At low salinities in the Zaire estuary, van Bennekom et al. (1978) found that PO_4 concentrations reached a maximum twice as high as the river concentration, accompanied by a decrease in the P concentration of suspended matter. They attribute this to inorganic PO_4 desorption from suspended matter.

Coastal upwelling of nutrient-enriched marine bottom water also can be an important source of nutrients for phytoplankton in some estuaries. In the Zaire estuary (van Bennekom 1978), upwelling is a greater NO_3^- source than the river water and about an equal PO_4 source. Here decomposition (to dissolved N and P) of *terrestrial* organic matter, deposited on the bottom in deep water, is the nutrient source in the upwelled water, and upwelling essentially represents an increase in the dissolved river nutrient flux.

Besides outside sources (rivers, upwelling) of nutrients in estuaries, recycling of nutrients from *internally* produced organic debris (phytoplankton debris, etc.) is also important as a phytoplankton nutrient source. (In some estuaries it is the dominant nutrient source.) *Regeneration* (return to dissolved forms) of nutrients from organic debris can occur by bacterial action within the water column as the debris settles out and on the bottom (benthic regeneration) either at the surface of the sediments or from within the sediments. In *shallow* estuaries, particularly those with long residence times, some 25%–50% of the organic matter that is fixed in the estuary is remineralized in the bottom sediments (Nixon 1981). Thus, benthic regeneration rates can play an important role in shallow estuaries where organic matter is decomposed on the bottom.

Nutrient nitrogen can also be added to estuaries by means of the *fixation* of N_2 from the atmosphere by blue-green algae and photosynthetic bacteria. Although Ryther and Dunstan (1971) feel that this process is unimportant in polluted eastern U.S. coastal estuaries, Wollast (1983) suggests it may be involved in the nitrogen balance of the Southern Bight of the North Sea.

The importance in estuaries of nutrient sources other than rivers on a worldwide basis can be calculated from flux data. For coastal regions in general (not only estuaries) it has been estimated by Wollast (1983) that, worldwide, less than 2% of N or P are provided to phytoplankton by dissolved N and P carried in rivers. More specifically for estuaries alone we calculate that approximately 25% of the phosphorus requirement of phytoplankton and 10% of the nitrogen requirement are provided *on the average* by rivers. This result is based on the total worldwide area of estuaries (1.4×10^6 km^2, from Williams 1980), the total river input to the ocean (including pollution) of inorganic dissolved P and N (Meybeck 1982; see also Chapter 5), the assumption that all river nutrient input is removed in estuaries, the assumption that the average primary productivity for coastal waters (270 g C/m^2/yr; Wollast 1983) is representative of estuaries, and the assumption of a standard Redfield ratio for plankton (see below) of C:N:P = 106:16:1. These results show that the behavior of nutrients in estuaries is highly complex and that simple input from rivers with planktonic uptake constitutes only part of the story.

Permanent removal of dissolved nutrients from estuaries can occur by (1) sedimentation and burial of biogenic debris in bottom sediments (as discussed above), (2) passage out to sea, (3) denitrification with loss of N_2 and N_2O to the atmosphere (see below), and (4) inorganic adsorption of phosphate on particles plus burial in sediments (see previous section).

Physical conditions are complex in those estuaries that have vertical stratification plus a bottom salt wedge and only local or partial mixing. In such cases nutrients removed from surface water may be regenerated in the bottom water, carried landward in the estuarine circulation, and isolated at least partly from the surface water. Redfield, Ketchum, and Richards (1963) refer to this as a *nutrient trap*. Since nutrients must reach surface water to be used by organisms, these complex physical conditions often make calculation of nutrient balances difficult or impossible.

Nitrogen, Phosphorus, and Limiting Nutrients

The nutrients nitrogen and phosphorus are used by phytoplankton, in forming organic matter, in definite ratios to carbon. An average composition for marine plankton, given in terms of the classic *Redfield ratio*, is $C_{106}N_{16}P_1$ (Redfield, Ketchum, and Richards 1963).   This is an idealized ratio and the actual marine phytoplankton nutrient utilization ratios can vary from 5N:1P to 16N:1P (Ryther and Dunstan 1971) depending upon the availability of nutrients in the water and the kind of phytoplankton growth.   Ryther and Dunstan (1971) estimate that in North American coastal waters the average phytoplankton N:P utilization ratio is about 10:1.   Apparently phytoplankton will use more phosphorus relative to nitrogen if more phosphorus is available—the so-called *luxury P consumption* (Redfield, Ketchum, and Richards 1963).

By analogy with lakes, P might be expected to be limiting in estuaries as it is in lakes.   (For a discussion of limiting nutrients see Chapter 6.)   However, *nitrogen* is generally found to be the limiting nutrient in coastal waters, not phosphorus (Ryther and Dunstan 1971).   This situation has three principal causes. First, the ratio of nitrogen to phosphorus (N:P) in many rivers, especially polluted rivers, is lower than that of estuarine plankton so that there is excess P left over in the estuary upon consumption of all N.   Second, due to denitrification (the bacterial reduction of dissolved NO_3^- to N_2 and N_2O), nutrient nitrogen can be selectively lost from the estuary.   Finally, upon deposition in sediments, N is regenerated much more slowly than P, such that the N:P ratio of the regenerated nutrients is lower than that used by plankton.

Pollution is a probable contributing factor to nitrogen deficiency, and to nitrogen becoming a limiting nutrient, in many estuaries.   Polluted river water along the U.S. East Coast is generally enriched in phosphorus over nitrogen with an average ratio of 5N:1P (Ryther and Dunstan 1971).   This ratio is decidedly less than that taken up by plankton in the same area, 10N:1P.   (However, nitrogen is not universally limiting in polluted estuaries as evidenced by the South Bight of the North Sea where P is the limiting nutrient in the spring; see van Bennekom, Gieskes, and Tijesen 1975.)

There is no doubt that loss of nutrient nitrogen via denitrification to N_2 is an important process in estuaries.   (A much smaller loss of nitrogen as N_2O, less than 10% of the total, is also found both in the water column and in bottom sediments; see McElroy et al. 1978; and Seitzinger, Nixon, and Pilson 1984.)   Seitzinger et al. (1980) and Seitzinger, Nixon, and Pilson (1984) have measured a considerable flux of N_2 from estuarine bottom sediments in Narragansett Bay due to denitrification.   The nitrogen loss to the atmosphere amounts to 35% of the organic nitrogen mineralized in the bottom sediments and almost half of the river input of fixed inorganic N in this polluted area.   The ratio of the dissolved N:P flux (excluding N_2) from the bottom sediments is about 8N:1P, somewhat nitrogen-deficient compared to the average phytoplankton consumption ratio in northeastern U.S. estuaries of around 10N:1P.   Likewise, Rowe et al. (1975) in other polluted U.S. East Coast waters found that some 20% of the N reaching the bottom was not returned to the water column as nutrient N (NO_3^-, NO_2^-, or NH_4^+) and the benthic dissolved nutrient regeneration ratio averaged only 4N:1P.   Thus, in these relatively polluted East Coast areas, around 20%–35% of the nitrogen reaching the bottom is being lost from the system (and not reaching the ocean), and the dissolved benthic nutrient fluxes range in N:P ratio from 4:1 to 8:1, enough to account for the observed N limitation in coastal waters, particularly when coupled with pollution (see above).

Another factor in causing N to be a limiting nutrient in estuaries involves the selective regeneration of phosphorus before burial in bottom sediments. In Long Island Sound, Krom and Berner (1981) found that sedimenting dead planktonic organic matter became depleted in P relative to both C and N as compared to the original plankton. This was deduced by comparing the elemental ratio of organic matter buried in the top few millimeters of the bottom sediments (C:N:P = 106:11:0.3) to that for the fresh plankton (C:N:P = 106:15:1). The rise in the ratio of N:P in organic matter upon sedimentation to the bottom means that P was regenerated during organic matter decomposition (while settling and while sitting on the bottom) faster than N. This preferential release of P (and burial of N) should help to explain why there is a deficiency of N in Long Island Sound water and why N is the limiting nutrient.

Krom and Berner (1981) also found that there was a considerable flux from sediments to the water column of dissolved PO_4 which had been adsorbed on ferric oxhydroxides. This P release accompanied the reduction of the iron minerals by H_2S in anoxic portions of the sediment. Thus, the flux of PO_4 from the bottom sediment was due to both the microbial breakdown of organic P and the release of adsorbed inorganic P.

Studies of N and P cycling have also been done in several unpolluted estuaries of different types which might be more representative of the natural situation. In the well-mixed Ochlockonee estuary in Florida, Kaul and Froehlich (1984) found that nitrogen is occasionally limiting. Some 20% of the river NO_3-N flux to the estuary does not reach the ocean because all nitrate removed by diatoms in the estuary is not regenerated to solution. Loss of nitrate could be due to denitrification to N_2, incomplete nitrification of NH_4^+, (which is released from organic decay) to NO_3^-, or burial of organic nitrogen derived from NO_3^-. Phosphate removed by diatoms, by contrast, is completely regenerated and the dissolved phosphate flux from the bottom is actually greater than that from diatom removal, due to bottom regeneration of riverborne particulate P. Thus, the low measured sediment regeneration ratio of $3NO_3$-N:$1PO_4$-P is due to both nitrate losses and phosphate gains. Edmond et al. (1980) also found a slight N limitation in the surface water of the Amazon estuary. Phosphate removed by diatoms from the surface water is almost completely remineralized in the salt wedge as compared to only 50% of the NO_3-N, resulting in a salt-wedge regeneration ratio of $5NO_3$-N:$1PO_4$-P. From anomalous O_2 versus NO_3^- plots one can speculate that nitrogen loss in the Amazon estuary may take place via denitrification, since little organic matter (and therefore, organic nitrogen) is buried in bottom sediments. In the

Zaire estuary, van Bennekom et al. (1978) found considerably less biogenic N and P removal than in the Amazon estuary, but where removal takes place, the limiting nutrient again is nitrogen.

Biogenic Silica Removal in Estuaries

Some planktonic organisms remove silica from solution as well as P and N. For example, diatoms use Si in forming their opaline-silica tests or shells. Silica removal associated with diatom blooms has been observed in the Amazon River estuary (Milliman and Boyle 1975) and also in a number of other estuaries (Ocklochonee in Florida, Scheldt in Netherlands, San Francisco Bay, Connecticut River, etc.). Milliman and Boyle (1975) found that some 25% of the Amazon River silica load is removed by diatoms. Likewise, Knapp, DeMaster, and Nitrouer (1981) found that Amazon silica removal varies from 20%–35% depending upon the part of the estuary studied. However, siliceous diatom tests do not seem to be accumulating to any degree in the Amazon estuarine or shelf sediments. There are several possible explanations for this: (1) biogenic sliceous debris is masked by the large Amazon suspended load; or (2) diatom tests are carried landward by the estuarine circulation; or (3) biogenic silica is redissolving in the bottom waters (salt wedge). While Edmond et al. (1981) found Si enrichment in the bottom salt wedge equal to only 20% of the biogenic silica removal, Knapp, DeMaster, and Nitrouer (1981), from bottom cores, found that 50%–80% of the Si removed by diatoms redissolves in the Amazon bottom water or at the sediment-water interface before burial. The latter means that *net* biogenic silica removal would amount to only 10% or less of the total Amazon silica load supplied by rivers. (Remember only 20%–35% of river-borne silica is taken up by diatoms.) The estuarine circulation of the Amazon is complex and part of the silica redissolved in bottom water may be transported by longshore currents or toward the river mouth before it returns to the surface water by upwelling or vertical mixing.

Dion (1983) concluded from silica concentrations in Long Island Sound sediments (DeMaster 1981) that only about 10% of the total Connecticut River silica load appears in the estuarine sediment. Since some 50% of the river silica flux is removed biogenically in the river itself, most of the siliceous biogenic detritus carried into the well-mixed estuary redissolves. In the small well-mixed Ochlockonee estuary in Florida, Kaul and Froelich (1984) found that about 20% of the river Si flux is removed biogenically, but that siliceous diatoms are almost entirely redissolved and thus essentially all the riverine silica flux reaches the ocean; that is, Si is *conservative* overall. Van Bennekom et al. (1978) found that silica is very nearly conservative in the Zaire River estuary where there is considerably less primary productivity. DeMaster (1981) has estimated that, on the average, 20% of riverine silica is removed (both biogenically and inorganically) in estuaries. However, because of evidence for redissolution of silica in both the Amazon and

Yangtze bottom waters and sediments, this should be a *maximum* value and the actual *net* Si removal is probably considerably less than 20% of the river flux.

Thus, to generalize from the limited number of estuaries studied, when there is biogenic silica removal taking place in estuaries (and it does not occur in all estuaries), diatoms remove 20%–25% of the total river-dissolved Si load. However, because of the redissolution of diatom tests on the bottom, the net removal of dissolved river silica by deposition of biogenic silica in natural estuarine sediments is probably far less than 20% and, in some well-mixed estuaries at least, silica is conservative.

Eutrophication and Organic Matter Pollution of Estuaries

Humans introduce a variety of dissolved pollutants into estuaries, for example hydrocarbons, heavy metals, and bacteria. But in this section, we shall confine ourselves only to estuarine pollution arising from (1) organic matter enrichment and (2) nutrient enrichment. Addition of nutrients (which bring about the artificial enhancement of planktonic production) and organic wastes can be referred to together as cultural *eutrophication* (Likens 1972) and we shall adopt this definition here. (For a discussion of definitions and further details on eutrophication, as applied to lakes, consult Chapter 6.)

Direct organic matter enrichment results from the addition to the estuary of large quantities of dissolved and particulate organic carbon and organic nitrogen, mainly from sewage. As we have discussed in previous chapters, respiration attending bacterial decomposition of this organic matter consumes dissolved O_2, so that the immediate result of organic matter enrichment in estuaries is the depletion of dissolved oxygen, particularly if there is not rapid resupply of O_2 by the estuarine circulation, and air-water exchange. Oxygen depletion is greatest in bottom waters, leading in some extreme cases to totally anoxic conditions at depth. An example of the latter is the Scheldt estuary in Belgium-Holland (Wollast 1983) where anoxic conditions extend over a length of over 30 km along the bottom. Some U.S. estuaries that exhibit oxygen depletion due to organic matter enrichment from sewage include the Delaware estuary, the Houston ship channel, the Hudson River estuary, and New York Harbor (O'Connor, Thomann, and Di Toro 1975; Simpson et al. 1975). Under unusual circumstances, a high degree of oxygen depletion can occur even in unpolluted estuaries which receive large quantities of natural organic matter—for example, in the Zaire estuary (van Bennekom et al. 1978).

In water quality papers, the amount of decomposable organic matter in water is often referred to in terms of its biochemical oxygen demand or BOD, which is a measure of the amount of dissolved oxygen consumed during decomposition of the organic matter in the water by microorganisms. Sewage treatment plants can reduce the biochemical oxygen demand of waste waters. With secondary sewage treatment, two-thirds of the oxygen demand of sewage waters entering New York

Harbor is removed.   However, such sewage treatment has a far smaller effect on the level of nutrients than it does on dissolved and suspended organic carbon (Simpson et al. 1975).

Nutrient enrichment, primarily increases in dissolved inorganic nitrogen and phosphorus, in estuaries leads to excessive phytoplankton (or algae) growth and high biological productivity.   Algae in water can be a nuisance in themselves, since they represent *internal* production of organic matter in estuaries.   This organic matter is broken down by bacterial respiration, and, thus, excessive phytoplankton growth causes increased O_2 depletion at depth.   (Additional O_2 is used up by the oxidation of NH_4^+ to NO_3^- and NO_2^- in estuaries receiving pollution-derived ammonia.)   In most estuarine studies, the term *eutrophication* refers only to excessive algal growth due to nutrient enrichment and not to direct organic matter addition.   However, externally produced organic matter from rivers and sewage outlets, which is added to an estuary, is broken down by bacterial respiration to dissolved inorganic N and P, producing additional nutrient enrichment.   Thus, excessive organic matter enrichment and nutrient enrichment are linked and both ultimately cause O_2 depletion.   This is why we use here an expanded definition of eutrophication that includes direct organic addition.

Nutrient enrichment usually involves the dissolved inorganic nitrogen forms NH_4^+, NO_3^-, and NO_2^-- and dissolved PO_4^{---} (and occasionally Si).   Sewage is a particularly rich source of ammonia and phosphate, even after secondary waste treatment.   The response of phytoplankton to increased pollution-derived nutrients (N, P, and Si) is more complex than bacterial response to increased organic matter and varies from estuary to estuary depending upon the estuarine circulation and original nutrient supply and balance.   (See earlier discussion on estuarine nutrients.)   In many eastern U.S. estuaries and coastal waters, where nitrogen tends to be limiting and the waters have lower N:P ratios than normally required by phytoplankton, nitrogen from pollution is often quickly consumed by phytoplankton leaving excess PO_4 in the water (Ryther and Dunstan 1971; Nixon 1981).   For this reason, excess PO_4 in coastal waters off the eastern United States tends to be a tracer of nutrient and organic pollution.   In this case, reductions in pollutive P (e.g., in detergents) will not greatly decrease algal growth, but reductions in pollution-derived N will (Ryther and Dunstan 1971).

Natural mechanisms, primarily denitrification, exist for removing excess N from estuarine water that do not exist for phosphorous.   Seitzinger and Nixon (1985) studied experimentally the effects of excess dissolved nutrient loading on the N_2 and N_2O release rates from marine bottom sediments.   They added excess nutrients to the water in a ratio roughly equal to organism nutrient utilization with the largest amount of nitrogen input equal to 65 times the usual anthropogenic nitrogen loading in Narragansett Bay (a nearby polluted estuary).   The denitrification rates did increase with increased nitrogen loading, but a constant or progressively smaller part of the nitrogen input was removed.   Thus, in polluted estuaries, denitrification cannot completely compensate for excessive nitrogen loading.   (The N_2O production, although still a small part of the gaseous nitrogen release, increased greatly with increased eutrophication during Seitzinger and Nixon's experiment which might be of concern because of atmospheric problems with N_2O; see Chapter 3.)

Other factors are also important in determining the extent of algal growth besides the concentrations of nitrogen and phosphorus. For example, in New York Harbor there is a low algal population, despite large pollutive concentrations of nitrogen and phosphorus, because there are high suspended matter concentrations which limit light and also because of rapid nutrient flushing times by the estuarine circulation. Phosphate, for example, has a residence time of only a week or less (Simpson et al. 1975).

In Europe, large-scale algal blooms occur in the South Bight of the North Sea along the Dutch coast where some 50% of the water from the polluted Rhine, Meuse, and Scheldt rivers is transported. Van Bennekom, Gieskes, and Tijesen (1975) found that for the summer diatom blooms, silica is the limiting nutrient. During the short spring algal blooms, phosphorus is limiting, but on the average there is excess phosphorus and nitrogen relative to silica, in these waters. The reason for the contrast with North American coastal waters, where N is limiting, is the high N:P mole ratio and low silica concentrations in the Rhine River (the dominant input). In several polluted (nitrogen- or phosphorus-enriched) eastern U.S. estuaries, all of the fluvial silica is also removed by diatoms and 80% of the fluvial Si flux is permanently deposited within the estuary (Froelich et al. 1985). This contrast with the nitrogen-limited estuaries discussed earlier points out the difficulties in making generalizations about pollution-derived nutrient overloading in estuaries and the necessity for studies of individual estuaries.

SUSPENDED SEDIMENT DEPOSITION IN MARGINAL MARINE ENVIRONMENTS

Most river-borne suspended sediment is deposited in deltas, estuaries, and other coastal marine environments (Gibbs 1981). Thus, from the point of view of geochemical cycling, it is important to study how this deposition comes about. River-borne suspended sediment consists of both inorganic particles (clays, iron oxide aggregates, etc.) and organic particles. We have already discussed the cycling of Fe, Al, Si, and N and P in estuaries and will confine the discussion here to the removal of the bulk of fine-grained river-borne inorganic and organic particulate material. (A number of heavy trace metals, such as Cu, Zn, and Pb, tend to be associated with fine suspended sediments and thus will be removed with them; see Turekian et al. 1980).

There are a number of sources of coastal and estuarine suspended sediment besides river-borne material (Bokuniewicz and Gordon 1980). Coastal erosion is a sediment source, as is oceanic suspended sediment carried to the coasts. In addition, organisms produce suspended organic particulate material. However, we shall mainly discuss the removal of river-borne suspended sediment since it is most interesting from the point of view of overall geochemical cycling.

Large-scale deposition of fine river-borne suspended sediment occurs in many deltas and estuaries and on the continental shelves. There are several factors that influence the transport and deposition of such suspended sediment: (1) physico-chemical processes that cause flocculation and aggregation of river-borne suspended particles on the transition from fresh water to seawater, (2) estuarine circulation and other hydrologic processes such as variations in current velocity due to alternating tides and changes in cross-sectional area, and (3) agglomeration by organisms as fecal pellets. The relative importance of these processes varies depending upon conditions and has been the subject of some debate (Dyer 1972; Meade 1972; Kranck 1973; Krone 1978; Burton 1976; Aston 1978; Bokuniewicz and Gordon 1980).

Flocculation (aggregation) of river-borne suspended particles involves the formation of large particle aggregates that have a greater settling rate than the original suspended particles. Flocculation and aggregation are tied up with two important factors: *cohesion* of the particles and *collision* of the particles (Kranck 1973; Krone 1978).

Clay minerals (kaolinite, illite, and smectite) that are carried in colloidal suspension in river water have a net negative charge on their faces due to a variety of causes (see Van Olphen 1977 for details). Each clay particle attracts a layer (Gouy layer) of positively charged cations around it to balance this negative charge. This results in excessive cation concentration around each clay particle relative to the bulk fresh water between particles. Thus, colliding clay particles repel one another and they do not aggregate. There is also an attractive force between the clay particles (Van der Waals force), but it is less strong in fresh water than the repulsive force due to the excess cation concentration around each particle. Since saline water has a greater concentration of dissolved charged ions (ionic strength) than fresh water, when river-borne clay particles encounter greater salinity, the concentration of ions in the water between particles increases. This brings about a thinning, or collapse, of the Gouy layer so that the charged particles may come closer together before any repulsion occurs. Thus, in saline water, the attractive or *cohesive* (Van der Waals) forces become stronger than the repulsive forces due to positive charge, and cohesion between clay particles can occur. (The preceding discussion is largely from Van Olphen 1977 to which the reader is referred for further details.) Since clay particles are always cohesive at a salinity greater than 1‰–3‰, increases in salinity beyond the initial mixing of fresh and saline water are not important in cohesion (Krone 1978).

The stability of clay suspensions against flocculation, when subjected to saline water, is dependent upon not only the salinity of the water but also on the type of clay involved. In experiments with brackish estuarine water, Edzwald, Upchurch, and O'Melia (1974) found that, at the same salinity, kaolinite was less stable towards flocculation than illite. In the Pamlico River estuary (North Carolina), Edzwald, Upchurch, and O'Melia attributed the distribution of kaolinite and illite in the bottom sediments to their relative stability. Kaolinite, being less stable, flocculated more rapidly than illite and was found to be more concentrated than illite in the upstream sediments near the river mouth. Krone (1978) quotes work that shows that the flocculation sequence with increasing salinity is kaolinite before illite before smectite and all flocculate at low salinities (<3‰).

Gibbs (1977) found a sequence of clay minerals deposited off the Amazon River mouth reminiscent of what would be expected for differential flocculation. Smectite increased with distance offshore while illite decreased greatly and kaolinite decreased less so. However, Gibbs attributed the laterally changing concentrations of clay minerals not to clay flocculation but to physical sorting of sediments by size. Differential clay flocculation was felt to be less important because possible natural organic and metal oxyhydroxide coatings on clay particles might modify the surface properties of the clays. Organic coatings (particularly of humic acid) probably also show surface effects (Burton 1976) since organics have a net negative charge in fresh water similar to clay minerals. (See earlier, under Inorganic Removal for a discussion of the estuarine removal of iron and organics.)

In order for clay particles to fully flocculate they must also collide with other clay particles. Collisions between suspended particles can occur due to three mechanisms (Krone 1978): (1) Brownian motion due to thermally induced water molecule motion, (2) differential settling velocities of different suspended particles, and (3) velocity gradients that cause particles of different speeds to collide with one another. With very high suspended sediment concentrations, the first two mechanisms for particle collisions, Brownian motion and differential settling, are important and cause the formation of floccules or aggregates of cohesive clay particles. Particles settle out in restricted quiet areas or at times in the tidal cycle when there is little motion. Since these aggregates are weak, they tend to be broken up by high velocities (Krone 1978). *Fluid mud*, a dense fluid layer with very high concentrations of suspended clay floccules (2,000–20,000 mg/ℓ; Dyer 1972) often forms on the bottom from collisions due to Brownian motion and differential settling. Fluid mud forms when there is a very large sediment supply (and low population of benthic animals) such as in the estuaries of the Chao Phraya, Thames, and Severn rivers (Dyer 1972; Meade 1972).

The velocity gradient is very high in the mixing zone of estuaries, where fresh water first meets saline water under tidal conditions. This can cause collisions and the rapid formation of strong aggregates when the suspended sediment concentration is lower than that needed for floccule growth by Brownian motion and differential settling. These strong aggregates, formed in areas of large velocity gradients, are resistant to being broken down as long as they do not become too large (Krone 1978). When the suspended sediment concentration is very low (<300 mg/ℓ; Dyer, 1972), collisions become too infrequent for appreciable flocculation via any of the above-mentioned processes, and sedimentation occurs instead by means of additional processes such as biogenic removal.

The importance of estuarine circulation processes in the nearshore transport of suspended sediment has been emphasized by Meade (1972). In moderately stratified estuaries (Figure 7.1) where there is both vertical mixing and a net landward flow at the bottom of the estuary, sediment while settling out is carried landward and temporarily trapped in the estuary. As a result, in many moderately stratified estuaries there is a *turbidity maximum,* or a zone of maximum suspended sediment concentration bounded by lower concentrations both landward and seaward (see Figure 7.5). This turbidity maximum occurs near the furthest (landward) extent of the bottom landward-flowing saline water, where it meets seaward-flowing fresh water (see Figure 7.6). The turbidity maximum is often accompanied by a zone of maximum sediment accumulation which is also a place of high velocity gradients. Since these high velocity gradients tend to promote greater collisions between suspended clay particles, the high sediment deposition may be a result of *both* estuarine circulation and velocity-gradient-induced flocculation and aggregation.

In salt-wedge estuaries, where river flow is dominant over tidal forces, little mixing occurs between the fresh, seaward-flowing river water in the upper layer, and the inward-flowing salt wedge below. Meade (1972) points out that when Mississippi River water, carrying a large suspended sediment load, meets the salt wedge, most of the suspended sediment is carried over the salt wedge with the river water. Because of the steep density gradient and turbulence at the fresh-

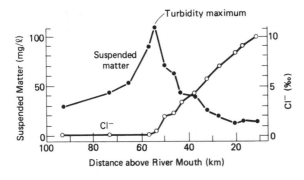

Figure 7.5 Concentration of suspended matter versus chlorinity in York River, Virginia, showing a *turbidity maximum* at the landward limit of sea salt penetration into the estuary; e.g., where the chloride concentration drops to nearly zero. (After R. H. Meade, 1972. "Transport and Deposition of Sediment in Estuaries," *Geol. Soc. Amer. Memoir 133*, p. 100, and B. N. Nelson, 1960. "Recent Sediment Studies in 1960," *Va. Polytech. Inst. J.*, 7(4), pp. 1–4.)

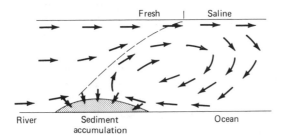

Figure 7.6 Suspended matter transport (arrows) and sediment accumulation near landward limit of seawater mixing in an estuary. Dashed line is fresh-saline water boundary. (After R. H. Meade, 1972. "Transport and Deposition of Sediment in Estuaries," *Geol. Soc. America Memoir 133*, p. 112.)

saline boundary, the river suspended sediment does not settle out and is carried out to sea to be deposited in the Mississippi Delta and on the Gulf of Mexico shelf. This is also true of the Zaire River (Eisma, Kalf, and van der Gaast 1978; van Bennekom et al. 1978), where most of the large particulate organic load is carried through the estuary in the surface layer and deposited at the head of the submarine canyon offshore.

The ability of moving water to transport sediment is dependent upon its velocity. As river water flows into an estuary and the cross section for flow widens, the water velocity decreases, its sediment transport ability diminishes, and sediment deposition is likely to occur. The reverse occurs when the flow cross section narrows. These changes affect both river-borne suspended sediment carried into estuaries, and sediment carried by tidal flow in and out of an estuary (Dyer 1972).

Extreme hydrologic conditions are often very important in near shore sedimentation (Dyer 1972). In river floods, there is much greater sediment transport than usual; the sediment load transported in several days may be greater than the total load for an average year. In addition, during river floods, the river flow dominates any estuarine circulation so that, in some estuaries like the Mississippi River, the salt wedge moves out of the estuary altogether and there is no bottom flow inward (Meade 1972). This means that the coarser suspended sediment is carried out of the estuary, and that sediment previously deposited in the estuary is eroded and flushed out by the dominant river flow (Dyer 1972). Instead of accumulating in the estuary, this sediment is deposited in deltas and on the continental shelf.

Another major process that results in sediment accumulation in coastal regions is *biogenic agglomeration*. A number of bottom-living (*benthic*) organisms feed on suspended particulate matter and agglomerate fine-grained suspended matter into fecal pellets that are denser and thus have a higher settling velocity than the original suspended particles (Rhoads 1974). Fecal pellets deposited on the bottom subsequently may be broken down by deposit feeders and either resuspended or agglutinated to the bottom by invertebrates (or plants). The benthic organisms involved include various bivalves (oysters, clams, mussels, scallops), copepods, tunicates, and barnacles (Rhoads 1974). Bottom feeders are capable of reworking and pelletizing the bottom sediment at a rate greater than the sediment deposition rate, as in Buzzard's Bay (Massachusetts) and Long Island Sound, so that the top centimeter of the bottom sediment may consist entirely of a layer of fecal pellets (Rhoads 1974). The pellet mantle in Long Island Sound is resuspended several times a year, and the rate-limiting step for permanent sediment accumulation is the conversion of fecal pellets to cohesive mud, which occurs at the bottom of the pellet layer (Bokuniewicz and Gordon 1980). In Long Island Sound, the thickness of sediment deposits is not apparently related to estuarine circulation (Bokuniewicz, Gebert, and Gordon 1976). This is because the biological processing rate is 50 times the rate of supply of silt-clay sediment (Bokuniewicz and Gordon 1980). Also, large suspended sediment concentrations, which would lead to the formation of fluid muds, do not occur. Thus, the reasons for the importance of biological agglomeration in areas such as Long Island Sound include a low sediment supply by rivers combined with a large benthic organism population capable of reworking the sediment rapidly.

Over the long run, most sediment brought by rivers passes out of estuaries due to their small size and small storage capacity, and this is evidenced by the paucity of estuarine sediments in the geological record. However, at present, relatively deep, unfilled estuaries are rather common as a result of the drowning of many coastlines by the rapid postglacial rise of sea level. As a result there are numerous instances of estuaries that trap a large proportion of the sediments brought to them by rivers. An example is the polluted Scheldt estuary in Belgium and the Netherlands, which receives a heavy river-borne particulate organic load. Seventy-five percent of the suspended particulate organic matter carried by fresh water is deposited in the estuary in a restricted area 30 km long of 1‰–10‰ salinity (Wollast and Peters 1978). Similarly in the Gironde estuary in France, 75% of the river-borne suspended load is trapped in the estuary (Allen, Sauzay, and Castaing 1976). In Long Island Sound, the suspended sediment retention has been estimated as nearly 100% (Gordon 1980; Bokuniewicz and Gordon 1980). The factors leading to such high sediment retention in Long Island Sound include a low river-borne suspended sediment load, a high biological processing rate relative to sediment supply rate (cited above), intermediate tidal power which does not resuspend sediment sufficiently to prevent deposition, and a large storage volume in the incompletely filled estuary. Some of the factors important in Long Island Sound sediment retention probably are also important elsewhere (Bokuniewicz and Gordon 1980).

In well-stratified (salt-wedge) estuaries there is considerable variation in the amount of sediment deposited within the estuary. The Amazon River carries a considerable terrigenous sediment load into the Amazon estuary. Some 95% of the suspended sediment in the surface water settles out before the salinity reaches 3‰ (Milliman, Summerhayes, and Bareto 1975; Milliman and Boyle 1975). There is a shoal just seaward of the river mouth, where there is strong vertical mixing and a turbidity maximum, and most of the surface suspended sediment settles out there. However, the sediment does not remain at this location, but is resuspended near the bottom of the water column by tidal and wave action (Milliman, Summerhayes, and Bareto 1975). Some of the suspended sediment is carried, by the northwest flowing longshore Guiana Current, onto the outer shelf where a gradual deposition process occurs. The remaining majority of the sediment is carried shoreward by landward-moving bottom water (Gibbs 1976), thus preventing escape of sediment to the ocean. By contrast to the Amazon, in other salt-wedge estuaries like the Zaire and the Mississippi rivers, discussed earlier, the river flow is so dominant that most river-borne sediment is carried out of the estuary in the surface layer and deposited elsewhere.

Human changes that affect the transport of river-borne suspended sediment through estuaries include dredging or dumping of sediment, construction of piers or barriers, diversion of freshwater flow from the river by dams or reservoirs, and addition of fresh water to the river (Dyer 1972; Meade 1972; Simpson et al. 1975). Dredging of river channels to aid navigation is common and often alters estuarine circulations by allowing the bottom landward-flowing salt wedge, in stratified estuaries, to penetrate further into the estuary.   Since sediment deposition often occurs where the saline water meets the fresh river flow, this moves the maximum sediment deposition zone further inland (Dyer 1972; Meade 1972).   Conversely, dumping of sediment (as in the Hudson River estuary) interferes with the inward flow of the bottom layer and results in localized sediment deposition (Simpson et al. 1975).   The construction of piers and barriers results in the reduction of current velocity and consequent sediment deposition.

Flood control upstream on a river, through the use of dams and reservoirs, can severely alter coastal sedimentation patterns.   The construction of a dam on the Savannah River, for example, eliminated the natural three-year flood pattern and allowed river-borne sediment that would normally have been flushed out of the Savannah River estuary during each flood to accumulate in the estuary (Meade 1972).   Also, suspended sediment tends to accumulate behind river dams, thus reducing the amount of suspended sediment reaching the sea.   Overall, however, as pointed out in Chapter 5, human activities have resulted in increased river-borne suspended loads and this has resulted in increased coastal sedimentation.

ANTIESTUARIES AND EVAPORITE FORMATION

The term *antiestuary* refers to an estuary that has a circulation that is the *reverse* of the typical estuarine circulation; that is, seawater flows into the estuary in the *surface* layer instead of in the bottom layer (see Figure 7.7).   Such a reverse flow develops in a coastal embayment or basin because of certain conditions: (1) circulation with the open ocean is *restricted* by the presence of a sill, sand bar, reef or other barrier at the entrance to the basin; (2) the basin has a high *net* rate of evaporation; that is, evaporation of water from the basin is greater than the combined input of fresh water from river runoff and rainfall.   This requires a warm, arid climate. Excess evaporation in the basin results in water loss which is greater than the input of fresh water from rainfall and/or river runoff.   This net water loss is made up for by surface seawater flow into the basin from the open ocean in order to maintain constant water volume in the basin (note the inward-sloping water surface in Figure 7.7.)   The evaporation of seawater in the surface layer of the basin leaves the excess salts behind and since the barrier prevents mixing with the open ocean, the surface water becomes denser and tends to settle toward the bottom of the basin. If the basin is to maintain a constant salinity, there must be a reverse flow at depth of denser, more saline water out over the sill to remove the excess salts.   In this situation, the combination of evaporative water loss and deep return flow of water to the ocean is equal to surface inflow of seawater.   The Mediterranean Sea is the best known present-day example of this type of reverse flow.   The salinity of the

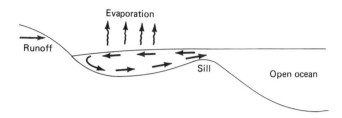

Figure 7.7   Antiestuarine circulation. Arrows indicate direction of water flow.

Mediterranean Sea is maintained at a somewhat greater value (37‰–38‰; Pickard and Emery 1982) than that of average ocean water (35‰).

The preceding (antiestuary) model for a restricted basin with net evaporation assumes that both a constant volume of water and constant salinity are maintained. If, however, such a restricted evaporative basin becomes completely isolated from the ocean so that the net water lost by evaporation cannot be replaced, its water volume decreases and the salinity accordingly increases. Ultimately the salinity becomes great enough for the precipitation of salts to take place. The sequence of salts that precipitate from the evaporation of seawater was determined originally by Usiglio in 1849. When normal seawater has been evaporated to 19% of its original volume gypsum ($CaSO_4 \cdot 2H_2O$) precipitates and halite ($NaCl$) precipitates at around 10% of the original volume. Further evaporation will also produce Mg salts ($MgSO_4$ and $MgCl_2$) and KCl; these are known as *bittern salts*. The relative thicknesses of gypsum:halite:Mg and K salts produced by complete seawater evaporation are approximately 1:20:5 (Hsü 1972). (Some $CaCO_3$ may form via evaporation but it is practically negligible compared to that removed from seawater by other means such as skeletons of organisms and much of this skeletal debris can be converted to dolomite $CaMg(CO_3)_2$ by the Mg-rich brines.) The formation of the evaporite minerals, $CaSO_4 \cdot 2H_2O$ (gypsum) and $NaCl$ (halite) provides an important mechanism for the removal of Na, Cl, and SO_4 from seawater (see Chapter 8).

One problem with the complete evaporation of seawater in a closed basin is that in order to deposit a 1-m thickness of $CaSO_4$, some 1700 m of seawater must be evaporated (Hsü 1972). In addition, for every meter of $CaSO_4$ deposited, one would expect 20 times as much halite and 5 times as much bitterns. In order to provide for greater evaporite deposition without the need to evaporate such great quantities of water, and for the formation of thick monomineral deposits of gypsum which are commonly encountered in the geological record, the *barred-basin* theory of evaporate deposition with its various modifications was proposed (Ochsenius 1877; King 1947; Scruton 1953). (For a discussion of the historical development of theories of evaporite deposition, see Hsü 1972.)

The theoretical *barred basin* has reverse circulation of the type described for an antiestuary. Evaporite deposition can occur, even though the water body is not completely cut off from the ocean, because the salinity of the basin water builds up to saturation with saline minerals faster than the water can be replaced by fresh seawater. Water balance is maintained by seawater influx, a very high evaporation rate, and subsurface return to the ocean of very saline brines (which are missing that part of the salts that has precipitated). In this way, a constant concentration of salts necessary for continual deposition of $CaSO_4$, for example, is maintained. As a result, large thicknesses of a single evaporite mineral can be deposited over a period of time without having to evaporite an extremely large thickness of seawater. Variations on this model also permit the deposition of different evaporite mineral facies in different parts of the basin where the salinity is different (see Fig. 7.8). For example, as the entering seawater becomes more and more saline, the mineral sequence $CaCO_3$ (calcite) or $CaMg(CO_3)_2$ (dolomite), $CaSO_4 \cdot 2H_2O$ (gypsum), $NaCl$ (halite) develops along the length of the basin with increasing distance from the mouth.

Thick, areally extensive evaporite deposits were formed at various times in the geologic past, for example, the late Miocene evaporites under the Mediterranean basin (Hsü 1972), the well-known Permian Zechstein in Europe (Schmalz 1970), and the Silurian Salina in the Michigan Basin (Briggs 1958). However, most present-day marine evaporite formation is rather limited both in areal extent and in the total removal of marine ions from seawater.

Present-day depositional environments where evaporite minerals are forming from marine waters include the following (Friedman 1978): (1) restricted basins, estuaries, or lagoons; (2) small isolated pools with occasional marine influx; and (3) coastal *sabkhas* or *salinas*, which are supratidal (above tide level) salt flats found particularly in the Persian Gulf.

Large-scale marine barred basins (or antiestuaries) precipitating evaporites do not exist at present. One small occurrence of the barred-basin type has been described. The Bocana de Virrilá on the Peruvian coast is a narrow, winding marine estuary with very restricted circulation (although not with an actual bar) which is precipitating evaporites in lateral zones along its length (Morris and Dickey 1957; Brantley et al. 1984). Seawater enters from the ocean, evaporates, and becomes more saline as it moves inland. Calcite deposition occurs first, followed by gypsum precipitation (at salinities greater than 160‰) and then halite as surface water becomes progressively more saline toward the estuary head. The inflow of marine water at least partly balances the water loss by evaporation. Since the estuarine circulation pattern has not been studied, it is not known whether bottom reverse flow of more saline estuarine water also occurs. No Mg and K salts seem

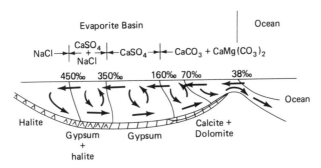

Figure 7.8 Restricted evaporite basin showing zoned deposition of evaporite minerals. Arrows indicate water flow. Contours are for approximate salinities (in ‰). (Modified from Scruton 1953.)

to be precipitating in the estuary, which means these seawater components must be removed somehow.

Small, shallow sea-marginal or tidal-flat pools, which are periodically flooded with seawater, then isolated and evaporated to dryness forming halite and gypsum, are known in several areas such as the Red Sea and Baja California. In the Red Sea pools, gypsum forms in the summer when the evaporation is particularly intense and the salinity reaches 330‰ (Friedman 1978). In the Baja California area, deposits of halite and gypsum form in pools or salt pans on tidal flats (Kinsman 1969). However, the total amount of NaCl and $CaSO_4$ removed from seawater is not great and reexposure to fresh seawater can result in loss by dissolution.

The best-known environment of modern evaporite formation is the coastal *sabhka* or *salina* which is particularly well developed along the Trucial Coast of the Persian Gulf (e.g., Kinsman 1969; Butler 1969; Shearman 1966, 1978). Here the main evaporite mineral forming is $CaSO_4$ (anhydrite or gypsum) and deposition occurs on vast mud flats above the reach of normal tides. Persian Gulf sabhkas occur in areas of very low relief in a very arid climate. The water table is never more than 1–2 m below the surface and consists primarily of seawater-derived brines near the coast grading into continentally derived brines inland. Seawater occasionally floods the sabhkas during storms, enters the sediments, and then evaporates, resulting in the formation of gypsum within the sediments. Dolomite, and additional gypsum is formed by reactions between $CaCO_3$ in the sediment and the residual brines. However, due to redissolution, halite does not accumulate.

Although large thicknesses of marine evaporites formed in the geologic past, removing considerable quantities of NaCl, and $CaSO_4$ from seawater, these salts are not being removed from seawater to any extent at present, and, thus, may be presumed to be building up in seawater. (For further discussion see Chapter 8.) The Mediterranean Sea exhibits antiestuarine circulation of the type associated with the barred-basin theory of evaporite formation, but it is not forming evaporites at present. However, this was not the case during the Miocene period (12–25 million years ago) when the Mediterranean underwent extensive evaporation and evaporite mineral deposition (Hsü 1972).

REFERENCES

ALLEN, G. P., G. SAUZAY, and J. H. CASTAING. 1976. Transport and deposition of suspended sediment in the Gironde estuary, France. In *Estuarine processes*, ed. M. Wiley, Vol. 2:63–81. New York: Academic Press.

ASTON, S. R. 1978. Estuarine chemistry. In *Chemical oceanography*, 2nd ed., ed. J. P. Riley and R. Chester, pp. 361–440. New York: Academic Press.

ASTON, S. R., and R. CHESTER. 1973. The influence of suspended particles on the precipitation of iron in natural waters, *Est. Coast. Mar. Sci.* 1:225–231.

BALE, A. J., and A. W. MORRIS. 1981. Laboratory simulation of chemical processes induced by estuarine mixing: The behavior of iron and phosphate in estuaries, *Est. Coast. Shelf Sci.* 13:1–10.

BERNER, R. A. 1971. *Principles of chemical sedimentology*, New York: McGraw-Hill. 240 pp.

BEWERS, J. M., and P. A. YEATS. 1980. Behavior of trace metals during estuarine mixing. In *Proceedings of the review and workshop on river inputs to ocean systems*, ed. J.-M. Martin, J. D. Burton, and D. Eisma, p. 103–115. Rome: FAO.

BOKUNIEWICZ, H. J., J. A. GEBERT, and R. L. GORDON. 1976. Sediment mass balance of a large estuary (Long Island Sound), *Est. Coast. Mar. Sci.* 4:523–536.

BOKUNIEWICZ, H. J., and R. B. GORDON. 1980. Sediment transport and deposition in Long Island Sound. In *Estuarine physics and chemistry: Studies in Long Island Sound*, ed. B. Saltzman, pp. 69–106. Adv. in Geophysics, vol. 22. New York: Academic Press.

BOWDEN, K. F. 1967. Circulation and diffusion. In *Estuaries*, ed. G. H. Lauff, pp. 15–36. AAAS Publ. no. 83. Washington, D.C.

BOYLE, E. A., R. COLLIER, A. T. DENGLER, J. M. EDMOND, A. C. NG, and R. F. STALLARD. 1974. On the chemical mass-balance in estuaries, *Geochim. Cosmochim. Acta* 38:1719–1728.

BOYLE, E. A., J. M. EDMOND, and E. R. SHOLKOVITZ. 1977. The mechanism of iron removal in estuaries, *Geochim. Cosmochim. Acta* 41:1313–1324.

BRIGGS, L. I. 1958. Evaporite facies, *J. Sed. Petrol.* 28:46–56.

BRANTLEY, S. L., N. E. MOLLER, D. A. CRERAR, and T. H. WEARE. 1984. Geochemistry of a modern marine evaporite: Bocana de Virrilá, Peru, *J. Sed. Petrol.* 54 (2): 0447–0462.

BURTON, J. D. 1976. Basic properties and processes in estuarine chemistry. In *Estuarine chemistry*, ed. J. D. Burton and P. S. Liss, pp. 1–35. London: Academic Press.

BUTLER, G. P. 1969. Modern evaporite deposition and geochemistry of coexisting brines, the sabkha, Trucial Coast, Arabian Gulf, *J. Sed. Petrol.* 39:70–89.

CAMERON, W. M., and D. W. PRITCHARD. 1963. Estuaries. In *The sea*, ed. M. N. Hill, vol. 2:306–324. New York: John Wiley.

COONLEY, L. S., JR., E. B. BAKER, and H. D. HOLLAND. 1971. Iron in the Mullica River and in Great Bay, New Jersey, *Chem. Geol.* 7:51–63.

CRERAR, D. A., J. L. MEANS, R. F. YURETICH, M. P. BORCSIK, J. L. AMSTER, D. W. HASTINGS, G. W. KNOX, K. E. LYON, and R. F. QUIETT. 1981. Hydrogeochemistry of the New Jersey coastal plain, 2. Transport and deposition of iron, aluminum, dissolved organic matter, and selected trace elements in stream, ground, and estuary water, *Chem. Geol.* 33:23–44.

DEMASTER, D. J. 1981. The supply and accumulation of silica in the marine environment, *Geochim. Cosmochim. Acta* 45:1715–1732.

DEUSER, W. G. 1975. Reducing environments. In *Chemical oceanography*, 2nd ed., ed. J. P. Riley and R. Chester, vol. 3, Chap. 16:1–37. New York: Academic Press.

DION, E. P. 1983. Trace elements and radionuclides in the Connecticut River and Amazon River Estuary. Ph.D. dissertation, Yale University, New Haven, Conn. 233 pp.

DYER, K. R. 1972. Sedimentation in estuaries. In *Estuarine environments*, ed. R. S. K. Barnes and J. Green, pp. 10–32. London: Applied Science.

EDMOND, J. M., E. A. BOYLE, B. GRANT, and R. F. STALLARD. 1981. The chemical mass balance in the Amazon plume, I: The nutrients. *Deep-Sea Research* 28A(11):1339–1374.

EDZWALD, J. K., T. B. UPCHURCH, and C. R. O'MELIA. 1974. Coagulation in estuaries, *Environ. Sci. Technol.* 8(1):58–63.

EISMA, D., J. KALF, and S. J. VAN DER GAAST. 1978. Suspended matter in the Zaire estuary and adjacent Atlantic Ocean, *Neth. J. Sea Res.* 12:382–406.

FAXI, LI. 1980. An analysis of the mechanisms of removal of reactive silicate in the estuarine zone. In *Proceedings of the review and workshop on river inputs to ocean systems*, ed. J. M. Martin, J. D. Burton, and D. Eisma, pp. 200–210. Rome: FAO.

FRIEDMAN, G. M. 1978. Depositional environments of evaporite deposits. In *Marine evaporites*, ed. W. E. Dean and B. C. Schrieiber, pp. 177–184. SEPM Short Course no. 4. Oklahoma City.

FROELICH, P. N., L. W. KAUL, J. T. BYRD, M. O. ANDRAE, and K. K. ROE. 1985. Arsenic, barium, germanium, tin, dimethyl sulfide and nutrient biogeochemistry in Charlotte Harbor, Florida, a phosphorus-enriched estuary, *Est. Coast. Shelf Sci.* 20:239–264.

GIBBS, R. J. 1976. Amazon River sediment transport in the Atlantic Ocean, *Geology* 4:45–58.

———. 1977. Clay mineral segregation in the marine environment, *J. Sed. Petrol.* 47 (1):237–243.

———. 1981. Sites of river-derived sedimentation in the ocean, *Geology* 9:77–80.

GORDON, R. B. 1980. The sedimentary system of Long Island Sound. In *Estuarine physics and chemistry: Studies in Long Island Sound*, ed. B. Saltzman, pp. 1–39. Adv. in Geophysics, vol. 22. New York: Academic Press.

HAHN, H. H., and W. STUMM. 1970. The role of coagulation in natural waters, *Amer. J. Sci.* 268:354–368.

HSÜ, K. T. 1972. Origin of saline giants: A critical review after the discovery of the Mediterranean evaporite, *Earth-Sci. Rev.* 8:371–396.

HYDES, D. J., and P. S. LISS. 1977. The behavior of dissolved aluminum in estuarine and coastal waters, *Est. Coast. Mar. Sci.* 5:755–769.

KAUL, L. W., and P. N. FROELICH, JR. 1984. Modeling estuarine nutrient geochemistry in a simple system, *Geochim. Cosmochim. Acta* 48:1417–1433.

KING, R. H. 1947. Sedimentation in Permian Castile Sea, *Amer. Assoc. Petrol. Geologists Bull.* 31:470–477.

KINSMAN, D. J. J. 1969. Modes of formation, sedimentary associations, and diagnostic features of shallow-water and supratidal evaporites, *Bull. Amer. Assoc. Petrol. Geologists* 53:830–840.

KLUMP, J. VAL, and C. S. MARTENS. 1981. Biogeochemical cycling in an organic-rich coastal marine basin, 2: Nutrient sediment-water exchange processes, *Geochim. Cosmochim. Acta* 45:101–121.

KNAPP, G. B., D. J. DEMASTER, and C. A. NITROUER. 1981. Processes affecting the uptake and accumulation of silica on the Amazon continental shelf, *GSA Abstracts* 13:488–489.

KRANCK, K. 1973. Flocculation of suspended sediment in the sea, *Nature* 246:348–350.

KROM, M. D., and R. A. BERNER. 1981. The diagenesis of phosphorus in a nearshore marine sediment, *Geochim. Cosmochim. Acta* 45:207–216.

KRONE, R. B. 1978. Aggregation of suspended particles in estuaries. In *Estuarine transport processes,* ed. B. Kjerfve, pp. 177–190. Columbia, S.C.: Univ. of South Carolina Press.

LIKENS, G. E. 1972. Eutrophication and aquatic ecosystems. In *Nutrients and eutrophication,* ed. G. E. Likens, pp. 3–13. Amer. Soc. of Limnology and Oceanography Spec. Sympos. Vol. 1. pp. 3–13.

LISS, P. S. 1976. Conservative and non-conservative behavior of dissolved constituents during estuarine mixing. In *Estuarine chemistry,* ed. J. D. Burton, and P. S. Liss, pp. 93–130. New York: Academic Press.

LODER, T. C., and R. P. REICHARD. 1981. The dynamics of conservative mixing in estuaries, *Estuaries* 4(1): 64–69.

McCARTHY, J. J., W. R. TAYLOR, and J. L. TAFT. 1975. The dynamics of nitrogen and phosphorus cycling in the open waters of Chesapeake Bay. In *Chemistry in the coastal environment,* ed. T. M. Church, pp. 664–681. ACS Sympos. ser. 18.

McELROY, M. B., J. W. ELKINS, S. C. WOFSY, C. E. KOLB, A. P. DURDIN, and W. A. KAPLAN. 1978. Production and release of N_2O from Potomac Estuary, *Limnol. Oceanogr.* 23(6):1168–1182.

MACKENZIE, F. T., and R. M. GARRELS. 1966. Chemical mass balance between rivers and oceans, *Amer. J. Sci.* 264:507–525.

MACKIN, J. E., and R. C. ALLER. 1984. Processes affecting the behavior of dissolved Al in estuarine waters, *Marine Chemistry* 14:213–232.

MEADE, R. H. 1972. Transport and deposition of sediment in estuaries, *GSA Memoir* 133:91–120.

MEYBECK, M. 1979. Concentrations des eaux fluviales en éléments majeurs et apports en solution aux oceans, *Rev. Géol. Dyn. Géogr. Phys.* 21(3):215–246.

———. 1982. Carbon, nitrogen and phosphorus transport by world rivers, *Amer. J. Sci.* 282:401–450.

MILLIMAN, J. D., and E. BOYLE. 1975. Biological uptake of dissolved silica in the Amazon River Estuary, *Science* 189:995–997.

MILLIMAN, J. D., C. P. SUMMERHAYES, and H. T. BARETO. 1975. Oceanography and suspended matter off the Amazon River, *J. Sed. Petrol.* 45:189–206.

MORRIS, A., A. J. BALE, and R. J. M. HOWLAND. 1981. Nutrient distributions in an estuary: Evidence of chemical precipitation of dissolved silicate and phosphate, *Est. Coast. Shelf Sci.* 12:205–217.

MORRIS, R. C., and P. A. DICKEY. 1975. Modern evaporite deposition in Peru, *Amer. Assoc. Petrol. Geologists Bull.* 41:2467–2474.

NELSON, B. N. 1960. Recent sediment studies in 1960, *Va. Polytech. Inst. J.* 7: n. 4, 1–4.

NIXON, S. W. 1981. Remineralization and nutrient cycling in coastal marine ecosystems. In *Estuaries and nutrients,* ed. B. J. Neilson and L. E. Cronin, pp. 111–138. Clifton, N.J.: Humana Press.

OCHSENIUS, K. 1877. *Die Bildung der Steinsalzlager und ihrer Mutterlaugensalze.* Halle, Germany: Pfeffer, 172 pp.

O'CONNER, D. J., R. V. THOMANN, and D. M. DI TORO. 1975. Water-quality analyses of estuarine systems. In *Estuaries, geophysics and the environment,* Natl. Res. Council Geophysics of Estuaries Panel, Natl. Acad. of Sciences, pp. 71–83. Washington, D.C.

OFFICER, C. B. 1979. Discussion of the behavior of nonconservative constituents in estuaries, *Est. Coast. Mar. Sci.* 9:91–94.

OFFICER, C. B., and D. R. LYNCH. 1981. Dynamics of mixing in estuaries, *Est. Coast. Shelf Sci.* 12:525–533.

PICKARD, G. L., and W. J. EMERY. 1982. *Descriptive physical oceanography,* 4th ed., New York: Pergamon Press. 249 pp.

PRITCHARD, D. W. 1967. What is an estuary: Physical viewpoint. In *Estuaries,* ed. G. H. Lauff, pp. 3–5. AAAS Publ. no. 83. Washington, D.C.

PRITCHARD, D. W., and H. H. CARTER. 1973. Estuarine circulation patterns. In *The estuarine environment: Estuaries and estuarine sedimentation*, J. R. Schubel, convener, pp. iv, 1–7. AGI Short Course. Washington, D.C.: AGI.

REDFIELD, A. C., B. H. KETCHUM, and R. A. RICHARDS. 1963. The influence of organisms on the composition of sea-water. In *The sea,* ed. M. N. Hill. vol. 2:26–77. New York: Wiley-Interscience.

RHOADS, D. C. 1974. Organism-sediment relations on the muddy sea floor, *Oceanogr. Mar. Biol. Ann. Rev.* (ed. H. Barnes) 1–2:263–300.

RICHARDS, F. 1965. Anoxic basins and fjords. In *Chemical oceanography,* ed. J. Riley and G. Skirrow, vol. 1:611–645. New York: Academic Press.

ROWE, G. T., C. M. CLIFFORD, K. L. SMITH, JR., and P. L. HAMILTON. 1975. Benthic nutrient regeneration and its coupling to primary production in coastal waters, *Nature* 255:215–217.

RYTHER, J. H., and W. M. DUNSTAN. 1971. Nitrogen, phosphorus and eutrophication in the coastal marine environment, *Science* 171:1008–1013.

SCHMALZ, R. F. 1970. Environment of marine evaporite deposition, *Miner. Ind.* 35(8): 1–7.

SCRUTON, P. C. 1953. Deposition of evaporites, *Bull. Amer. Assoc. Petrol. Geologists* 37: 2498–2512.

SEITZINGER, S. P., and S. W. NIXON. 1985. Eutrophication and the rate of denitrification and N_2O production in coastal marine sediments, *Limnol. Oceanogr.* 30(6):1332–1339.

SEITZINGER, S. P., S. W. NIXON, and M. E. Q. PILSON. 1984. Dentrification and nitrous oxide production in a coastal marine ecosystem, *Limnol. Oceanogr.* 29:73–83.

SEITZINGER, S. P., S. NIXON, M. E. Q. PILSON, and S. BURKE. 1980. Denitrification and N_2O production in near-shore marine sediments, *Geochim. Cosmochim. Acta* 44:1853–1860.

SHEARMAN, D. J. 1966. Origin of marine evaporites by diagenesis, *Inst. Mining Met. Trans.* 375:207–215.

———. 1978. Evaporites of coastal sabhkas. In *Marine evaporites,* ed. W. F. Dean and B. C. Schreiber, pp. 6–20. SEPM Short Course no. 4. Oklahoma City, Okla.

SHOLKOVITZ, E. R. 1976. Flocculation of dissolved organic and inorganic matter during the mixing of river and seawater, *Geochim. Cosmochim. Acta* 40:831–845.

SHOLKOVITZ, E. R., E. A. BOYLE, and N. B. PRICE. 1978. The removal of dissolved humic acids and iron during estuarine mixing, *Earth Planet. Sci. Letters* 40:130–136.

SIMPSON, H. J., S. C. WILLIAMS, C. R. OLSEN, and D. R. HAMMOND. 1975. Nutrient and particulate matter budgets in urban estuaries. In *Estuaries, geophysics and the environment*, Natl. Res. Council Geophysics of Estuaries Panel, Natl. Acad. of Sciences, pp. 94–103. Washington, D.C.

STEFANSSON, J., and F. A. RICHARDS. 1963. Processes contributing to the nutrient distribution off the Columbia River and Strait of Juan de Fuca, *Limnol. Oceanogr.* 8:394–410.

STOMMEL, H., and H. G. FARMER. 1952. *On the nature of estuarine circulation.* Woods Hole Tech. Report no. 52–63 (Pt. 3, Chap. 7).

SUNDBY, B., N. SILVERBERG, and R. CHESSELET. 1980. Estuarine mobilization and export of manganese: An example from the St. Lawrence Estuary. In *Proceedings of the review and workshop on river inputs to ocean systems*, ed. J.-M. Martin, J. D. Burton, and D. Eisma, pp. 231–238. Rome: FAO.

TUREKIAN, K. K., J. K. COCHRAN, L. K. BENNINGER, and A. C. ALLER. 1980. The sources and sinks of nuclides in Long Island Sound. In *Estuarine physics and chemistry: Studies in Long Island Sound*, ed. B. Saltzman, pp. 129–164. Adv. in Geophysics, vol. 22. New York: Academic Press.

UPCHURCH, J. B., J. K. EDZWALD, and C. R. O'MELIA. 1974. Phosphates in sediments of Pamlico Estuary, *Environ. Sci. Technol.* 8:56–58.

VAN BENNEKOM, A. J., G. W. BERGER, W. HELDER, and R. T. P. DE VRIES. 1978. Nutrient distribution in the Zaire estuary and river plume, *Neth. J. Sea Res.* 12:296–323.

VAN BENNEKOM, A. J., W. C. GIESKES, and S. B. TIJESEN. 1975. Eutrophication of Dutch coastal waters, *Proc. Royal Soc. London B.* 189:359–374.

VAN OLPHEN, H. 1977. *An introduction to clay colloid chemistry,* 2nd ed. New York: John Wiley. 318 pp.

WILLIAMS, P. J. LE B. 1980. Primary productivity and heterotrophic activity in estuaries. In *Proceedings of the review and workshop on river inputs to ocean systems*, ed. J.-M. Martin, J. D. Burton, and D. Eisma, pp. 243–246. Rome: FAO.

WOLLAST, R. 1980. Redox processes in estuaries. In *Proceedings of the review and workshop on river inputs to ocean systems*, ed. J.-M. Martin, J. D. Burton, and D. Eisma, pp. 211–222. Rome: FAO.

———. 1983. Interactions in estuaries and coastal waters. In *The major biogeochemical cycles and their interactions*, ed. B. Bolin and R. B. Cook, pp. 385–407. Chichester: John Wiley.

WOLLAST, R., and G. BILLEN. 1981. The fate of terrestrial organic carbon. In *The flux of carbon from the rivers to the oceans*, U.S. Dept. of Energy, CONF-8009/40/UC. Washington, D.C.

WOLLAST, R., and J. J. PETERS. 1978. Biogeochemical properties of an estuarine system: The river Scheldt. In *Biogeochemistry of estuarine sediments*, ed. E. D. Goldberg, pp. 279–293. Paris: UNESCO.

8

The Oceans

INTRODUCTION

In this chapter we shall discuss the largest portion of the world's water, that contained in the oceans. The principal and defining characteristic of the open oceans separating them from the marginal marine environments discussed in the previous chapter is their relatively uniform chemical composition. Seawater, as compared to all other natural waters, is amazingly constant in composition. Variations in total dissolved solids are small and, for well over 95% of the oceans, the salinity (total dissolved solids) ranges no more than ± 7% from its mean value of 35 parts per thousand (Sverdrup, Johnson, and Fleming 1942). In addition, the ratios of major ions to one another are even more constant. This uniformity of composition allows one to discuss many aspects of seawater chemistry in terms of average properties and still have the discussion pertain to most of the world's oceans. This is what is done here. We shall first present chemical data on the composition of seawater and then discuss how physical, biological, and geological factors both disturb and maintain this composition.

To acquaint the reader with seafloor morphology and, thereby, to provide a frame of reference during our discussion of seawater chemistry, a schematic diagram is presented in Figure 8.1, showing the principal physiographic divisions of the ocean floor.

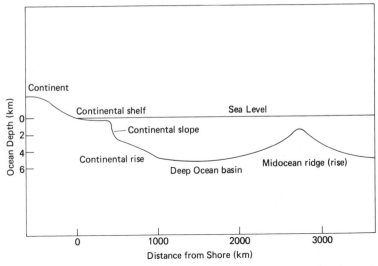

Figure 8.1 Generalized schematic cross section of the oceans showing major physiographic features. Note large differences in vertical and horizontal scales.

CHEMICAL COMPOSITION OF SEAWATER

The major dissolved constituents of seawater are almost the same ones encountered in continental waters as discussed in earlier chapters: Na^+, Ca^{++}, Mg^{++}, K^+, Cl^-, SO_4^{--}, and HCO_3^-. Concentrations of major components in average seawater of 35‰ salinity, where ‰ represents parts per thousand or grams per kilogram, are given in Table 8.1. Although the concentrations shown can range by

TABLE 8.1 Major Dissolved Components of Seawater for a Salinity of 35‰

Ion	Concentration g/kg	mM[a]	Percent Free Ion
Cl^-	19.354	558	100
Na^+	10.77	479	98
Mg^{++}	1.290	54.3	89
SO_4^{--}	2.712	28.9	39
Ca^{++}	0.412	10.5	88
K^+	0.399	10.4	98
HCO_3^{-}[b]	0.12	2.0	80

Sources: Wilson 1975; Skirrow 1975; and Millero and Schreiber 1982.

[a] mM = millimoles per liter at 25° C.

[b] For pH = 8.1, P = 1 atm., T = 25° C.

roughly $\pm 10\%$ due to changes in total salinity, the ratios of one ion to another vary by less than 1% (Wilson 1975). (The concentration of bicarbonate varies somewhat more, 5% to 10%, due to carbonate reactions; see later in this chapter). This constancy of seawater composition was first well documented by W. Dittmar (1884) as a part of the pioneering and world-famous Challenger Expedition and has since been thoroughly corroborated. Because of such constancy, the total salinity of seawater has historically been determined by measuring the concentration of the most abundant species, Cl^-, and multiplying this value by appropriate constants representing the ratios of the other major ions to chloride. (Salinity today, however, is normally determined by measuring electrical conductivity; see, e.g., Wilson 1975 for details.)

The major components listed in Table 8.1 are commonly assumed to be present in the ionic forms shown. However, according to a prominent theory of seawater composition (e.g., Whitfield 1975; Millero and Schreiber 1982), these "ions" can be viewed as being made up of a number of different individual species whose concentrations add up to the total concentration given for each element. The individual species consist of the actual ion shown (*free ion*) plus *ion pairs* formed with ions of opposite charge. For example, for SO_4^{--} the concentration shown represents the sum of concentrations of the following actual species: SO_4^{--}, $NaSO_4^-$, $CaSO_4^0$, and $MgSO_4^0$ (the superscript ^{0} represents an uncharged ion pair). Some idea of the degree of ion pairing for each element, taken from the results of Millero and Schreiber (1982), is also shown in Table 8.1. In addition to the high degree of ion-pairing exhibited by sulfate, Millero and Schreiber show that 15% of CO_3^{--}, 29% of HPO_4^{--}, and only 0.15% of PO_4^{---} are present as free, or unpaired, ions. This high degree of ion pairing helps to partly explain the different chemical behavior of highly paired ions as compared to those that are present mainly as free ions.

Many dissolved components of seawater of lesser concentration, in contrast to the major elements, do show variation, from place to place, in concentration ratios to chloride; in other words, they are *nonconservative* as compared to the *conservative* major elements. Some minor constituents (>1 μM) are shown along with concentration ranges in Table 8.2. Note that the extent of variability changes from element to element. Many other elements are dissolved in seawater in trace quantities (<1 μM) but they are not listed here because discussion of them is outside the scope of this book.

The reason why some elements are conservative and others are nonconservative is twofold. First of all, the major elements are all conservative because there is so much of them in seawater. Fluctuations resulting from river inputs or reactions in seawater are undetectably small due to the relatively large masses of these elements and the fact that potential variations are dissipated by mixing of the oceans accompanying the general circulation of seawater. In other words, the major elements have very long replacement times in seawater relative to the time scale of oceanic homogenization via mixing of 1000–2000 years (Broecker and Peng 1982). (Replacement time is the time, at the present rate of addition by rivers,

TABLE 8.2 Minor Dissolved Components of Seawater
(excluding trace components < 1 μM) Showing Ranges
in Concentration

Component	Concentration Range	
	(μg/kg or ppb)	μM[a]
Br^-	66,000–68,000[b]	840–880
H_3BO_3	24,000–27,000[b]	400–440
Sr^{++}	7700– 8100[b]	88– 92
F^-	1000– 1600[b]	50– 85
CO_3^{--}	3000–18,000	50–300
O_2	320– 9600	10–300
N_2	9500–19,000	300–600
CO_2	440– 3520	10– 80
Ar	360– 680	9– 17
H_4SiO_4-Si	<30– 5000	<0.5–180
NO_3^-	60– 2400	1– 40
NO_2^-	<4– 170	<0.1– 4
NH_4^+	<2– 40	<0.1– 2
Orthophosphate[c]	<10– 280	<0.1– 3
Organic carbon	300– 2000	—
Organic nitrogen	15– 200	—
Li^+	180– 185[b]	26– 27
Rb^+	115– 123[b]	1.3– 1.4

Sources: Wilson 1975; Kester 1975; Spencer 1975; Brewer 1975;
Skirrow 1975; and Williams 1975.

[a] μM = micromoles per liter.

[b] For a salinity of 35‰.

[c] Includes PO_4^{---}, HPO_4^{--}, and $H_2PO_4^-$; concentration range
expressed as μg P/kg.

necessary to build up to average concentration levels; see Chapter 6.) This allows
the major elements to be throughly mixed throughout the oceans. By contrast,
some trace elements (such as Fe) have such short replacement (residence) times,
relative to oceanic mixing, that they can exhibit spatially varying concentrations.
A list of replacement times for oceanic constituents of interest in the present chapter
is shown in Table 8.3.

The other principal reason for nonconservative behavior is that biological
processes acting within the oceans tend to deplete certain nutrient elements in
surface waters by biological uptake and to return these elements to solution at
depth due to death, settling out, decomposition, and dissolution. Some of the
chief elements involved are nitrogen, phosphorus, and silicon. Due to the rapidity
of these biological processes compared to the rate of vertical mixing of seawater,

TABLE 8.3 Replacement Time with Respect to River Addition, τ_r, for Some Major and Minor Dissolved Species in Seawater

Component	Concentration (μM)		$\tau_r{}^a$ (1000 yr)
	River Water	Seawater	
Cl^-	230	558,000	87,000
Na^+	315	479,000	55,000
Mg^{++}	150	54,300	13,000
SO_4^{--}	120	28,900	8700
Ca^{++}	367	10,500	1000
K^+	36	10,400	10,000
HCO_3^-	870	2000	83
H_4SiO_4	170	100	21
NO_3^-	10	20	72
Orthophosphate	0.7	1	50

Sources: Based on Tables 8.1 and 8.2 and data of Meybeck 1979, 1982 for world average river water.

[a] $\tau_r = ([SW]/[RW])\tau_w$ where τ_w = replacement (residence) time of H_2O = 36,000 yr; RW = river water; SW = seawater, and [ ] = concentration in μmoles per liter = μM.

strong vertical concentration gradients of the nutrient elements result. They are accompanied by corresponding gradients in dissolved O_2 and CO_3^{--} and together provide striking testimony to the efficacy of biological activity as a major process for the alteration of seawater composition. Much more on this important subject will be presented later in the present chapter.

Pressure and temperature also exert some effect on the composition of seawater. Because the oceans are deep, compared to other water bodies, they are subjected to much higher pressures. The mean depth of the oceans (excluding adjacent seas) is 4100 m (Sverdrup, Johnson, and Fleming 1942) which, at the bottom, corresponds to a pressure of about 400 atmospheres. High pressure exerts an influence on the composition of seawater, mainly by bringing about the dissolution of biogenic calcium carbonate falling to the bottom. Temperatures in the ocean, due to its thermohaline circulation (Chapter 2), in general, decrease with depth and this too exerts an influence on seawater composition by

1. Restricting vertical mixing due to thermally induced density stratification
2. Increasing the solubility of biogenic calcium carbonate
3. Decreasing the rate of bacterial decomposition of organic matter

The pH of seawater is controlled, on an oceanic time scale (thousands of years), by the dissolved bicarbonate-carbonate buffer system. (Variations in the ratio $[HCO_3^-]/[CO_3^{--}]$, however, are not large and, as a result, the pH of the open oceans ranges only between approximately 7.8 and 8.4 (see Skirrow 1975). The oxidation potential (Eh, pe) of seawater cannot be accurately measured (e.g., Stumm and Morgan 1981), but there is no doubt that it is controlled by dissolved O_2. The open ocean practically everywhere contains enough dissolved oxygen to enable O_2 control of its redox state.

MODELLING SEAWATER COMPOSITION

Sillén's Equilibrium Model

Several approaches to explaining the chemical composition of seawater have been used by previous workers. One approach, adopted by Sillén (1967) and others, assumes that the ocean represents a simple chemical equilibrium between seawater, the atmosphere, and solids deposited on the ocean floor. In other words, the oceans plus atmosphere plus sediments are thought of as a giant closed reaction vessel (Figure 8.2) where the aqueous solution portion has attained equilibrium with the included minerals via dissolution and precipitation, and with the overlying air space via gas exchange. The resulting solution composition is calculated by combining a large number of equilibrium expressions (e.g., the solubility product of $CaCO_3$) with an equation expressing the charge balance between cations and anions, and by specifying a given chloride content. Sillén was hindered by a lack of sufficiently accurate data for the various equilibrium constants, but he presupposed that once proper data were obtained, the resulting calculated major ion concentrations of seawater would approximate those actually found and would represent equilibrium with the atmosphere and the following minerals: quartz, calcite, kaolinite, illite (muscovite), smectite, chlorite, and either dolomite or phillipsite. (For compositions of these minerals see Tables 4.3 and 4.4; phillipsite is a hydrous Ca,Na,K aluminosilicate found occasionally in deep-sea sediments.)

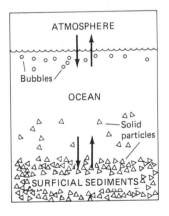

Figure 8.2 Sillén equilibrium model for the ocean. Large arrows represent exchange between gases, solids, and seawater. Exchange and mixing of the oceans are assumed to be sufficiently rapid that chemical equilibrium between all phases is attained.

Sillén's model is readily amenable to testing. If minerals, especially silicate minerals, readily react with seawater to maintain an equilibrium composition, then one would expect to find evidence of this reaction. For example, if a given detrital mineral carried into the oceans by rivers were not in equilibrium with seawater, it should dissolve and its components reprecipitate to form an equilibrium mineral. Thus, the equilibrium mineral should contain some evidence that it was formed from seawater. In addition, nonequilibrium minerals should not be present.

Most tests of these predictions have proven negative. The isotopic composition of clay minerals taken from the Atlantic sea floor do not indicate levels of either ^{87}Sr (Dasch 1969) or ^{18}O (Savin and Epstein 1970) expected for formation from seawater. Rather the minerals record the isotopic composition of the original rocks and weathering solutions from which they were formed on the continents. In addition, nonequilibrium silicates, such as K-feldspar, opaline silica, and gibbsite are often found in surficial marine sediments. Apparently, reaction times with seawater of most silicate minerals are so slow that an appreciable interconversion of minerals does not occur before depositional burial and removal from contact with the oceans. (An exception is the relatively rapid reaction of volcanic ash with seawater to form smectite; see section below on volcanic-seawater reaction.) Eventually upon burial to elevated temperatures, silicate-seawater reactions do take place with, for example, the diagenetic conversion of smectite to illite (Hower et al. 1976). However, these reactions occur at too great a depth to be considered as having a direct effect on seawater composition.

A further problem with the Sillén equilibrium model is that the biological activity of marine organisms is ignored. Because of the impact of solar energy resulting in photosynthesis, organisms can produce compounds that are not in equilibrium with seawater. A good example, for present purposes, is the secretion by planktonic organisms of opaline (amorphous) silica. The concentration of dissolved H_4SiO_4 in seawater closely mirrors opaline silica secretion and dissolution (upon death of the organisms) and bears no relation to the value expected for quartz saturation as predicted by the Sillén model. In addition DeMaster (1981) has shown that essentially all dissolved silica removal from seawater can be accounted for by opaline silica burial in sediments, with little left over to form equilibrium-type silicate minerals. Besides H_4SiO_4, other seawater components that are strongly affected by biological activity, and not explained by the Sillén model, include SO_4^{--}, NO_3^-, and phosphate.

Even though most tests of the Sillén model have been negative, this has not diminished the usefulness of the model as an idealized concept. Given sufficient time, unstable minerals should react with seawater to form stable minerals. Thus, the Sillén model points to the importance of silicate reactions on seawater composition, even though total chemical equilibrium at any one time may not be attained. In fact, as will be shown below, much silicate-seawater reaction may occur as a result of the contact of volcanic material, including volcanic ash, with ocean water. Also, there is little doubt that the precipitation and dissolution of $CaCO_3$ result in a reasonable approach to equilibrium between calcite and seawater.

In constructing more accurate models of seawater composition control, one should always keep in mind that for a number of reactions there may be a reasonable approach to equilibrium.

Oceanic Box Models

Since an equilibrium model does not work well to explain oceanic composition, recourse must be made to other models. The most widely used approach (e.g., Broecker 1971; Broecker and Peng 1982; Garrels, Mackenzie, and Hunt 1975) is that of box modelling, which has already been applied to lakes in Chapter 6. In box modelling the oceans are divided into several regions ("boxes") of uniform composition, and rates of change of concentration within each box are computed as the difference between input and output rates. Except for human perturbations, such as fossil fuel burning, the input and output rates are generally assumed to balance one another so that there is no change with time of concentrations within each box. This is known as steady-state box modelling.

In box modelling of the major, or conservative, elements the entire ocean is treated as a single, well-mixed, and therefore compositionally homogeneous box. This is illustrated in Figure 8.3. The difference between inputs, mainly from rivers, and outputs, mainly by sedimentation to the bottom and reaction with volcanics, results in changes in oceanic composition. If there is a steady state, then inputs and outputs balance one another and replacement times, as shown in Table 8.3, can be viewed as mean residence times. (For a detailed discussion of residence time see Chapter 6.)

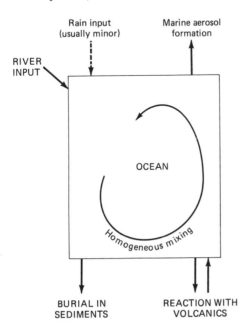

Figure 8.3 Simple box model appropriate for conservative elements in seawater. Note that, compared with the Sillén model of Figure 8.2, fluxes enter and leave the box, and the atmosphere and sediments are considered to be outside the box. Also note that, in contrast to lakes, there is no outlet, so that dissolved materials carried in by rivers can be removed only by sea-air transfer (marine aerosol formation), burial in sediments, or reaction with volcanics.

For nonconservative elements, especially those heavily involved in biological processes, the oceans are divided into a number of boxes, each representing a region of more or less homogeneous concentration. The simplest approach is that of a two-box ocean divided into surface seawater and deep seawater with the dividing line being the *permanent thermocline* or narrow zone of rapid vertical temperature change. This is similar to the model normally applied to lakes. Because of density stratification, the thermocline acts as a barrier to rapid exchange between deep and surface seawater and therefore behaves as a good box boundary. Also, vertical concentration-versus-depth profiles of nonconservative biological elements in general parallel the temperature profile so that distinctly different concentrations exist in deep and surface waters (e.g., see Figure 8.5). This makes such elements particularly amenable to two-box modelling.

A closer look at "deep water" shows that, based on temperature and salinity, it can be subdivided into a number of separate, internally homogeneous water masses (Antarctic Bottom Water, North Atlantic Deep Water, etc.; see Chapter 2). Thus, a more realistic box model of the oceans requires a number of boxes, one for each water mass. However, the use of many boxes leads to difficulty in that the rate of water flow and material exchange between water masses is not accurately known. Box modelling generally strikes a compromise between multibox sophistication and a desire to describe the processes under consideration in as simple a manner as possible.

Studies of ancient sedimentary rocks suggest that the chemical composition of the oceans has not changed drastically over the past 200 million years (e.g., Holland 1978). Therefore, as a first approximation, in constructing simple box models one may assume that a steady state has been established and that long-term inputs and outputs are in balance. This helps to explain the prevalent use of steady-state models both in chemical oceanography (e.g., Broecker and Peng, 1982) and in geochemical cycling (Garrels and Mackenzie 1971; Holland 1978). However, the effects of worldwide pollution by humans constitute a sudden change in the rate of addition of certain substances to the ocean (e.g., CO_2, SO_4^{--}, pesticides), and modelling of consequent concentration changes of these components requires a non-steady-state approach. A good example of non-steady-state modelling is that for the rate of excess CO_2 addition to the ocean which has resulted from the burning of fossil fuels (see Broecker et al. 1979 for a summary). Because of anthropogenic perturbations, more such modelling will be needed in the future (Stumm 1977).

Continuum Models

The ultimate and most accurate models for both conservative and nonconservative elements in the oceans are the so-called *continuum* models. In effect, the continuum models divide the ocean into an infinite number of boxes and treat concentration changes as continuous variations from place to place. Such models may be one-dimensional whereby only depth variations in concentration are treated (see e.g., Craig 1969), two-dimensional (e.g, lateral variations only), or three-dimensional (see e.g., Fiadeiro and Craig 1978). Concentration changes at a given point are described in terms of a differential equation which expresses the effects of turbulent diffusion, advective transport, and chemical (biochemical, radioactive decay, etc.) processes on concentration variation. Mixing and advection rates can be obtained by such modelling by applying a differential equation to temperature and conservative species (for example Cl^-) which undergo no chemical reaction and whose concentration variations are due solely to salinity variations. These rates can then be used in solving additional differential equations for the accompanying nonconservative elements. Although important, these models are mathematically complex and beyond the scope of the present book.

ENERGY SOURCES FOR CHEMICAL REACTIONS

Chemical reactions take place in the oceans, like everywhere else, as a result of chemical disequilibrium. Creation of this disequilibrium is brought about by the input of energy from two sources: the sun and the interior of the earth. Solar energy, by means of photosynthesis, enables the formation by marine plants of organic matter and hard parts (e.g., opaline silica) which are out of equilibrium with seawater. In addition, marine animals consume photosynthetically produced organic matter and also produce chemically unstable skeletal hard parts. As a consequence of both plant and animal activity, surface waters become depleted in certain elements, and deep waters, upon death and settling out of organic remains, become enriched in the same elements in an attempt to reattain chemical equilibrium via dissolution. Photosynthetic activity is concentrated in surface waters (top ~200 m) because of the lack of penetration of sunlight to greater depths. (Synthesis of organic matter by nonphotosynthetic marine organisms living at depth near volcanic centers on midocean ridge crests also occurs, but it is extremely rare and has only recently been discovered; see, e.g., Corliss et al. 1979)

The other major source of energy for chemical disequilibrium in the oceans is heat contained within the earth. This heat manifests itself, among other ways, as submarine volcanism. Basaltic minerals and volcanic glass, orginally formed from hot magmas poor in H_2O, are suddenly thrust into contact with seawater in which they are chemically unstable. As a result, silicate-seawater reactions occur. Reactions may occur at high temperatures or at low, seafloor temperatures. Temperature gradients near oceanic ridge crests result in the convective circulation of seawater through the ridges and the extensive heating of this water at depth. Basalt-seawater reactions are rapidly accelerated by increased temperature and, consequently, the seawater that eventually exists from the ridges is extensively altered in chemical composition.

Solar energy and that contained within the earth also have important *indirect* effects on oceanic composition. Solar energy brings about the evaporation of seawater, the transport of water vapor to the continents, and the formation of rainwater which then falls on the continents to form soil water, groundwater, lake water, and river water. Meanwhile, the interior energy of the earth, manifested by tectonic uplift and volcanism, causes rocks formed at depth to be uplifted into the zone of weathering on the continents. Here the rocks are unstable and react with soil waters and groundwaters via the mediation of (photosynthetically driven) green plants, to bring about the formation of new minerals. These minerals are (crudely) in equilibrium with continental waters but not necessarily in equilibrium with seawater, which has a distinctly different composition. Consequently, terrestrially formed weathering products are carried by rivers into the oceans where they undergo a variety of chemical reactions with seawater. In addition, any chemically unstable igneous and metamorphic minerals that managed to escape chemical weathering are also added to the oceans and may also undergo chemical reactions. In these ways, processes taking place on the continents in response to solar and earth-interior energy inputs, can also affect the composition of seawater.

MAJOR PROCESSES OF SEAWATER MODIFICATION

Processes affecting the chemistry of the oceans, besides addition of elements by rivers, can be classified into six categories. The first three, alluded to in the previous section, are the most important and will be discussed in detail here. They are (1) principal biological processes (secretion of hard parts and the production and decomposition of organic matter), (2) volcanic-seawater reaction, and (3) interaction with solid materials transported from the continents. Two additional processes, sea-to-air transfer of cyclic salts and the precipitation of evaporite minerals, have already been discussed, in previous chapters on rainwater (Chapter 3) and on marginal marine environments (Chapter 7), respectively, and will not be discussed further here. The sixth category constitutes special processes generally affecting only one or two elements, and includes pore-water burial of chloride and sodium, and a number of processes unique to the nitrogen and phosphorus cycles (denitrification, nitrification, N_2-fixation, addition of NO_3^- in rain, adsorption of phosphate on $CaCO_3$ and volcanogenic ferric oxides, phosphorite formation, and the sedimentation of fish bones). These special processes are also not discussed in this section but instead are treated under the chemical budgets for each appropriate element. The goal of this section is to deepen the reader's understanding of the three major processes that are mentioned time and again when discussing chemical budgets of the elements.

Biological Processes

Chemical reactions in the ocean which are intimately intertwined with life processes constitute major controls on the concentrations of the following seawater constituents: Ca^{++}, HCO_3^-, SO_4^{--}, H_4SiO_4, CO_2, O_2, NO_3^-, and orthophosphate. Biological activity also strongly affects many trace elements, for example copper and nickel (Boyle, Sclater, and Edmond 1977; Sclater, Boyle, and Edmond 1976). Three principal processes can be recognized: (1) the synthesis of soft tissues or organic matter, (2) the bacterial decomposition of organic matter upon death, and (3) the secretion of skeletal hard parts.

All organic matter is ultimately formed in surface waters by the process of photosynthesis.[a] This process requires the presence of light and therefore takes place only at water depths where sunlight can penetrate, that is, in the top few hundred meters of the oceans. The organisms involved are known as phytoplankton, which are minute floating marine plants that all contain chlorophyll which is necessary for photosynthesis.

Redfield (1958) has shown that the average elemental composition of phytoplankton from the open ocean in terms of the major components carbon, nitrogen, and phosphorus can be represented by the molar ratio $C:N:P = 106:16:1$. Therefore, one can represent marine photosynthesis by the reaction

$$106CO_2 + 16NO_3^- + HPO_4^{--} + 122H_2O$$

$$+ 18H^+ \xrightarrow{\text{light}} C_{106}H_{263}O_{110}N_{16}P + 138O_2$$

This reaction shows that photosynthesis involves not only the removal of CO_2 from solution and the production of O_2, but also the uptake of nutrients such as nitrate and orthophosphate. (Other nutrient elements, such as trace metals, are also taken up but their exact role in marine photosynthesis is not well established.) Because of the omnipresence of CO_2, H_2O, and light in most surface waters, the limiting factors in how much photosynthesis can occur generally are nitrate and (ortho-)phosphate, both of which occur at low concentrations. (Strictly speaking, other forms of nitrogen and phosphorus are also used for photosynthesis, e.g., NH_4^+, NO_2^-, and dissolved organic phosphorus, but they are normally present in seawater at concentrations even lower than that of NO_3^- and orthophosphate.)

Because of highly varying nitrate and phosphate concentrations in surface waters, the rate of photosynthesis, known as planktonic productivity, can vary considerably. For example, productivity in central gyres of the ocean, such as the Sargasso Sea, may be as little as one-tenth as fast as that occurring in coastal upwelling regions such as in the Pacific off Peru. This is illustrated in Figure 8.4. High marine productivity is brought about in the open ocean by mixing processes that bring deep, nutrient-rich water to the surface (and, in nearshore water, by river flow to the sea; see Chapter 7). Since phosphate and nitrate are highly enriched in subsurface waters (see Figures 8.5 and 8.6), any process that brings these waters up into the zone of light penetration will aid in photosynthesis. Two major processes, discussed in Chapter 2, are coastal upwelling and high-latitude mixing attending the formation of deep water.

Phytoplankton are eaten by zooplankton, which are in turn eaten by fish, and so on up the food chain. All during this activity, respiration is taking place both

[a]Except the rare organic matter synthesized near volcanic vents (see previous section).

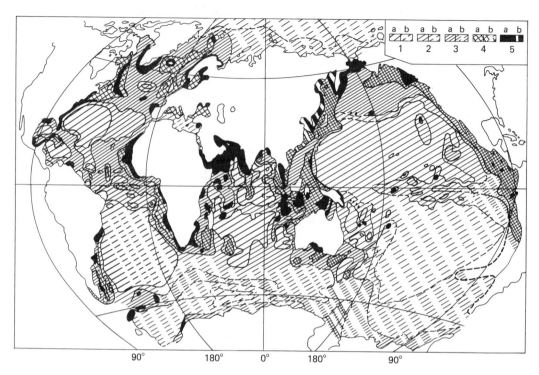

Figure 8.4 Rates of organic matter production (in mg C/m²/day) for the oceans: (1) less than 100; (2) 100–150; (3) 150–250; (4) 250–500; (5) more than 500. a = data from direct ^{14}C measurements; b = data from phytoplankton biomass, hydrogen, or oxygen saturation. (After O. J. Koblentz-Mishke, V. V. Volkovinsky, and J. G. Kabanova. "Plankton Primary Production of the World Ocean," In *Scientific Exploration of the South Pacific*, ed. W. S. Wooster, p. 185. Copyright © 1970 by the National Academy of Science, reprinted by permission of the publisher.)

by these higher organisms as well as by bacteria living on their dead remains. Respiration (used here in the strict sense of O_2 respiration) is the reverse of photosynthesis, and can be viewed essentially as the photosynthetic reaction given above written backwards. In other words, oxygen is consumed and CO_2, NO_3^-, and orthophosphate are liberated to solution and the rates of photosynthesis and respiration are so well adjusted that they almost balance one another in surface waters, but not quite. A small amount of dead organic matter sinks into deeper waters and this represents a net gain of photosynthesis over respiration. This organic matter is further oxidatively respired by bacteria in deep waters to produce CO_2, NO_3^-, and HPO_4^{--}, and to consume O_2, but in the absence of photosynthesis. As a result, there is a net input of nutrients and this helps to explain why higher concentrations of NO_3^-, HPO_4^{--}, and CO_2 and lower concentrations of O_2 are found at depth. Some typical depth distributions of these species are shown in

Figures 8.5 and 8.6. (Note in Figure 8.5 that higher phosphate concentrations are succeeded at even greater depths by somewhat lower concentrations, leading to the concept of the nutrient maximum and *oxygen minimum*. This concentration reversal results from the input of high-O_2, low-nutrient surface waters to great depths accompanying the general deep-water circulation; see Chapter 2.)

In the open ocean at all depths the ratio of dissolved NO_3^- to dissolved phosphate is almost the same as that found in average plankton (Redfield 1958). Therefore, during photosynthesis both nutrients are removed in the same proportions as their concentrations such that both become exhausted simultaneously; that is, both nutrients are limiting. This contrasts with the situation of most lakes where phosphate is limiting (Chapter 6) and many marginal marine areas where nitrate is limiting (Chapter 7). It is not obvious why the two nutrients should go to zero simultaneously in the open ocean, but it might represent the evolutionary adjustment of the elemental composition of seawater and marine plankton to one another over geological time (Redfield 1958).

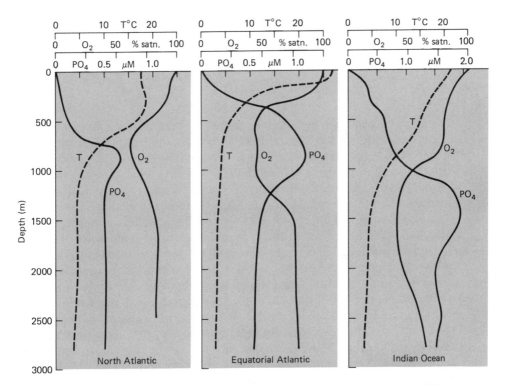

Figure 8.5 Depth profiles of dissolved oxygen (O_2) and phosphate (PO_4) and temperature (T) at stations in the north and equatorial Atlantic Ocean and in the Indian Ocean. Note the close anticorrelation between O_2 and PO_4^{-3}. (Adapted from J. P. Riley and R. Chester, *Introduction to Marine Chemistry*, p. 173. Copyright © 1971 by Academic Press, reprinted by permission of the publisher.)

Phosphate (μM)

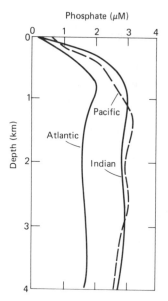

Nitrate (μM)

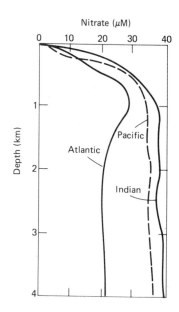

Figure 8.6 Average depth profiles of dissolved phosphate and nitrate in the major world oceans. (After H. V. Sverdrup, M. W. Johnson, and R. H. Fleming, *The Oceans*, pp. 241–242. Copyright © 1942, Renewed 1970 by Prentice-Hall, Inc., reprinted by permission of the publisher.)

Some organic matter survives bacterial decomposition during sinking and reaches the bottom where additional oxic decomposition occurs. Eventually a very small portion is buried by subsequent sedimentation—about 0.3% of that originally formed via photosynthesis (Holland 1978). The amount that is buried depends on the original amount falling (i.e., productivity) and the rate of burial via sedimentation of total solids. More rapid sedimentation allows burial of organic compounds that would be otherwise destroyed during prolonged exposure to the overlying water, and this helps to explain the generally lower contents of organic matter in deep-sea pelagic sediments, as compared to those deposited much more rapidly on the continental shelves. The relation between amounts of organic matter buried, sedimentation (burial) rate, and productivity has been studied extensively by Müller and Suess (1979) and the reader is referred to this publication for further details. It should be noted, however, that one of us (Berner 1982) has shown that most organic matter loss from the oceans to bottom sediments does not occur in upwelling and other classical oceanic areas, as studied by Müller and Suess, but rather in deltaic and other areas of very high sedimentation rate near river mouths where a simple relation between organic content, sedimentation rate, and productivity is not found.

The decomposition of organic matter by bacteria using dissolved O_2 continues after burial in sediments but only for a short time. In most sediments there is sufficient organic carbon burial that oxygen present in the interstitial waters is completely used up by this process within tens of centimeters of the sediment-water interface. In other words, most sediments are anoxic. (The widespread distribution of oxic, or red clay sediments in the deep sea gives a misleading impression of their quantitative importance. In fact, of the total fine-grained sediment carried each year to the oceans by rivers, less than 5% is deposited as red clay and similar types of deep-sea sediments; see Berner 1982 for a recent discussion.) Once sediments become anoxic, further organic matter decomposition is accomplished by bacteria, which use oxygen bound in a variety of other compounds. They are dissolved nitrate, manganese oxides, iron oxides, dissolved sulfate, and organic matter itself. The oxygen-containing substances become reduced while the organic matter is oxidized to CO_2. These compounds are attacked successively, until each is completely consumed in the order (see Table 8.4): nitrate reduction (denitrification), manganese reduction, iron reduction, sulfate reduction, and fermentation (methane formation). (For a discussion of anoxic bacterial processes and their succession consult Claypool and Kaplan 1974; Froelich et al. 1979; and Berner 1980.) Of these processes, the one of major interest to the present chapter is bacterial sulfate reduction.

In sediments that contain appreciable concentrations of organic matter, including most nearshore and continental margin sediments, the dominant process of organic matter decomposition is bacterial sulfate reduction (Goldhaber and Kaplan 1974; Westrich 1983). The bacteria that accomplish this are strict anaerobes (they are killed by even traces of dissolved O_2), and the overall reaction can be represented as

$$2CH_2O + SO_4^{--} \rightarrow H_2S + 2HCO_3^-$$

where CH_2O is a generalized representation of organic matter. Evidence for this

TABLE 8.4 Major Processes of Organic Matter Decomposition in Marine Sediments. Reactions succeed one another in the order written as each oxidant is completely consumed

Oxygenation (oxic)
$\quad CH_2O + O_2 \rightarrow CO_2 + H_2O$
Nitrate reduction (mainly anoxic)
$\quad 5CH_2O + 4NO_3^- \rightarrow 2N_2 + CO_2 + 4HCO_3^- + 3H_2O$
Manganese oxide reduction (mainly anoxic)
$\quad CH_2O + 2MnO_2 + 3CO_2 + H_2O \rightarrow 2Mn^{++} + 4HCO_3^-$
Ferric oxide (hydroxide) reduction (anoxic)
$\quad CH_2O + 4Fe(OH)_3 + 7CO_2 \rightarrow 4Fe^{++} + 8HCO_3^- + 3H_2O$
Sulfate reduction (anoxic)
$\quad 2CH_2O + SO_4^{--} \rightarrow H_2S + 2HCO_3^-$
Methane formation (anoxic)
$\quad 2CH_2O \rightarrow CH_4 + CO_2$

Note: Organic matter schematically represented as CH_2O.

reaction in sediments is provided by decreasing concentrations of interstitial dissolved sulfate with depth. Most hydrogen sulfide produced by sulfate reduction migrates out of the sediment and is subsequently oxidized, by O_2 in seawater, back to sulfate while the remainder reacts with the detrital iron minerals in the sediment to form a series of iron sulfides which are ultimately transformed to pyrite, FeS_2. (A minor amount of H_2S also reacts with organic matter.) The overall process of sedimentary pyrite formation is summarized in Figure 8.7. This process, once pyrite is permanently buried, constitutes a major mechanism for the removal of sulfate from seawater.

As shown in Figure 8.7, the amount of sulfur removed from the oceans as pyrite depends on the availability of organic matter, sulfate, and iron minerals. Organic matter is most important in that, without it, there can be no sulfate reduction. In addition, the reactivity or metabolizability of the organic matter dictates how fast sulfate is reduced, and therefore how fast pyrite can form. Abundant sulfate is present in seawater but it is available for pyrite formation only at relatively shallow depths in the sediment. This is because complete sulfate removal from the sediment pore water generally occurs within the top few meters so that at greater depths pyrite cannot form due to a lack of sulfate. Iron minerals are necessary for pyrite formation but, in terrigenous sediments (those derived from

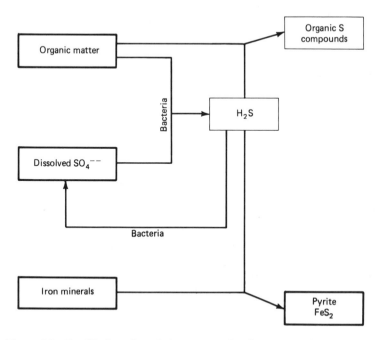

Figure 8.7 Simplified version of the process of sedimentary pyrite formation. (Intermediate steps involving elemental sulfur and iron monosulfides omitted; see also Berner 1984.)

weathering on the continents), the iron minerals are generally present to excess, as shown by the presence of abundant nonsulfidized iron. An example is shown in Figure 8.8. The form of iron most readily reactive toward H_2S is present as terrigenous fine-grained iron oxyhydroxides which enter the sediment as colloidal coatings on detrital clays, feldspars, and so forth. Lack of terrigenous iron in organic-rich $CaCO_3$ sediments helps to explain the relative absence of pyrite in these sediments which otherwise are high in H_2S. In summary, maximum pyrite formation is favored by the burial of high concentrations of fine-grained iron minerals and readily metabolized organic matter in sediments where rapid sulfate reduction occurs at shallow enough depths to enable replenishment of sulfate from the overlying water. In this way, an appreciable flux of sulfur into the sediment and its removal as pyrite is maintained. (For a further discussion of sedimentary pyrite formation consult Goldhaber and Kaplan 1974; and Berner 1984.)

So far we have dwelt upon biological activity only as it affects the synthesis and decomposition of organic matter. Another major biological process, important from both a chemical and a geological viewpoint, is the secretion of skeletal hard parts. Although a wide variety of minerals and mineraloids are known to be secreted by organisms (see, e.g., Lowenstam 1981), only quantitatively important ones will be discussed here. They are calcite ($CaCO_3$), aragonite ($CaCO_3$), magnesian calcite (defined here as calcite with more than 10 mole % $MgCO_3$ in solid solution), and opaline silica (SiO_2). A summary of the types of organisms that secrete each substance is listed in Table 8.5, and photographs of some common organisms are shown in Figure 8.9.

Besides mineralogy, an important distinction between organisms that secrete hard parts is that between benthos (bottom dwellers) and plankton (small, microscopic floating organisms). *All* important carbonate- and silica-secreting organisms

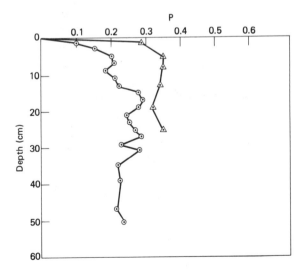

Figure 8.8 Plot of the degree of pyritization of iron, P, versus depth for two anoxic sediments of Long Island Sound, U.S.A. P is defined as iron contained in pyrite divided by total iron. Note that much of the iron is not converted to pyrite ($P \ll 1$) even though these sediments contain excess H_2S. (After R. A. Berner. "Sedimentary Pyrite Formation," *Amer. J. Science*, 268, p. 13. Copyright © 1970 by the Amer. J. Science, reprinted by permission of the publisher.)

TABLE 8.5 Quantitatively Important Plants and Animals that Secrete Calcite, Aragonite, Mg-calcite, and Opaline Silica

Mineral	Plants	Animals
Calcite	Coccolithophorids[a]	Foraminifera[a] Molluscs Bryozoans
Aragonite	Green algae	Molluscs Corals Pteropods[a] Bryozoans
Mg-calcite	Coralline (red) algae	Benthic foraminifera Echinoderms Serpulids (tubes)
Opaline silica	Diatoms[a]	Radiolaria[a] Sponges

Note: For further information on skeletal mineralogy consult Lowenstam 1981.

[a] Planktonic organisms.

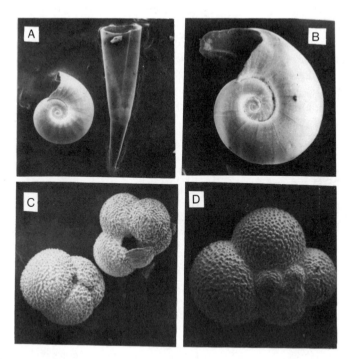

Figure 8.9 Photomicrographs of some common planktonic $CaCO_3$ secreting organisms: (A) pteropod shells (aragonite) × 20; (B) Pteropod shell (aragonite) × 10; (C) Foram tests (calcite) × 70; (D) Foram test (calcite) × 100.

(i.e., both benthos and plankton) live in surface waters (<200 m) where photosynthesis can occur and where abundant photosynthetically produced food is present. As a result, most benthic skeletal debris accumulates in sediments overlain by shallow waters, such as banks, atolls, and continental shelves, whereas plankton debris, because plankton can live over any depth of water, can also fall to the deep-sea floor. In this way deep-sea sediments differ from shallow-water sediments in containing far higher proportions of planktonic skeletal debris and this results in distinct mineralogical differences between the two sediment types. Deep-sea carbonate sediments contain mainly coccolith and planktonic foraminiferal calcite (pteropod aragonite is mainly dissolved away before burial) while shallow-water sediments are dominated by coral-algal-mollusc aragonite and magnesian calcite. (Strictly speaking, magnesian calcite with more than 10 mole % Mg and calcite are the same mineral but because of their different composition and different dissolution behavior, they are distinguished from one another.)

The fate of calcium carbonate is also different in the two environments. In the deep sea, much of the coccolith and foram calcite and almost all of the pteropod aragonite is dissolved prior to burial, whereas in shallow water (bank tops, etc.), little or no dissolution of $CaCO_3$ occurs. This is because the oceans are supersaturated with respect to calcite and aragonite in shallow water and undersaturated at depth. (For further discussion of $CaCO_3$ chemistry in the oceans, consult the section below on the oceanic budget for calcium.)

Opaline silica secreted by diatoms and radiolaria also undergoes dissolution upon death, but at all water depths. Opaline silica is distinctly more soluble than quartz and is undersaturated everywhere in the oceans. It is only because of the input of photosynthetic energy that plankton are able to remove opaline silica from surrounding seawater even though the water is undersaturated with respect to amorphous silica. As a result of the activity of diatoms and radiolaria, silica removal occurs in shallow water and, upon death and sinking, much of this silica is returned to solution at depth by dissolution, an overall process resembling the behavior of both organic matter and calcium carbonate.

The removal of biogenic elements in shallow water and their return in deep water can be described quantitatively in terms of box modelling. This has been done by Broecker (1971) (see also Broecker and Peng 1982) using a two-box model, illustrated here by Figure 8.10. Broecker divides the Atlantic and Pacific oceans each into two compartments, one representing shallow water, where photosynthesis and skeletal secretion occur, and the other deep water, which is dominated by organic matter decomposition and mineral dissolution. The deep reservoir is assumed to be 20 times the volume of the shallow one (see Figure 8.10). Water is exchanged between deep and shallow reservoirs by upwelling and downwelling accompanying normal oceanic circulation, and the average residence time of water at depth, which is also the time for a complete overturn of the oceans, is 1600 years. Combining this value with the volume of the deep reservoir and the rate of water flow by rivers to the sea, Broecker obtains a ratio of upwelling to river-water inflows to the shallow reservoir of 20:1. Finally, organic matter and hard

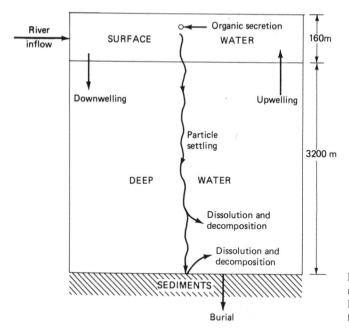

Figure 8.10 Two-compartment box model for the oceans as used by Broecker (1971). Arrows represent fluxes to and from each "box."

parts fall, by sinking of particles, from the shallow to the deep reservoir to complete the cycle.

By assuming a steady state for both shallow and deep reservoirs, Broecker was able to calculate the values for two useful parameters, denoted as f and g, from concentrations of each element in deep seawater, shallow seawater, and average river water. The parameters are defined as:

$$f = \frac{\text{particle flux into sediments}}{\text{particle flux from shallow to deep water}}$$

$$g = \frac{\text{particle flux from shallow to deep water}}{(\text{river input flux}) + (\text{upwelling input flux})}$$

The parameter f represents the fraction of a biogenic element falling into deep water that survives decomposition and dissolution to become buried in bottom sediments. In other words, it is a removal indicator. The parameter g represents the fraction of an element delivered to surface water by rivers and upwelling that is removed by biological secretion plus particle fallout. In other words, it is a measure of biogenic character. A g value of 1 means that all of the element carried into shallow water is removed by biological processes with none left over to be removed by downwelling.

Actual values of f and g were calculated by Broecker via the equations

$$f = \frac{1}{20 \left(\dfrac{[D] - [S]}{[R]} \right) + 1}$$

$$g = 1 - \left(\frac{20\,[S]/[R]}{20\,[D]/[R] + 1} \right)$$

where $[S]$, $[D]$, and $[R]$ represent concentration in shallow water, deep water, and average river water, respectively.

Values of f and g, calculated via the above equations for phosphorus, nitrogen, silicon, calcium, and inorganic carbon (CO_2, HCO_3^- and CO_3^{--}) at concentrations found in the Pacific Ocean are shown in Table 8.6. (Results for the Atlantic Ocean are similar.) As expected, high values of g for phosphorus and nitrogen are found, illustrating their highly biogenic character. High values are also found for silica. Lesser values are found for inorganic carbon, demonstrating its moderately biogenic character relative to downwelling, whereas a low value of $g = 0.01$ is found for Ca, demonstrating low, but still measurable, biogenicity. (Other nonbiogenic major elements, such as sodium, have g values unmeasurably different from zero.)

Values of f are all rather low, indicating that for these elements most of the biogenic material falling into deep water is returned to solution. Calcium and inorganic carbon do show a somewhat greater survival potential, however, and this may be the result of $CaCO_3$ being closer to equilibrium with deep seawater than the substances (organic matter, opaline silica) containing the other elements. At any rate the Broecker model (which can be applied to many additional elements)

TABLE 8.6 Average Concentration in Shallow and Deep Ocean Reservoirs (Using Pacific Ocean Data) and Corresponding f and g Values for Some Biogenic Elements According to the Broecker (1971) Two-Box Steady-State Model

	Concentration (μM)				
Element	Shallow Water [S]	Deep Water [D]	River Water [R]	f	g
P	0.2	2.5	0.7	0.015	0.92
N	3	35	20	0.03	0.92
Si	2	180	170	0.05	0.99
C inorg.	2050	2480	870	0.09	0.19
Ca	10,000	10,090	367	0.17	0.01

Sources: Concentration values from Broecker 1971; Tables 8.1 and 8.2; and Meybeck 1979, 1982. See also Broecker and Peng 1982.

provides a useful way of characterizing the behavior of biogenic elements in the oceans.

Volcanic-Seawater Reaction

Volcanic activity in the oceans is extensive and its products consist of submarine basaltic lava flows as well as widespread ash falls. The most abundant products of this volcanism include volcanic glass, pyroxenes, calcium plagioclase, and olivine, all of which are chemically unstable in seawater. As a result of their instability, and widespread distribution, these substances react with seawater altering its composition and producing new minerals in a variety of high-temperature and low-temperature environments. Along with biological processes, volcanic-seawater reactions constitute the two best documented mechanisms by which the composition of the modern oceans is created and maintained.

Volcanic-seawater reactions occur primarily as a result of seafloor spreading (for a further discussion of seafloor spreading and associated plate tectonics consult, e.g., Condie 1976 or Kennett 1982). New sea floor, consisting mainly of basalt is created by igneous activity at midocean rises and ridges and then is conveyed away from the rises by lateral spreading. Because of heating from below, a convective circulation of seawater occurs along the axes and flanks of the rises. Cool water descends, becomes heated, rises, and then returns to the ocean. If sediment overlying the basalt is not too thick, and the depth in the basalt is not too great, the amount of circulating seawater can be appreciable. Because of chemical reactions between seawater and basalt, this circulation leads to major changes in the composition of both the basalt and the circulating seawater.

Quantification of the effects of the reaction between basalt and seawater is very difficult because there is no general agreement as to the amount and location of the hydrothermal circulation. Location is important because high-temperature reactions, which occur near the axes of the rises, result, for many elements, in different changes in seawater composition from those reactions taking place at the lower temperatures encountered on the rise flanks (e.g., Thompson 1983). From geophysical modelling, Sleep et al. (1983) have estimated that nine-tenths of the total heat flow due to hydrothermal convection on the rises occur along the flanks with only one-tenth along the axes; this is in strong disagreement with the calculation of Edmond et al. (1979), who attribute essentially all heat flow, and therefore all chemical reaction, to the higher-temperature axial region.

The high-temperature (200°–400° C) reactions between basalt and seawater at rise axes have been well documented, from the viewpoint of both basalt composition and seawater composition. Evidence for seawater compositional alteration is derived from studies of heated water obtained from geothermal drill holes on Iceland (Holland 1978), from natural submarine vents in the eastern Pacific Ocean (Edmond et al. 1979; Von Damm et al. 1985), and from laboratory studies of high-temperature basalt-seawater reaction (for a summary, consult Mottl 1983). The results of Edmond et al. (1979) for water emerging from submarine vents atop the Galapagos spreading center are instructive in this respect. Although they sampled waters that were mixtures of hydrothermal solutions with ambient seawater, Edmond et al. were able to calculate, via an extrapolation scheme, the temperature (350° C) and composition of the end-member hot water derived from depth. Their results, in terms of compositional differences from ordinary seawater, agree reasonably well with those for the Icelandic waters and are shown in Table 8.7. Also, these results are in good agreement with those found at other hot springs sites in the eastern Pacific (Von Damm et al. 1985; Von Damm, Grant, and Edmond 1983) and in the laboratory hydrothermal experiments. All results indicate that hot basalt-seawater reaction at depth involves the complete removal of Mg^{++} and SO_4^{--} from seawater and the addition of Ca^{++}, H_4SiO_4, and K^+. Based on experimental studies (Bischoff and Dickson 1975; Mottl, Holland, and Carr 1979) initially sulfate removal occurs almost entirely via $CaSO_4$ precipitation (later, at higher temperatures, some of the sulfate may be reduced to H_2S), and the source of calcium for this is from the basalt. Therefore, the total amount of calcium released to seawater is equivalent to the increase in seawater Ca^{++} concentration plus the drop in seawater SO_4^{--} concentration, in other words, ΔCa^{++} minus ΔSO_4^{--} (see Table 8.7). Note that this total calcium release is equivalent, on a molar basis, to Mg^{++} uptake.

Seawater also reacts with basalt at lower temperatures as the basalt spreads away from the rise axes. In this case there is much less data on water chemistry due both to a paucity of clearly documented springs in rise flanks and to the slowness of laboratory reactions at low temperatures. Evidence has been obtained mostly

TABLE 8.7 Concentration Changes of Some Major Seawater Constituents Upon Reacting with Basalt at High Temperatures

Constituent	Concentration (mM)		Δ^b (mM)
	Seawater	Galapagos[a]	
Mg^{++}	54	0	-54
Ca^{++}	10	35	25
K^+	10	19	9
SO_4^{--}	29	0	-29
H_4SiO_4	0.1	~20	~20
ΔCa^{++} minus ΔSO_4^{--}	—	—	54[c]

Note: mM = millimoles per liter.

[a] Data are for the Galapagos spreading center at 350° C and are taken from the extrapolation of Edmond et al. (1979).

[b] Δ = concentration difference between 350° C Galapagos water and seawater.

[c] ΔCa^{++} minus ΔSO_4^{--} = total Ca^{++} released to solution.

from the study of altered basalts obtained by deep drilling.   A summary of studies of low-temperature basalt alteration is given by Thompson (1983).   As far as the elements of Table 8.7 are concerned, Thompson concludes that, overall, off-axis reactions result in uptake of K^+ and release of Ca^{++}, Mg^{++}, and H_4SiO_4 to seawater solution (data on sulfate aren't given).   Thus, for K^+ and Mg^{++} there appears to be a reversal of the sense of reaction found for high-temperature basalt-seawater reaction.   This is also true for a number of other elements, such as Li^+, Rb^+, and Ba^{++}, which are released to solution at high temperatures and taken up by basalt at low temperatures.

Other studies are in general agreement, qualitatively, with the results summarized by Thompson, but not for Mg^{++}.   First of all, the low-temperature (50° C) laborabory experiments of Crovisier et al. (1983) provide evidence for the initial *uptake* of Mg^{++} from seawater, in the form of microscopic layers of Mg-containing minerals on the surfaces of altered basaltic glass.   Second, several workers (e.g., Perry, Gieskes, and Lawrence 1976; Gieskes and Lawrence 1981) have shown that during long-term burial to hundreds of meters, sediments containing volcanic ash (fine fragments of basalt; see below) show consistent decreases in dissolved Mg^{++} (and K^+) with depth and increases in dissolved Ca^{++} which are essentially equivalent with and which mirror Mg^{++} decreases.   Accompanying this is the formation of smectite and the alteration of the ^{18}O content of the interstitial water, indicating reactions with silicate minerals.   An example of these changes is shown in Figure 8.11.   These diagenetic depth changes are best explained in terms of the reaction of volcanic material with interstitial seawater to form smectite, and the reactions may occur both within the sediment or below it.   (Some of the interstitial water compositional changes can be explained in terms of low-temperature alteration of basalts underlying the sediments.)   In either case, low-temperature volcanic-seawater reaction again involves the *uptake* of dissolved Mg^{++}, a corresponding release to solution of Ca^{++}, and a lesser uptake of K^+.   Third, the experiments of Seyfried and Bischoff (1979) indicate uptake of Mg^{++} by reaction with basalt at 70° C.   Finally, the data of Staudigel and Hart (1983) indicate that failure to quantify the precipitation of Mg^{++} to form smectite along veins in altered basalt may lead to erroneous conclusions regarding Mg^{++} release.   Based on these points, we feel that it is more likely that Mg^{++} is taken up from solution rather than released during low-temperature basalt-seawater reactions.

Another problem with low-temperature volcanic-seawater reaction is that, as alluded to above, it doesn't always involve basalt layers associated with sea-floor spreading.   Volcanic ash, consisting of fine fragments of basaltic (and other volcanic rock) constituents, is formed by explosive subaerial volcanism on oceanic islands and is subsequently carried by winds and water currents and sedimented to the sea floor.   The ash sitting on the bottom reacts with seawater, with the formation of smectitic clay minerals as a major consequence.   This process is especially common in the South Pacific Ocean (Peterson and Griffin 1964), but its quantitative importance, on a worldwide basis, has not been estimated.   At the rate by which sediments are buried in the deep (eupelagic) portions of the South Pacific, the

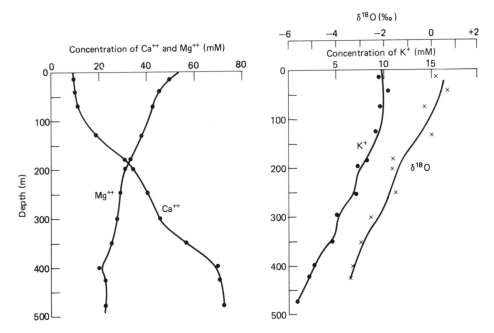

Figure 8.11   Depth distribution of concentrations of dissolved Ca^{++}, Mg^{++} and K^+ in interstitial water and the oxygen isotopic composition of the water (in terms of $\delta^{18}O$ relative to Standard Mean Ocean Water) for sediments from DSDP site 336. (Modified after J. M. Gieskes and J. R. Lawrence. "Alteration of Volcanic Matter in Deep Sea Sediments: Evidence from the Chemical Composition of Interstitial Waters from Deep Sea Drilling Cores," *Geochimica et Cosmochimica Acta*, 45, p. 1694. Copyright © 1981 by Pergamon Press, reprinted by permission of the publisher.)

process can be shown to be quantitatively unimportant, but for areas of both high explosive volcanic activity and high sedimentation rate, such as the East Indies, further quantitative estimates are needed.   At least, the chemical changes in seawater resulting from volcanic ash–seawater reaction are known from studies of interstitial water concentration in buried sediments (e.g., Gieskes and Lawrence 1981), as pointed out above.

Because of differences in uptake-versus-release behavior between high- and low-temperature (including volcanic ash) reaction, as well as lack of agreement as to the relative quantitative importance of axial (high-temperature) versus off-axial (lower-temperature) subsurface water circulation, it is very difficult to quantify the overall total effect of volcanic-seawater reaction on the chemical composition of seawater.   An attempt at quantification has been made by Edmond et al. (1979), but it is based solely on high-temperature reaction.   Their results are reproduced here in Table 8.8.   Thompson (1983) has attempted to include both high- and low-temperature reactions and one set of his results is also shown in Table 8.8.   (Thompson gives two sets of results depending upon whether all circulation, e.g., the approach

TABLE 8.8 Removal or Addition Fluxes of Some Major Seawater Constituents as a Result of Basalt-Seawater Reaction Near Midocean Ridges

Constituent	Flux (Tg/yr)		
	Edmond et al. (1979)	Thompson (1983)	Present Study
Mg^{++}	−187	−60	−119
Ca^{++}	140	73	191
K^+	51	−27	30
SO_4^{--} as S	−120	—	—[a]
H_4SiO_4 as Si	90	82	30

Note: Removal values shown as negative numbers. Tg = 10^{12}g.
[a]Less than 10% of Edmond et al. (1979) value.

of Edmond et al. [1979], or only 10% of the circulation, the approach of Sleep et al. [1983], is axial; here we report only the results for his 10% axial calculation.) Although there is considerable disagreement between the Edmond and Thompson results, one can still say that, overall, volcanic-seawater reaction results in the removal of Mg from the oceans and the addition of Ca and H_4SiO_4. The sign of the flux of K, however, must await further research.

In this book we have attempted our own crude quantitative estimates, based mainly on the assumption of a long-term steady-state composition for seawater; in other words, volcanic-seawater reaction is used to balance budgets. Details of the methods used to calculate fluxes are summarized later in the sections devoted to each of the elements; however, for the sake of comparison, results are also shown here in Table 8.8. Also, because of the possible importance of both biogenic and volcanic influences on the oceanic chemistry of dissolved sulfate and silica, additional discussion of these two species is presented here.

For sulfate there are serious problems with the removal value given by Edmond et al. (1979). Appreciable concentrations of $CaSO_4$ in altered basalt, as necessitated by the results of experimental studies, simply are not found when examining altered basalt samples obtained by drilling or dredging. Probably this $CaSO_4$ is redissolved later by other circulating water (Mottl, Holland, and Carr 1979). In addition, the sulfur budget of the oceans can be balanced (on a geologic time scale) by considering sulfur burial only in sediments; inclusion of such a large additional removal by reaction with basalt results in an unbalanced budget and unlikely changes in the sulfate concentration of seawater over time (see section in this chapter on the oceanic sulfur budget).

There are additional problems with the sulfur removal value of Edmond et al. (1979). Over geologic time, good agreement is found when one compares measured values of the carbon isotopic composition of the oceans (as recorded by limestones) with values calculated by a theoretical model (Garrels and Lerman

1984) that considers removal of sulfate from the oceans *only* as sedimentary pyrite and $CaSO_4$, and *not* by reaction with basalt.   Finally, mass balance calculations based on the isotopic composition of metal sulfides formed from H_2S exiting from submarine hydrothermal vents (e.g., Arnold and Sheppard 1981), indicates that sulfate reduction, via reaction with basalt, cannot be as important as envisaged by Edmond et al. (1979) (see also Von Damm et al. 1985).

Dissolved silica is also a problem.   Thompson (1983) suggests a large release of H_4SiO_4 to solution during low-temperature reaction, but the data of Staudigel and Hart (1983) suggest that most of this silica is reprecipitated to form veins of smectite within the basalt.   Furthermore, for Icelandic hydrothermal waters, flux values less than one-third that given for silica addition by Edmond et al. were obtained by Holland (1978).   For these reasons we use a worldwide silica flux, better in agreement with the Iceland results, of 30 Tg Si/yr.

Regardless of whose quantitative estimates are adopted, there is no doubt that the oceanic geochemical cycles of Mg, Ca, K, and many minor elements are appreciably affected by volcanic activity (see Edmond et al. 1979; Hart and Staudigel 1982; and Thompson 1983 for a discussion of Li^+, Rb^+, Ba^{++}, etc.).   In addition, the common occurrence of sodium-enriched basalts known as spilites, which are believed to form by the reaction of seawater at high temperatures with basalt, point to the possibility that Na^+ removal at high temperatures is also involved.   Compositional data from basalt-seawater experiments and from studies of altered basalts, and hydrothermal solutions, however, are inconclusive as to the quantitative effects on seawater sodium, and more research on this topic is needed.

Interaction with Detrital Solids

Detrital materials carried to the oceans by rivers consist to a large extent of silicate minerals, especially clay minerals, which are not in equilibrium with seawater. Therefore, upon their entering the oceans, chemical reactions take place.   Such reactions may involve the entire silicate mineral or only its surface.   In the former case we have formation of a new, generally more cation-rich mineral from the old detrital one and the process, because it resembles weathering, is referred to as *reverse weathering* (Mackenzie and Garrels 1966).   In the latter case, because of slowness of reaction, only chemical changes involving species on the mineral surface take place and this process is referred to as *adsorption-desorption* or, if ions are involved, *ion exchange*.   Together reverse weathering, adsorption-desorption, and ion exchange comprise all major reactions between river-borne silicate detritus and seawater.

The concept of reverse weathering was developed by Mackenzie and Garrels (1966) to provide a ready mechanism for the removal of several species, added to the oceans by rivers, for which at that time there were no well-documented removal processes.   This includes Na^+, K^+, Mg^{++}, HCO_3^-, and H_4SiO_4.   The reasoning goes as follows: aluminosilicate weathering products, which are depleted in cations and silica relative to the primary silicates from which they originally formed, upon entering seawater take up cations and silica and in the process convert HCO_3^- to CO_2. The overall reaction, written for example in terms of Na^+, can be expressed in a very generalized manner as

$$Na^+ + HCO_3^- + H_4SiO_4 + Al\text{-silicate} \rightarrow NaAl\text{-silicate} + CO_2 + H_2O$$

This reaction is very similar to those written for weathering reactions (see Chapter 4), except for going in the reverse direction, and this is how the term *reverse weathering* arose. However, one major difference from true reverse weathering is that the cation-enriched aluminosilicate formed is not the same as the primary silicates involved in weathering. In other words, feldspars, pyroxenes, and so forth, are not formed. (They are unstable in seawater.) Besides helping to remove cations and silica added to the oceans by rivers, reverse weathering is especially useful as a mechanism for the conversion of HCO_3^-, by the reaction with cation-depleted or "acid" aluminosilicates, to CO_2 so as to balance the opposite process which occurs during weathering.

Unfortunately, the concept of reverse weathering has not held up well under testing. Russell (1970), by closely examining changes in the chemical composition of highly degraded and cation-depleted clays carried by a Mexican river, found that upon addition to and deposition in seawater, there was no measurable *net uptake* of cations by the clay. The only change in clay composition observed was the *exchange* of river-borne cations for new seawater cations on the clay particle surfaces. This observation agrees with those cited earlier in this chapter, when discussing the Sillén equilibrium model, that very little evidence, whether mineralogical, compositional, or isotopic, exists for the formation of new clay minerals

TABLE 8.9 Change in Concentration in Interstitial Water for Various Ions versus Depth in a Sediment from the Brazil Basin, South Atlantic Ocean (Station CH 115-DD)

Sediment Depth (cm)	Concentration Change (Pore Water − Overlying Seawater) (mM)						
	pH	ΔNa^+	ΔMg^{++}	ΔCa^{++}	ΔK^+	ΔHCO_3^-	ΔSO_4^{--}
0	7.4	0.00	0.00	0.00	0.00	0.00	0.00
5	7.5	0.07	−0.04	0.17	−0.05	0.19	0.05
15	7.3	0.09	−0.35	0.45	−0.11	0.25	0.04
30	7.5	0.46	−0.42	0.50	−0.08	0.34	0.06
60	7.5	0.45	−0.58	0.76	−0.11	0.68	0.06
100	7.2	0.56	−0.78	0.97	−0.16	0.82	−0.01
195	7.4	0.95	−1.09	1.18	−0.26	1.12	−0.13

Source: Adapted from F. L. Sayles. "The Composition and Diagenesis of Interstitial Solutions. I. Fluxes Across the Seawater—Sediment Interface in the Atlantic Ocean,". *Geochimica et Cosmochimica Acta*, 43, p. 532. Copyright © 1979 by Pergamon Press, reprinted by permission of the publisher.

Note: Negative Δ values refer to uptake by the sediment (loss from pore water). mM = millimoles per liter.

from old detrital clay minerals in the marine environment. Furthermore, the results of DeMaster (1981) indicate that the riverine flux of silica can be removed from seawater entirely as opaline silica skeletal remains, without the necessity of invoking reverse weathering, as can the Mg^{++} flux be removed by volcanic-seawater reaction.

Some possible support for reverse weathering has been found more recently. Sayles (1979, 1981), using a very precise chemical technique, reports that small but measurable chemical concentration gradients exist in the concentration of dissolved Mg^{++}, K^+, Na^+, Ca^{++} and HCO_3^- across the sediment-water interface of many different pelagic (deep sea) sediments of the Atlantic Ocean. Some of his results are shown in Table 8.9 and Figure 8.12. Note that the gradients for Mg^{++}, K^+, and Ca^{++} are all in the same direction as those for volcanogenic sediments discussed earlier. In other words, there is evidence for Mg^{++} and K^+ uptake and Ca^{++} release from the sediment. Most of the Ca^{++} concentration gradient is due to $CaCO_3$ dissolution within the sediment, as witnessed by corresponding increases in HCO_3^-, but some excess Ca^{++} remains which cannot be explained by carbonate

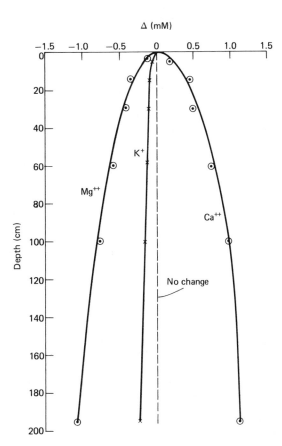

Figure 8.12 Concentration differences Δ between concentration of selected ions in pore waters and the concentration of each ion in the overlying seawater (in millimoles per liter = mM) as a function of depth in a sediment from the Brazil Basin, Atlantic Ocean. Negative Δ values refer to uptake by the sediment (loss from pore water) (Data from Sayles 1979, core CH-115DD.)

dissolution. (The sodium gradient is variable and problematical [Sayles 1981] and will be ignored here.) Since the Mg^{++}, K^+, and Ca^{++} gradients are all in the same direction as predicted for low-temperature volcanic ash–seawater reaction, it is tempting to ascribe these subtle changes at the seawater-sediment interface also to this process. However, most of the sediments are not obviously volcanogenic and it is possible that the observed concentration gradients in K^+ and Mg^{++} are, instead, brought about by reverse weathering. Also, some of the changes in the cations may be due simply to ion exchange (see below). Proper explanation of this data is important because worldwide uptake rates of K^+ and Mg^{++}, via diffusion, calculated by extrapolation from measured concentration gradients, approach the same level as rates of input of these elements from river water.

There is no doubt that the *surfaces* of clay minerals carried into the marine environment by rivers undergo reaction with seawater. From the viewpoint of the present chapter, the reaction of greatest interest is that of *cation exchange*. Clay minerals rapidly (hours to days) undergo exchanges of cations on their external surfaces and, in some clays such as smectites, in interlayer positions as well, when transferred from a water of one composition to another of a different composition. In this case of transfer from river water to seawater the degree of exchange is large because of the large difference in concentrations between the two waters.

An example of cation exchange upon transfer of a common clay mineral,

TABLE 8.10 Cation Exchange on Smectite upon Transfer from River Water to Seawater, and Oceanic Removal Rates

Ion	Surface Concentration[a] (mEq/100 g dry wt.)			Oceanic Removal (−) or Addition (+) Rate[b] (Tg/yr)
	Equilibrated with River Water	Equilibrated with Seawater	Change on Clay	
Ca^{++}	57.6	15.7	−41.9	+37
Mg^{++}	18.4	18.1	−0.3	+0.2
Na^+	2.3	44.3	+42.0	−42
K^+	0.6	2.7	+2.1	−4
H^+	2.0	0	−2.0	—
All ions (CEC)[c]	80.9	80.8	0	—

Source: Sayles and Mangelsdorf 1977.

[a] Values are based on smectite reacting for seven days with average seawater after being previously equilibrated for seven days with synthetic average river water (mEq = milliequivalents of +1 charge).

[b] Rates are based on a river flux of 14,000 Tg of sediment per year and an average cation exchange capacity of 25 mEq/100 g dry wt (Tg = teragram = 10^{12} g).

[c] CEC = cation exchange capacity.

smectite, from river water to seawater, taken from the work of Sayles and Mangelsdorf (1977), is shown in Table 8.10. (Later work of Sayles and Mangelsdorf using natural untreated river clay from the Amazon gave similar results.) Note that the principal change is the uptake of Na^+ and release of Ca^{++} from the clay, with a lesser exchange of K^+ for H^+. Holland (1978) has stated that the cationic exchange capacity of average river suspended sediment is approximately 25 mEq per 100 g. Using this value, the value of 14,000 Tg for the mass of suspended sediment carried to the sea by rivers each year (see Chapter 5), and assuming that the proportional changes for each element found by Sayles and Mangelsdorf for smectite can be extrapolated to all river-borne sediment, we have calculated present-day removal rates of Ca^{++}, Mg^{++}, Na^+, and K^+ from seawater as a result of cation exchange. These values are also shown in Table 8.10, and, for sodium, the rate represents an appreciable fraction (16%) of the total input rate of sodium to the ocean.

The values of Table 8.10 are based on the rate of suspended sediment input by present-day rivers. As we have shown in Chapter 5, this rate is unusually high due to increased erosion resulting from human agricultural activity. It may be high also because of an abundance of easily erodible debris produced by the last Pleistocene glaciation. In fact a number as low as 6000 Tg of solids per year, roughly half of the present rate, has been used by Garrels and Perry (1974) for the pre-Pleistocene long-term input rate. On this basis, for constructing long-term budgets (millions of years) we shall assume a rate of one-half that of the present. In this case the numbers listed in Table 8.10 must be divided by two to obtain the proper values for the effects of cation exchange.

In addition to simple cation exchange above, a slightly different exchange process occurs upon the transfer of certain clays from fresh water to the marine environments. This is potassium fixation. Whereas simple cation exchange involves the replacement of one hydrated cation on the clay surfaces or in interlayer positions by another from solution, in potassium fixation the addition of potassium to the clay involves dehydration of the added potassium ion, replacement of hydrated cations by potassium only in interlayer positions, and a consequent loss of exchangeability. This process involves mainly micas which have previously lost some K^+ during weathering (*degraded mica*) and which take up K^+ readily upon addition to seawater. It is an irreversible process as compared to simple cation exchange, which is reversible (The K^+ taken up from seawater is not readily lost upon reexposure of the mica to low-K fresh water.)

Quantitative evaluation of the importance of K-fixation as a control of seawater composition is provided by the work of Hoffman (1979). It was found that micalike clay (illite) from the Mississippi River when deposited in seawater on the Mississippi Delta took up K^+ in measurable quantities. Extrapolating these results to total river-borne suspended sediment (14,000 Tg/yr), we obtain a worldwide K^+ uptake rate of about 4 Tg/yr. This represents an appreciable but minor portion (10%) of the K^+ input by rivers.

CHEMICAL BUDGETS FOR INDIVIDUAL ELEMENTS

Summary of Processes

Through the remainder of this chapter we shall attempt to estimate quantitatively, using the simple box model presented earlier (Figure 8.3), the various inputs and outputs for each of the major components (and some minor components) of seawater. Two time scales will be used: one for the present (past tens of years) and one for geologic time (past tens of millions of years). Where sufficient independent data are available, we shall check to see if there is a steady-state balance between inputs and outputs on the million-year time scale. Otherwise we shall assume a steady state in order to quantify long-term outputs from inputs. In the case of biological elements, we shall also discuss how areally varying concentration differences are attained and maintained. As an aid to the reader, a summary of the major processes affecting each element is represented in Table 8.11.

During the upcoming discussion of major seawater species, the amounts added by rivers and amounts removed via aerosol (sea-spray) transfer across the air-seawater interface and not returned by marine rainfall (i.e., that transferred via the atmosphere from the oceans to the continents) must often be known. As an additional aid to the reader, a summary of river addition and net sea-salt transfer is presented here (Table 8.12). Rates of sea-air transfer are derived from the rates given in Chapter 5 for the cyclic salt contributions to rivers.

Chloride

There is little doubt that the chloride budget in the present ocean is badly out of balance and that the concentration of chloride is increasing with time, albeit very slowly because of the long replacement time of Cl^- of 87 million years. This

TABLE 8.11 Major Processes Affecting the Concentration of Specific Components of Seawater Numbered in Order of Approximate Decreasing Importance

Component	Input Processes	Output Processes
Chloride (Cl^-)	1. River-water addition (including pollution)	1. Evaporative NaCl deposition (in past)
		2. Net sea-air transfer
		3. Pore-water burial
Sodium (Na^+)	1. River-water addition (including pollution)	1. Evaporative NaCl deposition (in past)
		2. Net sea-air transfer
		3. Cation exchange
		4. Basalt-seawater reaction
		5. Pore-water burial

TABLE 8.11 (*Continued*)

Component	Input Processes	Output Processes
Sulfate (SO_4^{--})	1. River-water addition (including pollution) 2. Polluted rain and dry deposition	1. Evaporative $CaSO_4$ deposition (in past) 2. Biogenic pyrite formation 3. Net sea-air transfer
Magnesium (Mg^{++})	1. River-water addition	1. Volcanic-seawater reaction 2. Biogenic Mg-calcite deposition 3. Net sea-air transfer
Potassium (K^+)	1. River-water addition 2. Volcanic-seawater reaction (high temperature)	1. Low-temperature volcanic-seawater reaction or slow K^+ fixation or reverse weathering 2. Fixation on clays near river mouths 3. Net sea-air transfer
Calcium (Ca^{++})	1. River-water addition 2. Volcanic-seawater reaction 3. Cation exchange	1. Biogenic $CaCO_3$ deposition 2. Evaporitic $CaSO_4$ deposition (in past)
Bicarbonate (HCO_3^-)	1. River-water addition 2. Biogenic pyrite formation	1. $CaCO_3$ deposition
Silica (H_4SiO_4)	1. River-water addition 2. Basalt-seawater reaction	1. Biogenic silica deposition
Phosphorus (HPO_4^{-2}, PO_4^{-3}, $H_2PO_4^-$, organic P)	1. River-water addition (including pollution) 2. Rain and dry fallout	1. Burial of organic P 2. $CaCO_3$ deposition 3. Adsorption on volcanogenic ferric oxides 4. Phosphorite formation
Nitrogen (NO_3^-, NO_2^-, NH_4^+, organic N)	1. N_2 fixation 2. River-water addition (including pollution) 3. Rain and dry deposition	1. Denitrification 2. Burial of organic N

TABLE 8.12 Rates of Addition via Rivers of Major Elements to the Ocean (as Dissolved Species) and Rates of Net Loss from the Ocean by Transfer of Sea Salt to the Continents via the Atmosphere

Species	Rate of Addition from Rivers[a] (Tg/yr)	Rate of Net Sea Salt Loss to Atmosphere (Tg/yr)
Cl^-	308	40
Na^+	269	21
SO_4^{--}-S	143	4
Mg^{++}	137	3
K^+	52	1
Ca^{++}	550	0.5
HCO_3^-	1980	—
H_4SiO_4-Si	180	—

Sources: River-water data from Meybeck 1979; cyclic salt data from Chapter 5.

Note: Tg = 10^{12} g.

[a]Based on river water input of 37,400 km³/yr; includes pollution.

comes about in two ways. First, there are essentially no modern evaporite basins where appreciable NaCl (halite) precipitation is occurring and, thus, no corresponding removal of the large riverine influx of chloride derived from the dissolution of old salt beds. Second, about a third of the input of Cl^- is due to pollution (see Chapter 5) and the oceans have had far too little time to adjust to this extra addition. Combined, these two factors cause a large chloride imbalance.

An idea of the present-day Cl^- imbalance is provided by a quantitative comparison between known inputs and outputs. This is shown in Table 8.13. The only important output processes are net sea-salt aerosol transfer and burial of interstitial water in fine-grained sediments. The sea-air transfer rate is given in Table 8.12. The value for the burial rate is based on the average Cl^- content of shales of 0.12% (Holland 1978) and the sedimentation rate (riverine delivery rate) of fine-grained solids of 14,000 Tg/yr for the present day and half that value for the geologic past. (Shales should represent the ultimate burial of salts contained in interstitial water because essentially all H_2O is squeezed out by compaction while some salts are trapped by selective filtration.)

Evaporite basins where NaCl (halite) is deposited can form only where there is both an arid climate and restriction of water exchange with the open ocean so that high salinities can be attained (see Chapter 7). This combination of conditions is fortuitous and, thus, evaporite basin formation is basically a stochastic process being dependent on both tectonic and climatological factors. It occurs sporadically and the present happens to be a time when there are no major basins. Also, precipitation of NaCl (and $CaSO_4$), once evaporite basins are formed, is very rapid (e.g., King 1947) so that at times of major evaporite deposition the concentration of Cl⁻ in seawater would be expected to decrease suddenly. These factors, combined, would be expected to produce considerable fluctuation in the Cl⁻ content, and thus in the salinity of seawater. However, the salinity of the oceans over the past 600 million years cannot have deviated too greatly from the value found today, as evidenced by the presence of marine organisms in rocks deposited over this time which could not have survived unusually high or low salinities. Furthermore, even if all evaporite deposition ceased over a long time and the present existing mass of halite were all transferred via weathering to the ocean, the result, which is a maximum effect, would only be a doubling of the present Cl⁻ concentration.

Over the long term the average chloride content and salinity of seawater probably represents a steady-state balance. But superimposed on this balance one would expect small fluctuations brought about by the formation, or lack of formation, of evaporite basins. Overall a chloride-content-versus-time curve should look something like that depicted in Figure 8.13. Assuming steady state, a time-averaged balanced budget, using the present-day natural riverine input rate, is shown in Table 8.13. Sodium chloride deposition is used to balance the budget. Thus, *averaged over geologic time*, the major process for Cl⁻ removal from the oceans would be evaporitic NaCl deposition with sea-air sea-salt transfer a distant second.

Sodium

By contrast to chloride, sodium is a more complex element. It is involved in the same processes as chloride, as NaCl, but because it is a major constituent of silicate minerals, it is also involved in rock weathering, cation exchange, and possibly

TABLE 8.13   The Oceanic Chloride Budget (Rates in Tg Cl⁻/yr)

Present-Day Budget			
Inputs		Outputs	
Rivers (natural)	215	Net sea-air transfer	40
Rivers (pollution)	93	Pore-water burial	17
Total	308	Total	57

Long-Term (Balanced) Budget			
Inputs		Outputs	
Rivers	215	NaCl evaporative deposition	166
		Net sea-air transfer	40
		Pore-water burial	9
		Total	215

Note: Tg = 10¹² g. Replacement time for Cl⁻ is 87 million years.

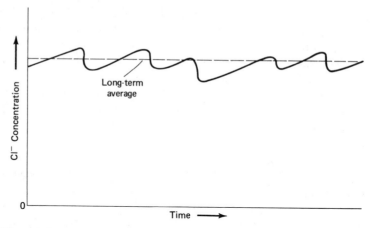

Figure 8.13 Schematic representation of change of chloride (Cl^-) concentration in seawater with geologic time. Sudden drops are due to the rapid precipitation of NaCl in evaporite basins.

volcanic-seawater reaction and reverse weathering. (The role of sodium in the latter two processes is not well established.) It provides a most interesting link between the geochemical cycles of silicates and evaporites.

The present-day oceanic budget of sodium is shown in Table 8.14 where only known, quantifiable removal mechanisms are listed. The cation exchange value is taken from Table 8.10 and the sea-air transfer value from Table 8.12. Pore-

TABLE 8.14 The Oceanic Sodium Budget (Rates in Tg Na^+/yr)

Present-Day Budget			
Inputs		**Outputs**	
Rivers (natural)	193	Cation exchange	42
Rivers (pollution)	76	Net sea-air transfer	21
		Pore-water burial	11
Total	269	Total	74

Long-Term Budget			
Inputs		**Outputs**	
Rivers	193	NaCl deposition	108
		Net sea-air transfer	21
		Cation exchange	21
		Pore-water burial	6
		Basalt-seawater reaction	37
		Total	193

Note: Tg = 10^{12} g. Replacement time for Na^+ is 55 million years.

water burial is that accompanying Cl^- burial as given in Table 8.13. Because of a lack of evaporative NaCl formation and a pollutional input of NaCl by rivers, it can be seen that the present ocean is badly out of balance with respect to Na^+, as it is with Cl^-. Also, over geological time, the fluctuations predicted for Cl^- concentration must also be mirrored by Na^+ concentration.

The long-term budget of sodium is also shown in Table 8.14. Here sodium removal, as evaporitic NaCl, is that corresponding to the rate of chloride removal used to balance the chloride budget (Table 8.13). Also, the rate of removal by cation exchange is one-half the value for the present ocean which would be expected if the present suspended sediment transport by rivers is about twice the long-term average. To balance the budget, a relatively minor amount (19%) of Na^+ is assumed to be removed by volcanic-seawater reaction. This volcanic removal must be considered a very tentative number, however, until independent data on the quantitative importance of this process is obtained.

Sulfate

Sulfate is another versatile element like sodium. It is removed from the oceans by several different processes. These include evaporative $CaSO_4$ precipitation as gypsum and anhydrite, biogenic pyrite (FeS_2) formation, net sea-air transfer, and possibly high-temperature basalt-seawater reaction. Like chloride and sodium, a large proportion of sulfate entering the oceans is derived from worldwide pollution. Unlike sodium and chloride, this pollutive addition occurs also by way of the atmosphere as a result of fossil fuel burning.

The sulfur budget for the present-day ocean is shown in Table 8.15. The value for biogenic pyrite formation is that derived by one of us (Berner 1982) from the well-documented constancy of the organic carbon to pyrite sulfur ratio in anoxic marine sediments and the integrated rate of organic carbon removal from the modern ocean. Most, by far, of the sulfur removal as pyrite occurs in continental margin sediments. (For a further discussion of biogenic pyrite formation, consult the section on biogenic processes.) The values for rain (and dry deposition) addition and net sea-air transfer are taken from the budget presented in Chapter 3. It is assumed (for the purposes of checking this assumption) that sulfate removal during basalt-seawater reaction is negligible. As can be seen, there is a very bad imbalance in the modern ocean budget, which is due to excessive inputs of pollutive sulfur and a lack of removal as $CaSO_4$ in evaporites. As will be shown, the imbalance is not likely to be caused by the omission of basalt-seawater reaction as a removal process.

The long-term budget for sulfate is also shown in Table 8.15. Pollution is excluded and the budget is balanced via removal of $CaSO_4$ in evaporite basins. The $CaSO_4$ removal rate, thus derived, is reasonable. The range of estimates by various workers for the ratio of $CaSO_4$ sulfur to pyrite sulfur in sedimentary rocks lies between 2:1 and 1:2 (R. M. Garrels, personal communication). If $CaSO_4$ weathers at about twice the rate for pyrite (see Chapter 5), then the two minerals would be expected to contribute fluxes to rivers in a ratio ranging from 1:1 to 4:1. To

TABLE 8.15   The Oceanic Sulfate Budget (Rates in Tg S/yr)

Present-Day Budget			
Inputs		Outputs	
Rivers (natural)	82	Biogenic pyrite formation	39
Rivers (pollution)	61	Net sea-air transfer (sea salt 4; H_2S 4)	8
Rain and dry deposition[a]	17	Total	47
Total	160		

Long-Term (Balanced) Budget			
Inputs		Outputs	
Rivers	82	Biogenic pyrite formation	39
Rain (unpolluted)	4	Evaporitic $CaSO_4$ deposition	39
Total	86	Net sea-air transfer	8
		Total	86

Note: Tg = 10^{12} g. To convert to sulfate, simply multiply by 3. The replacement time for SO_4^{--} (from river addition only) is 8.7 million years.

[a] Equal to input to marine air via transport from the continents 75% of which is due to pollution; see Table 3–12.

balance these fluxes, they would be expected to undergo removal from seawater in the same ratio. Thus, the ratio of 1 gypsum sulfur to 1 pyrite sulfur, obtained by balancing the long-term budget only with evaporitic $CaSO_4$ deposition, is not out of line with expectations.

Obtaining a reasonable long-term balance with evaporitic $CaSO_4$ deposition means that basalt-seawater reaction is not needed as a removal mechanism. If the total basalt-water removal value obtained by Edmond et al. (1979) were used (120 Tg S/yr), the long-term budget would be badly out of balance. For this reason and others given earlier in the section on volcanic-seawater reaction, it is believed that removal of SO_4^{--} by heated basalt is not a major mechanism for the removal of SO_4^{--} from seawater.

Magnesium

In contrast to sodium and sulfate, magnesium removal from the oceans is greatly affected by volcanic-seawater reaction. In fact, it most probably is the most important removal process. The only other significant processes of Mg removal are biogenic magnesian calcite secretion and net sea-air transfer and both are quantitatively much lower.

As pointed out earlier (see Table 8.8), there is a lack of agreement as to the rate by which Mg is removed from the oceans via volcanic-seawater reaction. All one can say is that the process is probably important. Here we assume it to be the dominant process and derive a removal rate by subtracting all other rates of removal from the total input so as to bring about steady state (see Table 8.16). The rate of removal in biogenic magnesian calcite was obtained from the approximate rate of deposition of $CaCO_3$ in shallow-water sediments, 1000 ± 300 Tg/yr (Milliman 1974; Hay and Southam 1977) and an average Mg content of 1.5 wt. % Mg in such sediments (Milliman 1974). The rate of net sea-air transfer was obtained from Table 8.12. The Mg-calcite and sea-air transfer values, when subtracted from the total input, give a steady-state value of 119 Tg/yr for the rate of removal of Mg by volcanic-seawater reaction (Table 8.16). Note that this value may include both high-temperature and low-temperature reactions with basalt if both involve uptake of seawater magnesium. At any rate, the Edmond et al. (1979) value of 187 Mt/yr for *high temperature* basalt-seawater reaction alone is distinctly higher than the value required for steady state and is believed by us to be too high.

In our budget we have not considered removal of Mg in authigenic clay minerals. Drever (1972) has made extreme maximum estimates of authigenic clay removal and even then he can account for only 27% of Mg added by rivers. We have also not considered removal of Mg in dolomite, $CaMg(CO_3)_3$, which is a common mineral in ancient sedimentary rocks. The reason for this is the scarcity of dolomite in sediments deposited over the past 100 million years, including those of the present ocean. Thus, our "long-term" budget does not extend to periods prior to 100 million years when dolomite formation was important.

Potassium

Of all the elements discussed in this chapter, potassium provides one of the biggest problems. This is because the chief removal mechanism has not been documented. The annual input of potassium to the oceans is 52 Tg from rivers and 30 Tg from high-temperature basalt-seawater reaction. (The latter value is derived by mul-

TABLE 8.16 The Oceanic Magnesium Budget (Rates in Tg Mg^{++}/yr)

(Balanced) Budget for Past 100 Million Years			
Inputs		Outputs	
Rivers	137	Volcanic-seawater reaction	119
		In biogenic $CaCO_3$	15
		Net sea-air transfer	3
		Total	137

Note: Tg $= 10^{12}$ g. Replacement time for Mg^{++} is 13 million years.

tiplying the Edmond et al. [1979] K^+ supply rate by the ratio of Mg uptake rate adopted here to the Edmond et al. value for Mg uptake. It is a probable maximum.) Of the 82 Tg delivered, only 1 Tg are removed by net sea-air transfer and ~4 Tg by fixation on deltaic clays (based on data of Hoffman [1979]). The remaining 77 Tg may be removed by low-temperature volcanic-seawater reaction. Several studies (e.g., Thompson 1983) have shown that, at low temperatures, basalts pick up K^+ during reaction with circulating seawater. In addition, slow steady depletion of K^+ in pore water with depth has been demonstrated for long columns of oceanic sediments (Gieskes and Lawrence 1981) in which Mg^{++} uptake via volcanic ash reaction with seawater is taking place. If similar, but more rapid uptake of K^+ occurs on volcanic ash near the sediment-water interface, then low-temperature volcanic-seawater reaction may represent the missing K^+ removal mechanism.

The data of Sayles (1979, 1981) (see also Table 8.9), however, suggest that slow potassium fixation and/or reverse weathering may constitute the missing mechanisms. On the basis of interstitial water concentration gradients for dissolved K^+, Sayles calculated that deep-sea sediments of the north, south, and central Atlantic type (nonvolcanogenic) on a worldwide basis could account for the removal of 60–75 Tg K^+/yr. Since there is no definitive evidence that K^+ removal in Sayles' sediment is due to volcanic-seawater reaction, it is possible that potassium fixation on illite and/or reverse weathering are the main processes of removal.

Until more definitive data is obtained we shall assume that the potassium budget is in balance, and that the major removal mechanism is either low-temperature volcanic-seawater reaction or deep-sea potassium fixation or reverse weathering or a combination of all three. Results are shown in Table 8.17.

TABLE 8.17   The Oceanic Potassium Budget (Rates in Tg K^+/yr)

Long-Term (Balanced) Budget			
Inputs		**Outputs**	
Rivers	52	Fixation on clay near river mouths	4
Volcanic-seawater reaction (high-temperature	30	Sea-air transfer	1
Total	82	Low-temperature volcanic-seawater reaction or slow fixation in deep sea or reverse weathering	77
		Total	82

Note: Tg = 10^{12} g.   Replacement time for K^+ is 10 million years.

Calcium

Calcium provides a refreshing contrast to the preceding elements in that only one mechanism of removal for the modern ocean is significant and it is quantitatively well known. That is the depositional burial in bottom sediments of biogenic $CaCO_3$ skeletal debris. Because of differences in mineralogy and types of organisms this deposition is divided into that occurring in shallow water (benthic aragonite and magnesium calcite) and that occurring in deep water (planktonic calcite). (For further discussion of skeletal mineralogy, consult the section on biogenic processes.) Shallow-water sediments today are receiving much higher quantities of $CaCO_3$ than normal because of the rapid postglacial (last 11,000 years) rise of sea level over the continental shelves (Hay and Southam 1977). Hay and Southam state that the present shallow-water (shelf and slope) deposition rate is approximately 1300 Tg of $CaCO_3$ per year as compared to the post-Miocene (last 25 million years) average value of 600 Tg/yr. The post-Miocene average deposition rate of planktonic $CaCO_3$ in the deep sea is 1100 Tg $CaCO_3$ per year. (Present-day deep-sea $CaCO_3$ sedimentation is assumed to be about the same.) These rates, in terms of Ca^{++} removal from the ocean, both for the present-day ocean and the average for the past 25 million years, are shown in Table 8.18.

Also listed in Table 8.18 are a variety of inputs. Besides rivers, there is calcium addition from cation exchange (taken from Table 8.10) and volcanic-sea-water reaction. The value for volcanic-seawater reaction is based on the rate of uptake of Mg^{++} by this process (Table 8.16) and the assumption, documented earlier, of an equivalent release of Ca^{++} (on a molar basis). The lower value for cation exchange for the past 25 million years is based on the assumption of a lower (by half) average sediment supply rate during this period (see section on interaction with detrital solids).

The remaining rate listed in Table 8.18 is that for Ca^{++} removal as $CaSO_4$ during periods of evaporite deposition. It is based on the rate of sulfate removal (Table 8.15) for this process and, as can be seen, is small when compared to removal as $CaCO_3$.

The data of Table 8.18 reveal two interesting observations. First, the removal of Ca^{++} from the present ocean considerably exceeds its input and this is caused by excessively rapid deposition of $CaCO_3$ on the continental shelves as a result of the rapid postglacial rise in sea level. By contrast over the long term (the past 25 million years), there is essential balance, well within errors of estimation, of Ca^{++} inputs and outputs. (Errors in estimating deposition rates can be as high as $\pm 50\%$.) This balance is what is to be expected and it demonstrates the essential validity of all the independent rate estimates used to construct the calcium budget. In this way it serves as a check on numbers used to balance other cycles (e.g., Mg).

The long-term balance also shows that removal of Ca^{++} as silicate minerals in the ocean, if it in fact occurs, is of negligible importance compared with removal as $CaCO_3$. Since Ca^{++} carried by rivers is derived partly from calcium silicate

TABLE 8.18 The Oceanic Calcium Budget (Rates in Tg Ca^{++}/yr)

Present-Day Budget			
Inputs		Outputs	
Rivers	550	CaCO$_3$ deposition:	
Volcanic-seawater		Shallow water	520
reaction	191		
Cation exchange	37	Deep sea	440
Total	778	Total	960

Budget for Past 25 Million Years			
Inputs		Outputs	
Rivers	550	CaCO$_3$ deposition:	
Volcanic-seawater		Shallow water	240
reaction	191	Deep sea	440
Cation exchange	19	Evaporitic CaSO$_4$	
		deposition	49
Total	760	Total	729

Note: Tg $= 10^{12}$ g. Replacement time (rivers only) for Ca^{++} is 1 million years.

(plagioclase) weathering (in Chapter 5 we suggest that 18% of river-borne Ca^{++} comes from silicates), removal from the oceans only as CaCO$_3$ infers an overall removal of CO$_2$ from the ocean-atmosphere system. However this CO$_2$ is returned by metamorphic and volcanic breakdown of CaCO$_3$ upon heating during deep burial (see Berner, Lasaga, and Garrels 1983).

Most of the CaCO$_3$ secreted by planktonic organisms (foraminifera, coccoliths, pteropods) does not become buried in deep-sea sediments. This is because seawater at great depth is undersaturated with respect to CaCO$_3$ and, as a result, material falling to such depths dissolves away. This gives rise to the concept of the *carbonate compensation depth*, which is abbreviated as CCD. (For a review discussion of the CCD and related phenomena, consult Berger 1976.) In the oceans as one goes downward the water passes from supersaturation to undersaturation with respect to CaCO$_3$, and, once undersaturation is attained, the rate of dissolution increases with increasing depth. The CCD is the depth where the rate of supply of calcareous skeletal debris from above is equalled by the downward increasing rate of dissolution. Below the CCD little or no CaCO$_3$ survives and, as a result, large areas of the deepest portions of the ocean (mainly in the Pacific) are relatively free of CaCO$_3$. A map of CaCO$_3$ distribution in the Atlantic Ocean is shown in Figure 8.14.

Although the matter is still somewhat controversial, most workers are in agreement that undersaturation and CaCO$_3$ dissolution occur above the carbonate compensation depth (see Figure 8.15) and that the CCD does not simply represent

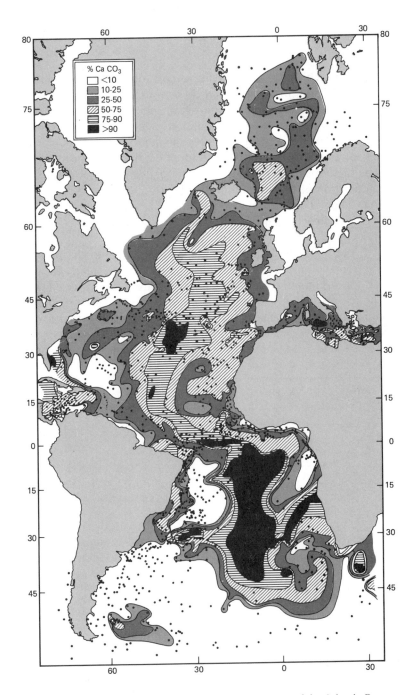

Figure 8.14   Distribution of CaCO₃ in deep sea sediments of the Atlantic Ocean. Note that the highest concentrations are located at the shallowest depths atop the Mid-Atlantic Ridge.   (After P. E. Biscaye, V. Kolla, and K. K. Turekian.   "Distribution of Calcium Carbonate in Surface Sediments of the Atlantic Ocean," *Journal of Geophysical Research*, 81, p. 2596.   Copyright © 1976 by the American Geophysical Union, reprinted by permission of the publisher.)

the boundary between overlying supersaturated and underlying undersaturated water. Dissolution occurs over a *range* of depths because it is not instantaneous and because different planktonic organisms dissolve at different rates. (The latter greatly complicates the interpretation of planktonic microfossils in ancient sediments.) The CCD is, then, simply the depth below which dissolution is complete. In fact, an additional depth shallower than the CCD has been created and referred to as the *lysocline* (Berger 1976), where a sudden downward change in the species composition of planktonic foraminifera occurs due to increased selective dissolution. This is shown along with the CCD, saturation depth, and R_0 depth (depth where microscopic evidence of foram dissolution first becomes detectable) in Figure 8.15.

Pteropods add another complicating factor to the problem of $CaCO_3$ preservation and burial. Although live pteropods are nearly as abundant as coccoliths and forams in surface water, their shells are virtually absent in bottom sediments. The reason for this is that pteropod shells are made of aragonite, which is distinctly more soluble than the calcite from which coccolith and foram (shells) are constructed. As a result, pteropod shells dissolve away at much shallower depths and are not found over most of the ocean floor, which is too deep for them to survive.

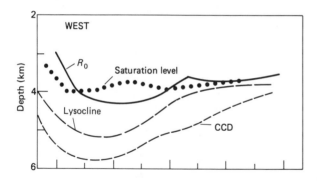

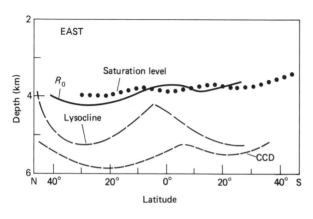

Figure 8.15 Plots of the depth of the carbonate compensation depth (CCD), lysocline, saturation level, and R_0 depth for surface sediments of the Atlantic Ocean as a function of latitude. The R_0 depth is that where evidence for dissolution is first encountered, and the saturation level is that depth in the water column below which calcite becomes undersaturated and can therefore dissolve. (After W. H. Berger. "Carbon Dioxide Excursions in the Deep Sea Record: Aspects of the Problem," In *The Fate of Fossil Fuel CO_2 in the Oceans,* ed. N. R. Andersen and A. Malahoff, Plenum Press, p. 512. Copyright © 1977 by Plenum Press, reprinted by permission of the publisher.)

The locus of $CaCO_3$ dissolution is a current topic of considerable controversy. Some workers believe that most dissolution occurs in the water column during the settling out of skeletal debris, while others emphasize dissolution within sediments from CO_2 produced in the sediment by organic matter decay. Most workers, by contrast, have focused on dissolution at the sediment-water interface, after settling out but before burial, as the major locus for dissolution. Probably all three locations—water column, sediment water interface, and within sediment—are important, but much more work is needed to clarify their relative importance.

$CaCO_3$ dissolves in the deep sea for two reasons. The first is that CO_2 is produced there by respiration without compensatory uptake via photosynthesis. As a result, the CO_2 accumulates, forms carbonic acid, and thereby brings about $CaCO_3$ dissolution. The overall reaction can be written

$$CO_2 + H_2O + CaCO_3 \rightarrow Ca^{++} + 2HCO_3^-$$

(This is simply the reverse of the reaction, occurring in surface water, representing $CaCO_3$ secretion by organisms.) The other reason for dissolution is that the solubility of $CaCO_3$ increases with increasing pressure so that at the deep-sea floor, with an average pressure of about 400 atmospheres, $CaCO_3$ is more than twice as soluble as at the ocean surface. In other words, the reaction

$$CaCO_3 \rightleftarrows Ca^{++} + CO_3^{--}$$

is driven to the right by increased pressure.

Regardless of cause or of location, the average degree of dissolution of all forms of $CaCO_3$ falling into the deep sea is large. From the model of Broecker (1971), discussed earlier in the section on biological processes (see Table 8.6) the f value (removal indicator) for Ca^{++} in the oceans is 0.17. This means that 83% of sedimented (or sedimenting) $CaCO_3$ is redissolved. This value includes all $CaCO_3$ falling below the carbonate compensation depth and almost all pteropod aragonite. It probably also includes an appreciable amount of aragonite and Mg-calcite benthic debris which has been dislodged by storm waves and currents from shallow-water resting places, such as bank tops and coral reefs, and carried out into the deep sea, where it dissolves (Berner and Honjo 1981). Dissolution of $CaCO_3$ injects extra Ca^{++} and HCO_3^- into deep water which is then ultimately carried into shallow water and removed via biological secretion. In this way the internal oceanic cycle of Ca^{++} (and HCO_3^-) is completed.

Bicarbonate

Dissolved HCO_3^- in the oceans is removed in two fundamentally different ways. Either it is neutralized by H^+ to form CO_2:

$$H^+ + HCO_3^- \rightarrow H_2O + CO_2 \qquad (8.1)$$

or it is precipitated to form a carbonate mineral:

$$Ca^{++} + HCO_3^- \rightarrow CaCO_3 + H^+ \tag{8.2}$$

In the first case, the CO_2 produced can be taken up by photosynthesis or lost to the atmosphere, but either way the consumption of H^+ ions, or of substances contributing H^+ ions, is required. (In the reverse weathering theory, H^+ ions are contributed by cation-free or "acid" clays.) In the second case H^+ ions are produced.

Since H^+ ions (or OH^- ions produced by the dissociation of H_2O to H^+) do not accumulate in seawater, which must remain close to neutrality (pH 7–9), the reactions written above must be balanced by other reactions so as to conserve H^+. The simplest way to do this is to add together the two reactions themselves. The resulting reaction is

$$Ca^{++} + 2HCO_3^- \rightarrow CaCO_3 + CO_2 + H_2O \tag{8.3}$$

This is the overall reaction occurring during the secretion of biogenic $CaCO_3$ and written in reverse it represents the dissolution of $CaCO_3$ (see previous section on the Ca^{++} budget).

Consideration of inputs and outputs (Table 8.19) shows that, over the long term (past 25 million years), the oceanic HCO_3^- budget is balanced simply by removing HCO_3^- as $CaCO_3$ according to reaction 8.3 above. There is no need to call on volcanic-seawater reaction. (Edmond et al. [1979], estimate a removal rate from high-temperature basalt-water reaction of 90 Tg HCO_3^- per year, which is small enough to lie well within errors of estimating the removal rates of $CaCO_3$.) More importantly there is no need for reverse weathering to supply H^+ ions for neutralization via reaction 8.1. One of the major reasons for originally developing the reverse weathering theory was to remove excess HCO_3^- added to the oceans by silicate weathering. This is no longer necessary if Ca^{++}, derived both from rivers and volcanic-seawater reaction, is capable of removing all the contributed HCO_3^- as $CaCO_3$.

As can be seen in Table 8.19, the ocean HCO_3^- budgets, both short and long term, very much resemble those of Ca^{++}. Over the short term, HCO_3^- is being depleted and, given no change in $CaCO_3$ removal rates, all HCO_3^- (and CO_3^{--}) in the oceans would be removed in ~200,000 years. However, since the $CaCO_3$ imbalance is a product of processes occurring only over the past 11,000 years, it is unlikely that the excessive $CaCO_3$ removal in the present-day budget will continue for another 200,000 years. Various feedback mechanisms such as the entire oceans becoming undersaturated with $CaCO_3$, act to prevent complete HCO_3^- exhaustion from coming about. However, small changes in HCO_3^- concentration can have a major bearing on the CO_2 budget of the atmosphere.

An interesting additional input of HCO_3^- to the oceans, besides delivery by rivers, is worth mentioning. As shown in Table 8.19, about 7% of the HCO_3^- is a consequence of the oxidation of organic matter accompanying the bacterial re-

TABLE 8.19 The Oceanic Bicarbonate Budget (Rates in Tg HCO_3^-/yr)

Present-Day Budget			
Inputs		Outputs	
Rivers	1980	CaCO₃ deposition:	
Biogenic pyrite formation	145	Shallow water	1580
		Deep sea	1340
Total	2125	Total	2920

Budget for Past 25 Million Years			
Inputs		Outputs	
Rivers	1980	CaCO₃ deposition:	
Biogenic pyrite formation	145	Shallow water	730
		Deep sea	1340
Total	2125	Total	2070

Note: $Tg = 10^{12}$ g. Replacement time for HCO_3^- (river input only) is 83,000 years.

duction of interstitial sulfate in sediments to H_2S:

$$2CH_2O + SO_4^{--} \rightarrow H_2S + 2HCO_3^-$$

For every mole of H_2S removed as iron sulfides, there are two moles of HCO_3^- given off to the interstitial and eventually to the overlying water. The removal of 39 Tg of S per year as pyrite implies an input of 145 Tg of HCO_3^- per year to the oceans.

The HCO_3^- budget presented here should not be confused with that of CO_2 in seawater even though the two species are readily interconverted to one another. Dissolved CO_2 is affected by the synthesis and destruction of organic matter as well as by the precipitation and dissolution of $CaCO_3$, whereas HCO_3^- is affected only by the latter process. Dissolved HCO_3^- is electrically charged and cannot *by itself* be taken up by photosynthetic organisms to form electrically neutral organic matter. Failure to take this simple fact into consideration has led to many incorrect statements in the scientific literature.

Silica

The oceanic budget for silica has been worked out in detail by DeMaster (1981) who has determined sedimentation rates, silica contents, and so forth for a large number of sediments, and it is his budget, basically, that is presented here. This is shown in Table 8.20. (H_4SiO_4 addition and SiO_2 removal are given in terms of Si.) The main thing to note is that all H_4SiO_4 added to the oceans can be removed

TABLE 8.20 The Oceanic Silica Budget (Rates in Tg Si/yr).

Present-Day Budget			
Inputs		Outputs	
Rivers	180	Biogenic silica deposition:	
Basalt-seawater		Antarctic Ocean	117
reaction	30	Bering Sea	13
Total	210	North Pacific Ocean	7
		Sea of Okhotsk	7
		Gulf of California	5
		Walvis Bay	3
		Estuaries	38
		Other areas	≤13
		Total	190–203

Source: Outputs from DeMaster 1981.

Notes: Tg = 10^{12} g. To convert to Tg of SiO_2, multiply by 2.14. The replacement time for river-borne H_4SiO_4 is 21,000 years. The removal value for estuaries may be a maximum . . . see Chap. 7.

as biogenic opaline silica ($SiO_2 \cdot nH_2O$) without the necessity of calling upon reverse weathering or any other process to remove an apparent excess. (The value for the volcanic addition of H_4SiO_4 is derived from Iceland borehole data; if the estimates of Thompson [1983] or Edmond et al. [1979] are used, an imbalance of 35% results.) Thus, silica in the oceans, like Ca^{++}, is almost totally biogenic.

Unlike $CaCO_3$, opaline silica is secreted almost entirely by planktonic organisms (radiolaria and diatoms) and it dissolves at all depths. The latter comes about because of the high degree of undersaturation of the oceans with respect to silica. In fact, surface water, because of biogenic removal, is more undersaturated than deep water. Apparently, the diatoms and radiolaria are able to defy the usual predictions for chemical reactions and to remove dissolved silica from undersaturated solution. This is allowed because excess energy is provided by sunlight for the biosynthetic process.

Once the diatoms and radiolaria die, their siliceous remains immediately begin to dissolve and, in contrast to $CaCO_3$, most dissolution occurs during settling of particles in the water column. Some, however, make it to the bottom and continue to dissolve during initial burial·as evidenced by elevated concentrations of H_4SiO_4 in the pore waters of most sediments. (Values of removal rate determined by DeMaster are for final burial at sediment depths below the top 2–20 cm where most dissolution in the sediment takes place.) Regardless of where it occurs, dissolution is very effective and 95% of the opaline silica falling to the bottom is dissolved. This is shown by an *f* value, according to the Broecker model, of 0.05 (see Table 8.6). In addition, biogenic silica removal is so efficient in surface water that the *g* value is 0.99, which is the highest value for any of the biogenic elements considered. It is remarkable that in the oceans silica is a more biogenic element than either nitrogen or phosphorus even though it is the basic constituent of completely abiological rocks, such as granite or basalt!

Opaline silica, because it dissolves so fast, accumulates only in areas where it is rapidly produced in overlying surface waters. Since it is biogenic, its production is dictated by nutrients such as phosphorus and nitrogen. As a result, the areas of appreciable biogenic silica deposition are those overlain by fertile surface waters, high in dissolved N and P, which owe their fertility to coastal upwelling and other ocean circulation processes. It isn't fortuitous that several of the silica depositional areas listed in Table 8.20 are also areas of high organic matter accumulation.

Dissolved silica has the lowest replacement time, 21,000 years, of any element discussed in this chapter. This suggests that imbalances in inputs and outputs, if maintained for short times, should result in appreciable changes in the average concentration of H_4SiO_4 in seawater. The geologic record for the past several hundred million years, however, provides no evidence for large excursions. There is no evidence for widespread nonbiogenic silica precipitation which would be effective if the average silica level ever attained values exceeding the solubility of opaline silica (about six times the present maximum level in the deep Pacific Ocean). In addition there is evidence that silica removal by planktonic organisms has been going on for at least 200 million years. Apparently the siliceous plankton are able to respond rapidly to changes in input and thereby maintain an overall level of dissolved silica far below the saturation value. Rapid response is obvious when one considers the very high g value for silica of 0.99.

Phosphorus

The oceanic phosphorus budget is considerably different from the budgets for the elements discussed so far in that the removal mechanisms are largely unique to this element and the effects of pollution are relatively more important. Because of the mining and use of phosphate for fertilizer, detergents, and so forth (see Chapters 5 and 6), much of the phosphorus delivered by rivers to the ocean is pollutional in origin and this results in a present-day delivery rate that is considerably higher than that experienced during the geologic, prehuman past. For example, Meybeck (1982) estimates that the present rate of delivery of dissolved phosphate alone is twice as high as that expected in the absence of pollution.

Another complicating factor is the role of suspended phosphorus compounds carried by rivers to the sea. This includes both particulate organic and inorganic phosphorus. The latter is comprised of phosphorus adsorbed onto soil clays and ferric oxides as well as detrital primary apatite (calcium phosphate) eroded from rocks. The basic problem is that some of this particulate suspended phosphorus can be solubilized in the oceans, thus providing an additional input to the sea (see Chapter 7). Some particulate organic phosphorus is decomposed by bacteria and some phosphorus adsorbed on ferric oxide is rapidly solubilized when the iron is reduced to Fe^{++} in anoxic marine sediments (Krom and Berner 1981). The question is how much. Since particulate phosphorus is delivered at a rate about ten times that of dissolved phosphorus, the problem of not knowing the proportion that is solubilized becomes critical.

In Table 8.21 are given our best estimates of phosphorus inputs for both the present ocean and for the geologic past (see also Chapter 5). River input rates for dissolved inorganic (ortho-P) and organic phosphorus are based on the data of Meybeck (1982). The rain (plus dry fallout) net input for the present ocean is based on the "reactive" or seawater-soluble portion of land-derived total fallout, as determined by Graham and Duce (1979). The long-term rain input is assumed to be half that of the present day, mirroring the change in the river-dissolved input. The value given for river-borne particulate reactive phosphorus is based on our crude assumption that 10% of the total particulate phosphorus given by Meybeck (1982) is solubilized in the oceans. This assumes that most of the particulate P is unreactive detrital apatite which is in keeping with the observations of Rittenberg,

TABLE 8.21 The Oceanic Phosphorus Budget (Rates in Tg P/yr)

Present-Day Budget			
Inputs		Outputs	
Rivers:		Organic P burial	2.0
Natural dissolved P (organic plus ortho-P)	1.0	$CaCO_3$ deposition	0.7
Dissolved P from pollution	1.0	Adsorption on volcanogenic Fe oxides	0.1[a]
Particulate reactive P (mostly pollution)	2.0	Phosphorite formation	<0.1[a]
Rain (plus dry fallout)	0.2	Fish debris deposition	<0.02[a]
Total	4.2	Total	2.8–2.9

Long-Term (Balanced) Budget			
Inputs		Outputs	
Rivers:		Organic P burial	0.5[a]
Dissolved ortho-P	0.4	$CaCO_3$ deposition	0.5
Dissolved organic P	0.6	Adsorption on volcanogenic Fe oxides	0.1[a]
Particulate reactive P	0.1	Phosphorite formation	0.1
Rain (plus dry fallout)	0.1	Total	1.2
Total	1.2		

Source: River input data from Meybeck 1982; [a]data from Froelich et al. 1982.

Note: Tg = 10^{12} g. The replacement time for phosphorus via river addition (of dissolved orthophosphate only) is 50,000 years.

Emery, and Orr (1955). Finally, we assume that 95% of the seawater-soluble particulate phosphorus is pollutional in origin, thus giving a much smaller long-term river-borne reactive particulate phosphorus input. This much smaller input is required to bring the long-term phosphorus budget into balance. For the present oceans, most of the imbalance is due to excess inputs of both reactive particulate and dissolved phosphate arising from pollution.

The output processes shown in Table 8.21 are rather different than those described earlier in this chapter for other elements and, thus, they deserve special discussion. Some organic matter, carried to the oceans by rivers and formed *in situ* by photosynthesis, survives bacterial destruction and eventually becomes buried in marine sediments. This material contains organic phosphorus compounds so that its burial constitutes a mechanism for the removal of phosphate from seawater. The removal rate given in Table 8.21 for the present-day oceans is based on an organic carbon removal rate of 126 Tg C/yr (Berner 1982) and an average C/P weight ratio in buried organic matter of 65. The latter value was obtained from a plot of C/P versus organic C content (Froelich et al. 1982) and the average organic C content for modern fine-grained sediments of 0.6% (Berner 1982). The long-term output rate for organic phosphorus is based on the data of Froelich et al. (1982). Phosphorus is also removed from seawater as a component of skeletal calcium carbonate. This phosphorus is taken up both during growth of the shell and by adsorption from interstitial water upon death and sedimentation to the sea floor. The output values given for this process in Table 8.21 are taken from the $CaCO_3$ output rates of Table 8.18 and the average phosphorus content of marine $CaCO_3$ of 300 ppm (Froelich et al. 1982). Note that, along with organic phosphorus burial, $CaCO_3$ deposition constitutes one of the two most important processes of phosphorus removal from the oceans.

The two additional removal mechanisms also deserve comment. It has been established (Berner 1973; see Froelich et al. 1982 for a review) that an appreciable proportion of phosphate is removed from seawater as an indirect result of volcanic-seawater reaction. During the reaction of basalt with heated seawater, an appreciable buildup of Fe^{++} in solution occurs as a result of the attack of hot acidic and oxygen-free water on iron-containing minerals. Upon travelling upward to the sea floor, the water mixes with oxygenated seawater and, as a result, the ferrous iron is oxidized and precipitated to form hydrous ferric oxides. This accounts for the high degree of enrichment of ferric oxides in sediments along the crests of midoceanic rises. These oxides readily adsorb and remove phosphate from seawater and, thus, their formation constitutes a process of oceanic phosphorus removal. Although of lesser quantitative importance, this process is unique in that it provides a link between volcanogenic processes associated with sea-floor spreading and biological processes that are intimately associated with the nutrient element phosphorus.

The remaining major phosphate removal process (removal as fish bones is insignificant) is phosphorite formation (Froelich et al. 1982). Phosphorite is calcium phosphate that has precipitated (inorganically) from seawater to form deposits which are often of economic importance.   Phosphorite formation is sporadic over geologic time, with long periods, including the present, where it is relatively unimportant, alternating with occasional bursts of rapid and massive calcium phosphate precipitation such as occurred in the Permian period ($\sim$200 million years ago) which is evidenced by the areally extensive Phosphoria formation of the western United States.   (There is more phosphorus in this formation than there is in the entire present-day oceans!)   However, because we do not fully understand the genesis of large phosphorite deposits, it is assumed in Table 8.21 that the *average* long-term (past $\sim$10 million years) rate of phosphorite removal is roughly the same as that occurring today.   This is probably a minimum and more work on this subject is badly needed.

Besides the overall input-output cycle discussed above, phosphorus also undergoes important internal cycling within the oceans.   It is a highly biogenic element ($g = 0.92$ from Table 8.6) and a key nutrient in photosynthetic production. As a result, there is extensive dissolved phosphorus uptake in surface waters, both as orthophosphate ($H_2PO_4^-$, HPO_4^{--}, and PO_4^{---}) and organic phosphorus, and return to solution at depth.   Over 98% of the phosphorus falling to the bottom, as a constituent of organic matter or $CaCO_3$, is returned to solution.   In other words, the f value (Table 8.6) is 0.015.   This means that the flux of phosphorus to surface water via upwelling is 66 times greater than its input from rivers.   As in lakes, phosphorus (and nitrogen) undergo many transfers between deep and surface waters before becoming buried permanently in sediments.

Nitrogen

In one respect the nitrogen cycle in the oceans is simpler than that for the other elements already discussed, but in all other respects it is more complicated.   The simplifying feature is that nitrogen is removed in bottom sediments almost entirely as a constituent of organic matter.   In other words, there are no important sedimentary nitrogen-containing minerals.   The total removal rate for nitrogen in sediments can thus be determined from a knowledge of the organic carbon removal rate and the average C/N ratio of marine sedimentary organic matter.

Nitrogen is more complicated than the other elements in that it can be lost to or gained from the atmosphere in the form of a gas, specifically N_2 (and to a lesser extent N_2O).   During photosynthesis, certain organisms, most prominently the blue-green algae, fix dissolved N_2 to form organic nitrogen compounds such as proteins.   This lost N_2 is then replaced by gas transfer from the atmosphere to the oceans.   The ability to fix N_2 enables the fixing organisms to continue to photosynthetically produce organic matter even in the absence of nitrate, or other dissolved nitrogen-containing nutrients.   The N_2 lost by this process is eventually returned to seawater and the atmosphere by the process of denitrification.   This process occurs mainly in sediments and occasionally in the water column (such as the eastern tropical Pacific) where dissolved oxygen concentrations are low.   It involves the reduction of nitrate to N_2 (see Table 8.4) by the reaction

$$5CH_2O + 4NO_3^- \rightarrow 2N_2 + CO_2 + 4HCO_3^- + 3H_2O$$

Denitrification also results in the formation of lesser amounts of N_2O.

Another complicating factor in the oceanic ntirogen cycle is the presence of inorganic nitrogen in several different oxidation states, all of which can be used as nutrients. Besides N_2, which is a special case, the major ones are nitrate (NO_3^-, nitrite (NO_2^-), and ammonium (NH_4^+) with nitrate predominating in abundance (see Table 8.2). This contrasts with the other two major nutrients, phosphorus and silica, each of which exists in only one oxidation state. Ammonium forms by the decomposition of organic nitrogen compounds and bacterial nitrate reduction whereas nitrite and nitrate form by the bacterial oxidation of NH_4^+ using O_2 in seawater. This is all shown, along with N_2 fixation, denitrification, and other processes, in the schematic nitrogen cycle diagram for the oceans, Figure 8.16.

Like phosphorus and silicon, nitrogen is also a highly biogenic element ($g =$

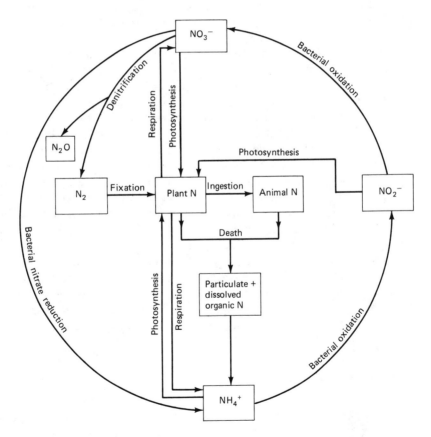

Figure 8.16 Schematic representation of the marine nitrogen cycle.

TABLE 8.22   The Oceanic Nitrogen Budget (Rates in Tg N/yr)

Present-Day Budget			
Inputs		Outputs	
Rivers:		Organic N burial in sediments	14
Natural dissolved inorganic N (88% as NO_3^- -N)	4.5	Denitrification	40–120
		Total	54–134
Natural dissolved organic N	10		
Pollutive dissolved N	7		
Particulate organic N	21		
Rain and dry deposition	20		
Fixation of N_2	10–90		
Total	73–153		

Note: $Tg = 10^{12}$ g.   The replacement time for NO_3^- added by rivers is 72,000 years.

0.92), with most of the organic nitrogen that sediments out being returned to solution at depth ($f = 0.03$) (see Table 8.6).   Thus, it also exhibits extensive internal cycling between deep and surface waters prior to burial.   Unlike phosphorus and silicon, nitrogen is not involved, to any appreciable extent, in mineral dissolution and precipitation.   In other words, it is strictly biogenic.

The overall input-output balance for nitrogen in the oceans is shown in Table 8.22.   Since many of the values shown in this table are poorly known, no attempt is made to obtain a long-term balanced budget.   Especially critical are the large values for N_2 fixation and denitrification, which are known only to about a factor of two (see, e.g., Simpson 1977).   Here we can only express a range of values.   River input is taken from Meybeck (1982), the net rain and dry fallout value is from Chapter 3 of the present book, and the removal rate as organic nitrogen is based on the organic carbon removal rate (Berner 1982) and an average C/N weight ratio of marine sedimentary organic matter (~9; see Meybeck 1982).   As can be seen, there may well be an imbalance for the present-day oceans and, even if poorly known values for N_2 fixation and denitrification were improved, this balance would probably still persist.   This all points to the importance of pollutive inputs, mainly from fertilizers and human and animal wastes, as the major cause of this imbalance.

REFERENCES

ARNOLD, M., and S. M. F. SHEPPARD.   1981.   East Pacific Rise at latitude 21°N: Isotopic composition and origin of the hydrothermal sulphur, *Earth Planet. Sci. Letters* 56:148–156.

BERGER, W. 1976. Biogenous deep sea sediments: Production, preservation, and interpretation. In *Chemical oceanography*, ed. J. P. Riley and R. Chester, vol. 5:265–387. New York: Academic Press.

———. 1977. Carbon dioxide excursions in the deep sea record: Aspects of the problem. In *The fate of fossil fuel CO₂ in the oceans*, ed. N. R. Andersen and A. Malahoff, pp. 502–542. New York: Plenum Press.

BERNER, R. A. 1970. Sedimentary pyrite formation, *Amer. J. Sci.* 268:1–23.

———. 1973. Phosphate removal from seawater by adsorption on volcanogenic ferric oxides, *Earth Planet. Sci. Letters* 18:77–86.

———. 1980. *Early diagenesis: A theoretical approach.* Princeton, N.J.: Princeton Univ. Press, 241 pp.

———. 1982. Burial of organic carbon and pyrite sulfur in the modern ocean: Its geochemical and environmental significance, *Amer. J. Sci.* 282:451–473.

———. 1984. Sedimentary pyrite formation: An update, *Geochim. Cosmochim. Acta* 48:605–615.

BERNER, R. A., and S. HONJO. 1981. Pelagic sedimentation of aragonite: Its geochemical significance, *Science* 211:940–942.

BERNER, R. A., A. C. LASAGA, and R. M. GARRELS. 1983. The carbonate-silicate geochemical cycle and its effect on atmospheric carbon dioxide over the past 100 million years, *Amer. J. Sci.* 283:641–683.

BISCAYE, P. E., V. KOLLA, and K. K. TUREKIAN. 1976. Distribution of calcium carbonate in surface sediments of the Atlantic Ocean, *J. Geophys. Res.* 81:2595–2603.

BISCHOFF, J. L., and F. W. DICKSON. 1975. Seawater-basalt interaction at 200° C and 500 bars: Implications for origin of sea-floor heavy metal deposits and regulation of seawater chemistry, *Earth Planet. Sci. Letters* 25:385–397.

BOYLE, E. A., F. R. SCLATER, and J. M. EDMOND. 1977. The distribution of dissolved copper in the Pacific, *Earth Planet. Sci. Letters* 37:38–54.

BREWER, P. G. 1975. Minor elements in sea water. In *Chemical Oceanography,* 2nd ed., ed. J. P. Riley and G. Skirrow, vol. 1:301–363. London: Academic Press.

BROECKER, W. S. 1971. A kinetic model for the chemical composition of seawater, *Quaternary Res.* 1:188–207.

BROECKER, W. S., and T.-H. PENG. 1982. *Tracers in the sea.* Palisades, N.Y.: Eldigio Press. 690 pp.

BROECKER, W. S., T. TAKAHASHI, H. J. SIMPSON, and T.-H. PENG. 1979. Fate of fossil fuel carbon dioxide and the global carbon budget, *Science* 206:409–418.

CLAYPOOL, G., and I. R. KAPLAN. 1974. The origin and distribution of methane in marine sediments. In *Natural gases in marine sediments*, ed. I. R. Kaplan, pp. 99–139. New York: Plenum Press.

CONDIE, K. C. 1976. *Plate tectonics and crustal evolution.* New York: Pergamon Press. 288 pp.

CORLISS, J. B., J. DYMOND, L. I. GORDON, J. M. EDMOND, R. P. VON HERZEN, R. D. BALLARD, K. GREEN, D. WILLIAMS, A. BAINBRIDGE, K. CRANE, and T. H. VAN ANDEL. 1979. Submarine thermal springs on the Galapagos Rift, *Science* 203:1073–1083.

CRAIG, H. 1969. Abyssal carbon and radiocarbon in the Pacific, *J. Geophys. Res.* 74: 5491–5506.

CROVISIER, J. L., J. H. THOMASSIN, T. JUTEAU, J. P. EBERHART, J. C. TOURAY, and P. BAILLIF. 1983. Experimental seawater-basaltic glass interaction at 50° C, *Geochim. Cosmochim. Acta* 47:377–388.

DASCH, E. J. 1969. Strontium isotopes in weathering profiles, deep-sea sediments, and sedimentary rocks, *Geochim. Cosmochim. Acta* 33:1521–1552.

DeMASTER, D. J. 1981. The supply and accumulation of silica in the marine environment, *Geochim. Cosmochim. Acta* 45:1715–1732.

DITTMAR, W. 1884. Report on researches into the composition of ocean water collected by H.M.S. Challenger. In *Challenger Reports*, vol. 1, *Physics and Chemistry*, pp. 1–251. London: H.M. Stationery Office.

DREVER, J. I. 1972. The magnesium problem. In *The sea*, ed. E. D. Goldberg, vol. 5: 337–357. New York: Wiley-Interscience.

EDMOND, J. M., C. MEASURES, R. E. McDUFF, L. H. CHAN, R. COLLIER, B. GRANT, L. J. GORDON, and J. B. CORLISS. 1979. Ridge crest hydrothermal activity and the balances of the major and minor elements in the ocean: The Galapagos data, *Earth Planet. Sci. Letters* 46:1–18.

FIADEIRO, M., and H. CRAIG. 1978. Three-dimensional modeling of tracers in the deep Pacific Ocean, I: Salinity and oxygen, *J. Marine Res.* 36:323–355.

FROELICH, P. N., M. L. BENDER, N. A. LUEDTKE, G. R. HEATH, and T. DeVRIES. 1982. The marine phosphorus cycle, *Amer. J. Sci.* 282:474–511.

FROELICH, P. N., G. P. KLINKHAMMER, M. L. BENDER, N. A. LUEDTKE, G. R. HEATH, D. CULLEN, P. DAUPHIN, D. HAMMOND, B. HARTMAN, and V. MAYNARD. 1979. Early oxidation of organic matter in pelagic sediments of the eastern equatorial Atlantic: Suboxic diagenesis, *Geochim. Cosmochim. Acta* 43:1075–1090.

GARRELS, R. M., and A. LERMAN. 1984. Coupling of the sedimentary sulfur and carbon cycles—an improved model, *Amer. J. Sci.* 284:989–1007.

GARRELS, R. M., and F. T. MACKENZIE. 1971. *Evolution of sedimentary rocks.* New York: W. W. Norton. 397 pp.

GARRELS, R. M., F. T. MACKENZIE, and C. HUNT. 1975. *Chemical cycles and the global environment.* Los Altos, Calif.: Wm. Kaufman, Inc. 206 pp.

GARRELS, R. M., and E. A. PERRY. 1974. Cycling of carbon, sulfur, and oxygen through geologic time. In *The sea*, ed. E. D. Goldberg, vol. 5: 303–336. New York: Wiley-Interscience.

GIESKES, J. M., and J. R. LAWRENCE. 1981. Alteration of volcanic matter in deep sea sediments: Evidence from the chemical composition of interstitial waters from deep sea drilling cores, *Geochim. Cosmochim. Acta* 45:1687–1703.

GOLDHABER, M. B., and I. R. KAPLAN. 1974. The sulfur cycle. In *The sea*, ed. E. D. Goldberg, vol. 5:569–655. New York: Wiley-Interscience.

GRAHAM, W. F., and R. A. DUCE. 1979. Atmospheric pathways of the phosphorus cycle, *Geochim. Cosmochim. Acta* 43:1195–1208.

HART, S. R., and H. STAUDIGEL. 1982. The control of alkalies and uranium in seawater by ocean crust alteration, *Earth Planet. Sci. Letters* 58:202–212.

HAY, W. W., and J. R. SOUTHAM. 1977. Modulation of marine sedimentation by the continental shelves. In *The fate of fossil fuel CO₂ in the oceans*, ed. N. R. Andersen and A. Malahoff, pp. 569–604. New York: Plenum Press.

HOFFMAN, J. C. 1979. An evaluation of potassium uptake by Mississippi River borne clays following deposition in the Gulf of Mexico. Ph.D. dissertation, Case-Western Reserve University, Cleveland, Ohio. 136 pp.

HOLLAND, H. D. 1978. *The chemistry of the atmosphere and oceans*, New York: Wiley-Interscience. 351 pp.

HOWER, J., E. V. ESLINGER, M. E. HOWER, and E. A. PERRY. 1976. Mechanism of burial metamorphism of argillaceous sediments, mineralogical and chemical evidence, *Bull. Geol. Soc. Amer.* 87:725–737.

KENNETT, J. 1982. *Marine geology*. Englewood Cliffs, N.J.: Prentice Hall. 813 pp.

KESTER, D. R. 1975. Dissolved gases other than CO₂. In *Chemical oceanography*, 2nd ed., vol. 1:498–556. ed. J. P. Riley and G. Skirrow, London: Acad. Press.

KING, R. H. 1947. Sedimentation in Permian Castile Sea, *Bull. Amer. Assn. Petrol. Geologists* 31:470–477.

KOBLENTZ-MISHKE, O. J., V. V. VOLKOVINSKY, and J. G. KABANOVA. 1970. Plankton primary production of the world ocean. In *Scientific exploration of the South Pacific*, ed. W. S. Wooster, pp. 183–193. Washington, D.C.: National Academy of Science.

KROM, M. D., and R. A. BERNER. 1981. The diagenesis of phosphorus in a nearshore marine sediment, *Geochim. Cosmochim. Acta* 45:207–216.

LOWENSTAM, H. A. 1981. Minerals formed by organisms, *Science* 211:1126–1130.

MACKENZIE, F. T., and R. M. GARRELS. 1966. Chemical mass balance between rivers and oceans, *Amer. J. Sci.* 264:507–525.

MEYBECK, M. 1979. Concentrations des eux fluviales en éléments majeurs et apprts en solution aux océans, *Rev. Géol. Dyn. Géogr. Phys.* 21:215–246.

MEYBECK, M. 1982. Carbon, nitrogen, and phosphorus transport by world rivers, *Amer. J. Sci.* 282:401–450.

MILLERO, F. J., and D. R. SCHREIBER. 1982. Use of the ion pairing model to estimate activity coefficients of the ionic components of natural waters, *Amer. J. Sci.* 282:1508–1540.

MILLIMAN, J. D. 1974. *Marine carbonates*. New York: Springer-Verlag. 375 pp.

MOTTL, M. J. 1983. Hydrothermal processes at seafloor spreading centers: Application of basalt-seawater experimental results. In *Hydrothermal processes at seafloor spreading centers*, ed. P. A. Rona, K. Bostrom, L. Laubier, and K. L. Smith, pp.225–278. NATO Conf. ser. 4, vol. 12. New York: Plenum Press.

MOTTL, M. J., H. D. HOLLAND, and R. F. CARR. 1979. Chemical exchange during hydrothermal alteration of basalt by seawater, II: Experimental results for Fe, Mn, and sulfur species, *Geochim. Cosmochim. Acta* 43:869–884.

MÜLLER, P. J., and E. SUESS. 1979. Productivity, sedimentation, and sedimentary organic matter in the oceans, I: Organic carbon perservation, *Deep Sea Research* 26A:1347–1362.

PERRY, E. A., J. M. GIESKES, and J. R. LAWRENCE. 1976. Mg, Ca, and ¹⁸O/¹⁶O exchange in the sediment-pore water system, hole 149, DSDP, *Geochim. Cosmochim. Acta* 40:413–423.

PETERSON, M. N. A., and J. J. GRIFFIN. 1964. Volcanism and clay minerals in the southeastern Pacific, *Marine Res.* 22:13–21.

REDFIELD, A. C.   1958.   The biological control of chemical factors in the environment, *American Scientist* 46:205–222.

RILEY, J. P., and R. CHESTER.   1971.   *Introduction to marine chemistry*.   New York: Academic Press, 465 pp.

RITTENBERG, S. C., K. O. EMERY, and W. L. ORR.   1955.   Regeneration of nutrients in sediments of marine basins, *Deep Sea Research* 3:23–45.

RUSSELL, K. L.   1970.   Geochemistry and halmyrolysis of clay minerals, Rio Ameca, Mexico, *Geochim. Cosmochim. Acta.* 34:893–907.

SAVIN, S. M., and S. EPSTEIN.   1970.   The oxygen and hydrogen isotope geochemistry of ocean sediments and shales, *Geochim. Cosmochim. Acta* 34:43–63.

SAYLES, F. L.   1979.   The composition and diagenesis of interstitial solutions, I: Fluxes across the seawater-sediment interface in the Atlantic Ocean, *Geochim. Cosmochim. Acta* 43:527–545.

————.   1981.   The composition and diagenesis of interstitial solutions, II: Fluxes and diagenesis at the water-sediment interface in the high latitude North and South Atlantic, *Geochim. Cosmochim. Acta* 45:1061–1086.

SAYLES, F. L., and P. C. MANGELSDORF.   1977.   The equilibration of clay minerals with seawater: Exchange reactions, *Geochim. Cosmochim. Acta* 41:951–960.

SCLATER, F. R., E. BOYLE, and J. M. EDMOND.   1976.   On the marine geochemistry of nickel, *Earth Planet. Sci. Letters* 31:119–128.

SEYFRIED, W. E., and J. L. BISCHOFF.   1979.   Low temperature basalt alteration by seawater: An experimental study at 70° and 150° C, *Geochim. Cosmochim. Acta* 43:1937–1947.

SILLÉN, L. G.   1967.   The ocean as a chemical system, *Science* 156:1189–1197.

SIMPSON, H. J.   1977.   Man and the global nitrogen cycle group report.   In *Global chemical cycles and their alteration by man*, ed. W. Stumm, pp. 253–274.   Berlin: Dahlem Konferenzen.

SKIRROW, G.   1975.   The dissolved gases—carbon dioxide.   In *Chemical oceanography*, 2nd ed., ed. J. P. Riley and G. Skirrow, vol. 2:245–300.   London: Academic Press.

SLEEP, N. H., J. L. MORTON, L. E. BURNS, and T. J. WOLERY.   1983.   Geophysical constraints on the volume of hydrothermal flow at ridge axes.   In *Hydrothermal processes at seafloor spreading centers*, ed. P. A. Rona, K. Bostrom, L. Laubier, and K. L. Smith, pp. 53–70.   NATO Conf. ser. 4, vol. 12.   New York: Plenum Press.

SPENCER, C. P.   1975.   The micronutrient elements.   In *Chemical oceanography*, 2nd ed., ed. J. P. Riley and G. Skirrow, vol. 2:245–300.   London: Academic Press.

STAUDIGEL, H., and S. R. HART.   1983.   Alteration of basaltic glass: Mechanisms and significance for the oceanic crust–seawater budget, *Geochim. Cosmochim. Acta* 47:337–350.

STUMM, W., ED.   1977.   *Global chemical cycles and their alterations by man*.   Berlin: Dahlem Konferenzen, 346 pp.

STUMM, W., and J. J. MORGAN.   1981.   *Aquatic chemistry*.   New York: John Wiley.   780 pp.

SVERDRUP, H. V., M. W. JOHNSON, and R. H. FLEMING.   1942.   *The oceans*.   Englewood Cliffs, N.J.: Prentice-Hall, 1087 pp.

THOMPSON, G. 1983. Basalt-seawater reaction. In *Hydrothermal processes at seafloor spreading centers,* ed. P. A. Rona, K. Bostrom, L. Laubier and K. L. Smith, pp. 225–278. NATO Conf. ser. 4, vol. 12. New York: Plenum Press.

VON DAMM. K. L., J. M. EDMOND, B. GRANT, C. I. MEASURES, B. WALDEN, and R. F. WEISS. 1985. Chemistry of submarine hydrothermal solutions at 21°N, East Pacific Rise, *Geochim. Cosmochim. Acta* 49:2197–2220.

VON DAMM, K. L., B. GRANT, and J. M. EDMOND. 1983. Preliminary report on the chemistry of hydrothermal solutions at 21° North, East Pacific Rise. In *Hydrothermal processes at seafloor spreading centers*, ed. P. A. Rona, K. Bostrom, L. Laubier, and K. L. Smith, pp. 369–389. NATO conf. ser. 4, vol. 12. New York: Plenum Press.

WESTRICH, J. T. 1983. The consequences and controls of bacterial sulfate reduction in marine sediments. Ph.D. Dissertation, Yale University. 530 pp.

WHITFIELD, M. 1975. Sea water as an electrolyte solution. In *Chemical oceanography, 2nd ed.*, vol. 1, ed. J. P. Riley and G. Skirrow. London: Academic Press. pp. 44–171.

WILLIAMS, P. J. 1975. Biological and chemical aspects of dissolved organic material in sea water. In *Chemical oceanography, 2nd ed.,* vol. 2, ed. J. P. Riley and G. Skirrow. London: Academic Press, pp. 301–363.

WILSON, T. R. S. 1975. Salinity and the major elements of sea water. In *Chemical oceanography, 2nd ed.,* vol. 1, ed. J. P. Riley and G. Skirrow. London: Academic Press. pp. 365–413.

Index